Organic Coatings:
Their Origin and Development

Organic Coatings:
Their Origin and Development

Proceedings of the International Symposium on the History of
Organic Coatings, held September 11–15, 1989, in Miami Beach,
Florida, USA

Editors

Raymond B. Seymour
University of Southern Mississippi
Hattiesburg, Mississippi

Herman F. Mark
Polytechnic University
Brooklyn, New York

Elsevier
New York • Amsterdam • London

Elsevier Science Publishing Co., Inc.
655 Avenue of the Americas
New York, New York 10010

Sole distributors outside the United States and Canada:
Elsevier Science Publishers B.V.
P.O. Box 211, 1000 AE Amsterdam, the Netherlands

© 1990 by Elsevier Science Publishing Co., Inc.

This book has been registered with the Copyright Clearance Center, Inc.
For further information, please contact the Copyright Clearance Center, Inc.,
Salem, Massachusetts.

This book is printed on acid-free paper.

Library of Congress Cataloging-in-Publication Data

International Symposium on the History of Organic Coating (1989 :
    Miami Beach, Fla.)
      Organic coatings : their origin and development : proceedings of
    the International Symposium on the History of Organic Coatings, held
    September 11–15, 1989, in Miami Beach, Florida, USA / editors,
    Raymond B. Seymour, Herman F. Mark.
        p.   cm.
      1. Plastic coatings – History – Congresses.   I. Seymour, Raymond
    Benedict, 1912–   . II. Mark, H.F. (Herman Francis), 1895–
    III. Title.
    TP1175.S6I59 1989   89-25769
    667'.9 – dc20   CIP
    ISBN: 0-444-01520-5

Current printing (last digit):
10 9 8 7 6 5 4 3 2 1

Manufactured in the United States of America

CONTENTS

Foreword

After the development of the art of painting, and the use of lacquers and oleoresinous paints, many centuries past before the introduction of a host of synthetic coatings. The Monk, Theophilous, in his "De Diversis Attibus," summarized coatings science, prior to the 12th century. In a series of books on paints, edited in the 1940's, Dr. Mattiello also supplied some up to date information on the History of Coatings. Fortunately, a group of modern science historians presented reports at the ACS Symposium in order to update all aspects of the history of coatings.

The book on the origin and development of high performance polymers, which was coedited with Dr. Kirshenbaum, served as a valuable addition to our knowledge of history of that phase of polymer science. Hopefully, this treatise on the history of coatings will be as well received by those scientists who are applying the knowledge developed by macromolecular giants in previous years.

Preface

Since the drying of oleoresinous paints involved man's first controlled polymerization, Nobel Laureate Lord Todd's Statement, "I am inclined to think that the development of polymerization is perhaps the biggest thing chemistry has done," is an appropriate opening sentence for a preface on the History of Coatings.

The statement made by Nobel Laureate Hermann Staudinger's antagonists to support their polymer science incredulity viz, "we are shocked, like zoologists would be, if they were told somewhere in Africa was found an elephant which was 1500 feet long and 300 feet high," might also be applied to some more modern scientist who overlook history and coatings science. Yet, the 10 billion dollar US coatings industry, which is the nation's major consumer of organic chemicals, also contributes to the prevention of degradation of our modern stationary and mobile structures and to our artistic lifestyle.

Through its establishment of the Center For History of Chemistry, the American Chemical Society has demonstrated the need to recognize past contributions to science. The funds supplied by firms such as, Arco, Dow, du Pont, Exxon and GE for the establishment of this center have been augmented by a generous gift from Dr. Arnold Beckman.

Fortunately, this center chose the history of polymer chemistry as its first project. This book on the proceedings of the International Symposium on History of Coatings, is published in support of the center's first project.

The world history of painting techniques, which will be published in 1990 by the Federations of the Associations of Technicians from Continental Europe, in the field of paint, varnish, enamel and printing inks and The Coatings Handbook to be published in 1990 by Elsevier will also be in tune with the first project selected by the Center for History of Chemistry.

ACKNOWLEDGMENTS

The editors wish to thank the many authors who presented reports at the International Symposium on The History of Organic Coatings at the National Meeting of The American Chemical Society at Miami Beach, FL, September 11-14, 1989. The editors appreciate the assistance of Ms. Machell Haynes who aided in the organization of the symposium and the typing of many of the chapters in the book.

Origin and Development of Polymeric Coatings

Raymond B. Seymour
University of Southern Mississippi
Hattiesburg, MS 39406

Abstract

Prior to the mid 19th Century, the only polymer technology available was in the field of organic coatings. The art of decorating the walls of caves was practiced many thousand years ago. The early materials were aqueous dispersions of earth colors and resinous additives, such as crushed berries, eggs, milk or tree sap. Aqueous dispersions, such as distemper, and white wash were used for many centuries but were replaced, in part, by oleoresinous paints. The recipe for the production of these primitive paints was published by Theophilus in the 12th Century. While vegetable oil-modified paints, such as alkyds continue to be used, they were replaced, to large extent, by the introduction of cellulose nitrate lacquers in the 19th Century. While all of the various types of coatings continue to be used, the trend is toward water-borne and high solids coatings which meet environmental specifications.

Waterborne Coatings

As a result of regulations, which limit the amount of volatile organic solvents in coatings, there has been a renewed interest in water borne coatings, which were the earliest types of paints used by the caveman. He used earth colorants as pigments and crushed berries, animal blood, egg whites, and sap from dandelions, milkweed and trees as adhesives in these crude paints.

The paintings of the grand bison at Altarmira, Spain and the chinese horse at Laucaux, France are believed to be 15,000 years old. The Obiri rock paintings in Arnhem Land in Northern Australia also date back to prehistoric times. Some 5000 years ago, the Egyptians improved their coatings by adding other adhesives, such as casein and the term "distemper" from the Latin temperare, meaning to mix, was used to describe these water borne paints.

Pigments

The principal pigments, used by the Palaeolithic artisans were charred wood (black), chalk (white) and iron and manganese oxide for red and yellow colors. About 3 or 4 thousand years ago, the Egyptians supplemented these basic pigments with lapis lazuli (blue), azurite (green), ochres (red and yellow), orpiment (yellow), malachite (green), gypsum (white) and lamp black.

The Egyptians also developed vegetable colorants such as those from the madder root. The ancient Romans used red lead as a pigment in coatings but the first synthetic pigments were Egyptian blue and white lead (cerussa). White lead was made by Pliny by the reaction of vinegar on lead sheets in the atmosphere over 2500 years ago. Egyptian blue was produced by the calcination of a mixture of lime, alumina, silica, soda ash and copper oxide at least 10 thousand years ago.

Other classic white pigments were zinc oxide, zinc sulfide, lithopone and basic white lead sulfate which were introduced in 1770, 1783, 1847, and 1855, respectively. Titanium dioxide, which was produced from black ilmenite in 1924, became the major white pigment and is the worlds most widely used pigment.

Published 1990 by Elsevier Science Publishing Co., Inc.
Organic Coatings: Their Origin and Development
R.B. Seymour and H.F. Mark, Editors

1

Ancient Binders

Shellac, which was erroneously called "Indian Amber" by Pliny is one of the few resins obtained from insects. This resin, is secreted by a coccid insect (Laccifer lacca) which feeds on the lac trees in India and Thailand. It was used for making lac sticks over 3000 years ago. These sticks were used to coat rotating objects on a lathe by pressing the lac stick which was softened by the friction developed during rotation.

Lac, which is the hard resinous secretion is dissolved in ethanol and the residue of the insects and twigs is removed by filtration. The color which ranges from light yellow to dark orange is dependent on the type of host tree used as a source of the shellac and the extent of refining. Since shellac is insoluble in aliphatic hydrocarbon solvent such as "mineral spirits" it continues to be used as a stain suppressant seeker under other solvent based coatings shellac is also used for coating pills, candy and fruit.

The word lacquer is derived form the word lac. However, the first true lacquer was Japan lacquer was obtained from the sap of a sumac tree (Rhus Vernicera) during the chou dynasty about 3000 years ago. The use of Japan lacquers was expanded and improved during the ming period (1362-1644). The resin which was obtained by thermal evaporation of the aqueous solution was mixed with pigments, and used as a high gloss coating. It was not uncommon to build a thick coating by the application of as many as 250 layers of lacquer.

Varnish based on solutions of amber were used as early as 250 BC but the formations were not until a monk named Theophilus described the production of an oil varnish by dissolving resin in hot oil. Amber was one of the resins used at that time. The name varnish is derived from vernix which is the latin word for amber. Subsequently, natural resins were obtained from trees in tropical regions. The resins were classified as ancient or fossil, semi-fossil and recent, depending on when they were separated from the trees. These resins were also named after their geographic source, such as Manila, Batu Dammer, Congo, and Kauri.

Oleoresinous Paints

The term paint is usually applied to a coating consisting of pigments dispersed in a drying oil such as linseed oil but through misuse, this term has been expanded to include many other coatings, including water based coatings. There is some evidence that drying oils were used in ancient Egypt and more definite reports that linseed oil was used for coating in the 4th Century A.D.

Linseed oil, which is obtained from the seeds of flax (Lininum usitatissimum) was the first vegetable oil binder for making paints. This oil is a glyceryl ester of unsaturated acids, such as linoleic and oleic acids. Linseed oil forms a film when exposed to the atmosphere and the rate of this "hardening" is accelerated when white lead is mixed with the oil. It has been suggested that a trace of free acid in the oil reacts with the lead salts to produce an oil soluble salt which is called a drier or siccative. Detailed recipes for producing oleoresinous paints were supplied by Theophilus.

Presbyter Theophilus , who was also called Rugerus, in cooperation with another monk, called Roger, wrote "De Diversis Artibus" which described painting and other practical arts used in church decoration. Neither of the coauthors showed any scientific talent but they did record unpublished recipes which had been used for centuries.

Linseed oil continues to account for more than 50 percent of the drying oils used in paints but its share of the market which was over 90 percent in 1900 has been decreased by the use of soybean oil, tung oil, dehydrated castor oil, oiticica oil and fish oil. The hardening (drying) of these glyceryl glesters in air is dependent on the extent of unsaturation and is accelerated by the addition of driers.

Driers

Lead salts of organic acids, which were the original driers, function as polymerization catalysts throughout the entire film depth. In contrast, cobalt salts function as surface driers. Hence, until recently, it has been the practice to use combinations of lead and cobalt driers in order to achieve a uniform drying rate. Manganese and zirconium salts may be used in place of lead. Many organic acids have been used to form these heavy metal salts but the most widely used salts are napthenates which are based on naphthenic acid which is a residue of petroleum resinous. Other acids used for the production of driers are octoic, tallic, rosin and linolenic acids.

Drying of Paints

It has been recognized for many centuries, that film formation from oleoresinous paints, was dependent on atmospheric exposure of the unsaturated oils. It is now recognized that crosslinking between the polymer chains occur and that an insoluble polymer network is produced after the absorption of oxygen in this "drying" reaction.

The mechanisms in the drying of nonconjugated acids involves the formation of hydroperoxide groups on the allylic carbons, i.e., carbon atoms of the methylene groups adjacent to the double bonds. Polymerization occurs via a radical chain mechanism, called autoxidation. This theory also applies to conjugated acids, such as linolenates, but the "drying reaction" is much faster with these polyunsaturated oils.

Cellulose Nitrate (Pyroxylene)

Solutions of shellac in ethanol and sumac sap were used as coatings centuries ago but the first lacquer from a manmade polymer was not available until the middle of the 19th Century. Braconnet nitrated starch in 1833 and Pelouze nitrated cellulose 5 years later. The names of xyloidine and pyroxylene were used by the inventors to describe their products.

In 1846, Schönbein improved the nitration process by using a mixture of nitric and sulfuric acids and patented this cellulose trinitrate, which he called guncotton. Schönbein observed that his guncotton was soluble in a 50:50 mixtures of ethanol and ethyl ether but credit is usually given to Maynard, who announced the availability of collodion as a waterproof coating for wounds in 1847. In 1882, Stephens used amyl acetate as a solvent for cellulose nitrate and Wilson and Green patented pyroxylene coatings for carriages and automobiles in 1884.

The availability of hugh stockpiles of surplus guncotton, after World War I, catalyzed peacetime uses for this polymer. In 1923, cellulose nitrate (CN) lacquers, erroneously called nitrocellulose lacquers, were used as automobile finishes, under the trade name of Duco. After the introduction of competitive resins for automotive coatings, catalyzed lacquer and a multicolored CN were developed in the 1950's. "Super lacquers based on cellulose nitrate-isocyanate resins were introduced in the 1960's when CN continued to hold second place, next to alkyds, among the industrial finishes. Alkyd-CN blends and CN-blends with copolymers of vinyl chloride and vinyl acetate are also used as commercial coatings. Additional information of the history of cellulosic coatings is provided in subsequent chapters by Dr. Fisher.

Alkyds

The first polyester resin was produced by Berzelius by the condensation of glyceryl tartrate in 1847. W. Smith made Glyptal coatings in 1901 by the controlled esterification of glycerol by phthalic anhydride. This technology was improved by Friedburg a few years later and by Kienle in 1921.

Kienle recognized that the condensation of difunctional

reactants produced linear polymers and that trifunctional reactants produced crosslinked, infusible network polymers. Accordingly, he reacted ethylene glycol and phthalic anhydride, in the presence of drying oils and obtained linear polymers, which could undergo autoxidation polymerization, like that described for oleoresinous paints. He used parts of the words of the reactants alcohol and acid to coin the word alkyd. A profile on Kienle is provided in a subsequent chapter by Dr. Kauffman and an entire subsequent chapter is also devoted to the history of alkyds.

Kienle also classified his alkyd resins as short, medium, and long oil on the basis of their having an oil content of less than 40, less than 50 and over 50 percent oil, respectively. The Kienle patent was declared invalid by the U.S. Patent Office in 1935 but the name alkyd is still used to describe the major coating resin used worldwide.

Phenolic Resins

Phenolic resins, which were noncrystalline and lacked a precise melting point, were produced and discarded as undesirable "goos and gunks" by the leading organic chemists of the late 19th Century. As a result, many of the first and second generation students of these professors, avoided these and other polymers and concentrated their efforts on crystallizable and distillable compounds.

In 1872, Bayer condensed phenol with methylal and, at the suggestion of Fischer, Kleeberg repeated this experiment, using formaldehyde, instead of methylal in 1891. Smith patented phenolic resins for use as electrical insulators in 1899 and Swinburne produced these products commercially in 1904.

Baekeland, a visitor from Belgium, who decided to remain in the USA, was aware of the importance of functionality and of the mistakes made by his predecessors in their investigations of phenolic resins. Fortunately, he had been awarded enough money by Eastman for his Velox photographic paper patent to enable him to establish his own laboratory and to chose his research projects.

In 1907, he produced an ethanol-soluble novolac, which was used as a substitute for shellac, by the condensation of an excess of phenol with formaldehyde, under acid conditions. In 1908, he produced resole prepolymers by the condensation of phenol and formaldehyde, under mild alkaline conditions. These prepolymers called A-stage, could be converted into infusible C-stage insoluble castings or coatings, in the presence of strong acids. Nevertheless, neither the novolac or resole resins could be used as substitutes for natural resins in oleoresinous varnishes.

However, Albert heated the phenolic resin (PF) with a large excess of rosin and produced a product which could be used with tung oil to produce "four hour enamels." In

1928, Turkington patented oil soluble varnishes which were produced by the alkaline condensation of formaldehyde with para-substituted phenols, such as p-phenylphenol. Additional information on the history of phenolic coatings is provided in a subsequent chapter.

Amino Coatings

Tollens described resins produced by the condensation of urea and formaldehyde in 1884 and John patented this polymer. (UF) in 1918. In 1936, Henkel patented comparable resins (MF) based on melamine instead of urea. Both resins, which were described by the term amino resins, are insoluble in common solvents and are used as additives for curing other coatings. For example, the addition of MF reduces the curing time of alkyd resins by 50%. Soluble amino resins, which can be used as coatings, are produced when they are modified by etherification with butanol. Additional information on amino resins is provided in a subsequent chapter by Dr. Albrecht.

Vinyl Chloride Polymers and Related Resins

Regnault described polyvinyl chloride (PVC) in 1835, but since no solvents were available for this polymer, no PVC coatings were produced. In 1920, Reid of Union Carbide, and Voss and Dickhauser of I. G. Farbenindustrie filed for patents on a more soluble copolymer of vinyl chloride and vinyl acetate. This copolymer, which was produced, under the trade name of Vinylite, was used for coating beer cans in 1936.

The versatility of this copolymer, copolymers of vinyl chloride and vinylidene chloride and PVC was increased, in 1944, by the development of plastisols by suspending a resin, produced by emulsion polymerization, in a liquid plasticizer and then fusing the mixture at elevated temperatures.

While polyvinyl alcohol (PVAL), obtained by the hydrolysis of polyvinyl acetate (PVAC) is insoluble in organic solvents, the acetal, produced by the condensation of butyraldehyde and PVAL is soluble and is used as a base for "wash Primers" or metal conditioners.

Polyvinyl acetate (PVAC) was patented by Klatte and Rollet in 1914. PVAC is used as an adhesive and the major resinous component of a widely used water borne coating. Additional information on the history of vinyl coatings is provided in a subsequent chapter.

Acrylic Esters

While acrylic acid was polymerized in 1847, its esters remained as laboratory curiosities until the 1900's when Rohm wrote his Ph.D dissertation on acrylic esters. He continued the investigation of these products for several decades. In 1956, General Motors Company replaced some of its cellulosic automotive finishes by acrylic coatings. Some of these were thermoset by the incorporation of melamine resins or by the introduction of additional functionality in the acrylic monomers. Additional information on acrylic latex coatings is provided in a subsequent chapter by Dr. Harren.

Silicones

Polysiloxane was synthesized by Kipping, in the early 1900's. Since he believed that these polymers were ketones, he called them silicones. While he was pessimistic about their commercial use, these water and temperature-resistant coatings were commercialized in the 1930's by Rochow of General Electric Company, McGregor of Union Carbide and Bass of

Dow Corning. Additional information on silicone coatings is provided in a subsequent chapter by Drs. Finzel and Plueddemann.

Epoxy Resins

Ethyoxyline resins, which are now called epoxy resins, were patented by Schlack in 1939. Many of the prepolymers were versatile resins produced by the condensation of a diphenol (bisphenol A) and epichlorohydrin. Since these resins possess terminal oxirane (epoxy) groups, they can be crosslinked by reaction with polyamines at ordinary temperatures. These resins may be also be crosslinked by esterification of the pendant hydroxyl groups by cyclic anhydrides at elevated temperatures. Additional information on epoxy resin coatings is provided in a subsequent chapter by Dr. Reddy.

Polyurethanes

The original resins, which Bayer produced in 1937, by the condensation of aliphatic diisocyanates and diols were elastomers and foams. However, both one and two component polyurethane coatings are available. The two component systems are produced by the addition of aromatic or aliphatic diisocyanates to an ester or ether with terminal hydroxy groups.

The room temperature-one component systems contain unreacted terminal isocyanate groups, which react with moisture in the atmosphere to produce an amine which undergoes an additional crosslinking reaction. Baked polyurethane coatings are produced by heating phenol-capped isocyanates to remove the phenol and permit the isocyanate group to react with diols present in the coating system. Additional information on the history of polyurethane coatings provided in a subsequent chapter by Dr. Edwards.

Other Coatings and Profiles

As described in a subsequent chapter by Dr. Skolnik, rosin and other naval stores have been used for many years in coating. Actually, turpentine was one of the first non-alcoholic solvents used by paint artisans.

As outlined in a subsequent chapter by Dr. Deanin, the advancement of the science of coatings has been impeded by a lack of adequate educational facilities. Hopefully, his report will encourage other universities to provide courses in coatings science.

In addition to the profile by Dr. Kienle, provided by Dr. Kauffman, Drs. Fisher and Mattson have provided profiles on Drs. Doolittle, Mattiello and Long. Dr. Tarbell, who was the recipient of the Dexter award at the Miami Beach meeting, has published many joint reports on history of chemistry, with his wife. The profile on Dr. Long is also written by a husband and wife team.

The chapter on History of Plasticizers is written by Dr. Craver, who could be called "the father of plasticizers." He, like the Tarbells and Mattsons is also a member of an energetic husband and wife team. The chapter on the history of block copolymers is provided by Dr. Hsieh who is also a pioneer in this field. Likewise, the chapter on the history of polyvinylidene chloride coatings is provided by Dr. Gaylord who is another well known polymer science pioneer.

The chapter on membrane coatings was provided by Dr. Fritsch who is also recognized as a pioneer in his field. Other chapters on the history of polystyrene coatings, power coatings, diagnostic coatings, container coatings, automotive coatings, photo-cured coatings and radiation-cured coatings have been provided by Drs. Stevens, Usmani, Robinson, and Wu, respectively.

Hopefully, these reports will prove the inaccuracy of the phrase, "ignorance in action" applied to coatings science by IUPAC.

Application of Coatings

Coatings were originally applied by hand but this technique was upgraded by the use of feathers and brushes. Brushing continues to be the most widely used technique for the application of paint. Rollers were used to apply paint in the 1940's and this technique is still used. Spray guns were developed in the 1920's and paints continued to be applied by high pressure spraying.

The spraying technique has been upgraded by hot spray application and by airless spray in which atomization occurs as a result of a pressure drop as the paint leaves the nozzle of the gun. More uniform thickness of coatings and higher efficiency are obtained when the paint is electrically charged or when the atomized spray passes through an electric field before it is deposited on the surface to be coated.

In the early 1970's, Brewer developed techniques for the electrodeposition of water borne coatings. In this technique, the resin particles carry positive or negative charges that are deposited by anionic or cathodic techniques. Coatings may also be deposited as powder which is then fused by heating and the deposited coatings may be cured by electron beam (EB) or ultraviolet radiation in the presence of a sensitizer, such as benzophenone. Information on powder coatings, electrodeposition, photocuring and electron beam curing is supplied in subsequent chapters by Drs. Gribble, Usmani, Hoyle and Kinstle, respectively.

Future Trends

The art of coatings, which is one of man's oldest arts, is gradually becoming transformed to a science. In spite of the application of many competitive materials, coatings continue to be used at an unprecedented rate for both decoration and protection. Over 1 billion (10^9) gallons of coatings are produced annually in the US at a selling price of $10 billion ($10^9$). This volume will continue to increase and may increase dramatically as new resins and new application techniques are developed.

WORLD WIDE HISTORY OF PAINT

Joseph H. Boatwright

To reproduce the history of paint from its inception to the present day in one or more chapters of a book is much like reducing a football field to a ping-pong table! Consequently this chapter will introduce only the high points and leave the details to other publications. References will be omitted, since they would consume almost as much room as the chapter!

The earliest paints we have been able to find are those in Europe and Australia. Those in Europe were produced by the Neanderthal man and/or the Cro-Magnon man; those in Australia were produced by the same Aborigines that live there today. Both types of paints date back, as nearly as we can tell, to approximately 20,000 years B.C. While many of the drawings (since that is essentially what they are) are monochromatic, others represent an ingenious palette of colors made from natural earthen materials. Many of the paints were applied with fingertips (more smudges than painting), while others were applied with primitive but useful brushes. These brushes were made by the simple expedient of chewing on the tips of fairly soft twigs until the ends frayed, and became rudimentary bristles. They were then dipped in the paint and brushed where desired. The early paints used many natural pigments which we still use -- iron oxide red and yellows, chalk, charcoal, terra verde, and whatever else was available. Binders again were rudimentary, but quite useful: animal fats, blood, egg whites and yolks, etc. While these prehistoric people are given credit for inventing paint, those in Europe can be considered to be most ingenious people: their work was deep in the caves - a half mile more inside - so they had to devise a system of illumination so that they could see to work. At the same time their paints were often applied to the ceilings of caves, so that they had to devise scaffolding of one kind or another to enable them to climb up to paint. For an apelike man (?) with no intelligence (?), this is nothing short of phenomenal!

While these caves have been sealed off, in many cases, for thousands of years, it is still remarkable that the paint has lasted for approximately 20,000 years. Those in Australia have much the same background, and the Aborigines must have had the same sort of ingenious approach when it came to ex-expressing their thoughts.

Discoveries in the Libyan Desert indicate that paints of this type were adopted by the Egyptians. From this they are said to have developed their system of hieroglyphics, and this led to the development of an alphabet by the Phoenicians. It is said that this type of drawing and the same techniques are used today by various tribes in the center of Africa.

As we move to the North American continent, we find that a race of people living west of the Pecos River in Texas about 9000 B.C. also learned how to make a primitive paint. These people, who were Caucasian and not American Indian, used their primitive paints much in the same manner as their European and Australian predecessors. They paints the rock walls of their living quarters with pictures of animals, people, etc. Their paints also were made of natural pigments, and whatever binders they could procure. It is intriguing that this information has just come to light, and we find that the American Indian had predecessors on this continent!

Perhaps the next step in the history of paint would be in the Middle East in Biblical times. we find that Noah was admonished to "pitch the Ark within and without with pitch" (Gen. 6:14). The word "pitch" does not have the meaning to us now as it did in those days; the pitch they used was nothing more than a liquid asphalt which seeped out of the ground. If you visit the Middle East now, or close to home, the island of Trinidad, you will find

Published 1990 by Elsevier Science Publishing Co., Inc.
Organic Coatings: Their Origin and Development
R.B. Seymour and H.F. Mark, Editors

these areas in which asphalt still seeps out of the ground. Under some conditions we would not think of asphalt as a paint, but we have to consider it as a coating, which brings it into the context of our history. Also, when one remembers that there are thousands of gallons of asphalt roofing coating sold in this country. Also, when I had a paint factory many years ago, one of our biggest movers was an asphalt aluminum paint - it still fits!

Several thousand years after Noah, we find that the Egyptians were beginning to make and use paint. They too used many natural pigments and materials in their paints and were the first to develop synthetic pigments. What was known as Egyptian Blue was composed of lime, alumina, silica, soda ash, and copper oxides. It was made, according to Vitruvius, by calcining a mixture of sand, soda and copper. Their natural colors were red and yellow ochres, cinnabar, hematite, orpiment, burnt yellow marl, gold leaf, malachite, azurite, charcoal, lampblack, and gypsum. The Egyptians also developed an organic pigment which was a madder lake upona gypsum base. They employed a variety of organic and inorganic materials as their binders: gum arabic, egg whites and yolks, gelatin, beeswax, etc. Lime plaster and plaster of Paris were also used. Their ships were coated with both asphalt and balsam exudations. Even though drying oils were known, there is no evidence that they knew how to use them as coatings. As the Egyptians were one of the earliest of civilizations, they progressed further into the manufacture and use of paints than did any of their predecessors.

The Egyptians made coffins of wood, linen and papyrus by using gypsum and glue as the adhesive. They were then decorated with lime paints. A coating of varnish similar to Venice turpentine of Canada balsam was used at one time as the final coating. The stones of temples were highly colored by rubbing pigments into the stones much in the same manner as stoves were "blackened" in the earlier part of this century.

Before I am challenged about my chronology, let me say that while I have placed Noah before the Egyptians, it just as well may have been the other way around! Our ability to establish specific dates for the use of any type of paint by any people is tenuous to say the least -- we do the best we can!

Concurrently with the Egyptians and/or Noah, the Japanese and Chinese were beginning to develop their famous lacquers. The basis for these lacquers is a latex extracted from the so-called "Varnish Tree", which is known in Japan as the "Urushi" tree. This latex is incredible, and, to my knowledge, has never been synthesized. Strangely though, the tree belongs to the same family as the poison ivy plant, and like it, all parts of the Urushi tree are toxic - tree, sap and latex. Those who tap the tree must wear gloves and protective clothing. This latex is not really the sap of the tree, but is a healing compound for wounds to the tree. Contrary to mose paints and coatings, the lacquer made from this latex will not cure in a dry atmosphere; it must be very humid for any curing to take place. The Japanese (and Chinese) have learned how to make fantastic lacquers with this latex, and have produced coatings that will last thousands of years under the most difficult of conditions. the story is told about a passenger liner plying the Pacific that had its lounge doors decorated with this lacquer. After many years of traveling through the tropics in all manner of weather, all the paint, interior and exterior, wore off the ship. The two doors into the lounge, however, were unscathed, and bore no degradation whatever!

While this may seem out of place in a history of paint, it is related to coatings, and may have more application in the future than it has had in the past. This is a natural fungicide contained by the Japanese Hinoki tree. The natural fungicide, which has been extracted, is known as "Hinokitol"; it is better known in this country as "Thujaplicin". The thujaplicin and its related compound, thujic acid, are currently being tested in this country by the Chicago Society for Coatings Technology. It has been, and may still be, in the process of being tested in Jamaica by some paint firms. For those chemists who are interested, thujaplicin belongs to the family of Tropolones, which are rings analagous to benzene, except that the ring contains seven carbons. Back to the hinokitol and the hinoki tree, timbers from this tree have been used in Japanese temples, and have lasted, again, thousands of years with no treatment, coating of any kind, or any other assistance.

The American Indian developed paint also, but just exactly when is unknown. Possibly he fits in with the time of the Greeks and Romans in their early uses of colors to make paints. The Indian, as did the other early peoples, used mixtures of various natural oxides, charcoal, chalk, etc. to make their paints. He also used animal fats, blood, egg white and yolks, etc. as his binders. The Indian did not decorate caves (at least none have been found), but he did decorate his face with "war paint". He also decorated the exterior of his teepees. When the first settlers came to this country, they found the Indians still using the same paints.

From the birth of Christ on, we find that the early artists were becoming adept at making and using paints, particularly the Greeks and Romans. Early Cretan and Etruscan were done in fresco, with glue and albumin as binders. While frescos abound, no Roman or Greek paints on panels or canvass have been found. Pliny the Elder describes the materials and techniques used in painting. Besides the pigments common to the Egyptians, the Romans made several artificial colors: white lead, litharge, red lead, yellow lead oxide, verdigris, and bone black. Lake pigments were made using woods, kermes, and madder. These were precipitated upon a white earth, or were mixed with honey and chalk for painting. Pitch was used during this period for sealing the timbers on ships, and a mixture of wax and pitch was used for ship bottoms. Resins and oils were used only for liniments, and there is no record of any varnishes being made, except, of course, the bituminous coatings made with pitch.

The Orientals developed their own techniques as time passed. The art of suspending pigments in water, with or without a binder, was common. The Persians used gum arabic as a binder, while the Chinese used a glue. In India crayons were made from boiled rice, and colors were applied by brushes and a crude stylus. East Indians used cinnabar, lac dye, red ochre, lampblack and lime for pigments. Lime was blended with the above colors to produce pastels of one kind or another. The Chinese and Japanese used as colors cinnabar, azurite, copper carbonate, malachite, ultramarine, indigo lake, orpiment, lime, red lead, litharge, lampblack, black ochre, gold powder, and brown coral powder. Gold powder was used only in drawings of Buddha, since yellow was the sacred color of the Orient.

The East Indians used shellac and sealing wax, incorporating the pigments in lac sticks and applying them to pieces of wood in a lathe. As the lathe turned, the friction created sufficient heat to melt the shellac, and thereby apply a coating to the wood. It is interesting to note that the art of applying paint by brush was known at that early date, but the manufacturing methods of these brushes is unknown. The Japanese were great craftsmen, and in their efforts to produce works in the greatest detail, some brushes were used containing only one bristle!

The Mayan civilization of Mexico and Central America relied upon natural organic pigments for their colors. These are very interesting, since they do not appear to be duplicated in other parts of the world. Whites were made by mixing lime with the juice of the chichebe plant. Natural earths (iron oxides) were used for browns and reds. The finest reds were made from splinters of heartwood, and blues were made from plants containing compounds of aniline. They also had an inorganic blue called Mayan Blue. Yellows were made from the fruit of the achiote, or from chips of the fustics. Brushes were made from either bird plumes or hair. Many of the Mayan paints had good durability.

As we advance into Medieval Europe, we find that the art of paint, as well as the art of paint manufacture was progressing. There is much literature on the subject of making various types of coatings, and the ingredients used. There is some problem for the unitiated, however, in trying to decipher just what the author meant when he discussed certain raw materials. The names were common in those days, but are, to a great extent, unknown in the present. However, it is known that the advent of oil as an ingredient in the manufacture of varnish occurred sometime around the sixth century. Still the preferred medium for the manufacture of paint was egg albumin.

Greater interest was shown in oils about the time of the Renaissance. During this period of time, the artist was not only the painter, but he also was the paint manufacturer. Since the artist required only a small amount of paint, his batches were only as large as he would need for a particular project. An ounce of paint was perhaps a normal batch, while a really large mural could consume a six ounce batch! Their manufacturing equipment, while crude, was still being used as lab equipment up to the period of World War I. Apothecaries used the same mortar and pestle even into the early part of this century.

There are treatises written during the fifteenth century (some are preserved in the Vatican) describing the running of rosin and sandarac, and the preparation of sandarac and mastic varnishes containing linseed oil. These varnishes were often used to coat armor, cross-bows, and other weapons. A Flemish artist of the 15th century was able to produce paintings with unusual durability, but no one is certain jsut how he accomplished this feat. Speculation has it that he painted with a tempera of an egg-yolk emulsion, and then coated the whole with a glaze made with an oil containing varnish. Many of the early artists were able to manufacture coatings of remarkable qualities. One of the most noted with Antonio Stradivari who coated his magnificient violins with a finish that is unknown to this day. It is fascinating to think that this craftsman created violins that are worth fortunes today -- approximately $250,000 each!

As time passed, more varnishes were made from a variety of resins, at least one oil (linseed) and possibly others. And soon thinners were used to allow the painter to utilize them at ambient temperatures. Again, the largest batch was six ounces! Around the middle of the 17th century, it became apparent that driers had a place in paint. Such materials had been suggested as early as the 2nd century, but they were not accepted until a millenium later. Bleaching of oils was known, but little practiced. Oils were usually purified by boiling with water.

Until the late 1600's very few recipes for paint or varnish were ever recorded. A few were written for various grinding media, but by and large, everything was kept secret by the families which made and used the paints. Some of the old varnish formulae, which had been publsihed, were intriguing, since they listed materials that are seldom ever heard of today. A good example is "dragon blood"; most people do not know what it is, let alone how to use it. Strangely, it can still be bought, if one just knows where to look!

Jumping to the early days of this country, the first settlers lived in log cabins, which precluded the use of paint. As their building techniques improved, and industrialization was beginning, they learned to saw logs into boards, and they built what we might consider primitive, conventional homes. Still, they were never painted! Slowly but surely tastes changed, and people began to decorate their homes, first on the inside and then on the outside.

At that time protection of the surface had not been considered, but appearances ruled. To paint a house, each painter bought his own raw materials (there were no stores that sold finished paint) and made his own paint according to his own formula.

Watin, in 1773, was the first to detail the technical preparation of paints and varnishes. Copals and ambers were the resins of choice for varnishes at the time of the Revolution. Slowly but surely varnish factories began to operate throughout Europe, and this may be termed the real beginning of the paint industry. About this same time, paint manufacturing itself progressed from the mortar and pestle type of operation to a rough stone trough with a ball muller enabling one to make much larger batches of paint with ease. The Boston Stone, which has been preserved by the Federation of Societies for Coatings Technology is a prime example of this device.

The 20th century was the beginning of real advances in paint technology, and from this point on, improvements in both machinery and technology were made at a geometric pace. The paint industry has advanced more in the past fifty years than it has in the previous 20,000 years! What will the next century bring?

For those who are interested in pursuing this subject in greater detail, I refer you to the book "World History of Painting Techniques" being edited by Jacques Roire of the AFTPV (Association Francaise des Techniciens des Peintures, Vernis, Encres d'Imprimerie, Colles et Adhesifs. The book is to be published in the near future, and notices will appear in various periodicals announcing its availability.

THE HISTORY OF NAVAL STORES IN COATINGS

HERMAN SKOLNIK
239 Waverly Rd., Wilmington, DE 19803

ABSTRACT
Oleoresins, their turpentine and rosin components, and
their fossilized resins, e.g., amber and copal, have played an
important role in the evolving history of coatings from the
Paleolithic period to the present. Over most of this history,
coatings was an art and used primarily for decorative purposes.
Coatings emerged slowly as a technology and industry in England
in the 18th century and in Europe and the U.S. in the early
19th century. With the blossoming of chemistry as a science in
the 19th century, constituents of oleoresins were separated
and identified, and in the 20th century, based on extensive
research and development, converted to a stable of products
that made possible the coatings industry of today.

INTRODUCTION

Varnish is defined as composition of resinous matter
in a volatile solvent. Conifers have been manufacturing varnish
compositions for the past 200 million years. Widely distributed
over the earth, there are over 500 species of conifers which
on being chipped (cut into) exude oleoresin (also called gum)
consisting chiefly of a mixture of resin acids and turpentine.
Oleoresin as a product of and from conifers is rooted in antiquity. Through their powers of observing, thinking, and then
innovating, our early ancestors in the Stone Ages fashioned
tools from stones and wood; mastered the making of fire, which
led to cooking then to pottery, metallurgy, and mining; domes-
ticated animals; and initiated civilization with agriculture.

EARLY PROCESS FOR OBTAINING OLEORESIN AND ITS PRODUCTS

On burning conifer wood for cooking and for making pottery, they observed oleoresin flowing out of the wood. Using
a process similar to the making of charcoal, they built a stone
furnace fitted with a conduit for the tar and pitch to flow out,
being careful to avoid flames in the heated wood. The oil (turpentine) was recovered by placing flocks of loose wool or
fleecy sheepskin above the heated wood to absorb the vapors.
On wringing out the sheepskin they obtained the oil (cedar oil,
a turpentine) called cedruim in the Near East.
They learned to obtain oleoresin from conifers by making
an incision with a stone axe in the living tree. By cutting
deeply into the tree to form a box into which the oleoresin
flowed, they were able to gather and transfer the exudate into
clay pots. The exudate in the open clay pot was cooked until a
thick pitch remained in the bottom. The oily vapors that arose
from the boiling exudate was absorbed in sheepskin placed over
the pot which was wrung out to recover the oil. Thus by the
innovation of a crude distillation process, our early ancestors
were able to separate oleoresin into three products: turpentine,
tar, and pitch.

UTILIZATION OF OLEORESIN PRODUCTS

In the eastern part of the Mediterranean where cedars
abounded, Phoenicians were able to build seaworthy boats with
cedar planks and using the tar and pitch from cedars to cauld
the boats. This occurred long before Noah was instructed to
caulk the ark "within and without with pitch." The Phoenician's

Published 1990 by Elsevier Science Publishing Co., Inc.
Organic Coatings: Their Origin and Development
R.B. Seymour and H.F. Mark, Editors

caulking art was diffused to Egypt where pitches and balsams
were used as a protective coating for their boats. Egypt obtained cedars of Lebanon timber and tar, pitch, and cedar oil
from Phoenicia for making and caulking their boats and for
using cedar oil to embalm their dead. By about 2500 B.C.,
Egyptians had developed varnishes and paints based on cedar
oil and applied them to buildings, sculpture, and coffins for
their mummies.

The Persian name for cedar oil was Terpentin or Turmentin
which became Terebinthine in Greece, Terebentine in France,
Turpentyne in England, and Turpentine in the U.S.

In ancient Egypt, amber dissolved in cedar oil was used
as a varnish. They named this fossilized resin Bernice which
then underwent a series of changes to Verenice, Vernix, and
in the 12th century Medieval Latin to Varnish. Thales in Greece
observed the attractive property of amber when rubbed and
named it Electron (the property he observed was static elec-
tricity). That amber is a fossilized resin from an extinct
species of conifer was first made known in 1767 by Friedrick
Samuel Bock in his book "Attempt at a Natural History of
Prussian Amber."

The production of tar, pitch, and oil in Greece from the
Aleppo pine also predates biblical times. They called the
crude exudate Colophony for its geographical source, Colophon
in Asia Minor. Colophony in the literature of terpenoid has
been used as a synonym for rosin.

THE FIRST LACQUER

Some time before 200 B.C., Chinese used the exudation
from the conifer Rhus Vernicifera (which became known as the
Varnish Tree) as a coating. The product became known as Japanese lacquer by 400 A.D. when the technology was advanced in
Japan. Japanese lacquers are a type of oleoresin which dries
by oxidation in a damp atmosphere. The word lacquer was derived from the French word Lacre for·resin, which was derived
from the Latin word Lac for milk.

The Chinese added carbon black to the lacquer and first
used it in the 2nd century B.C. to write on bamboo strips.
They also applied it as a paint for pottery and by the 2nd
century A.D. to buildings and musical instruments. They obtained carbon black by carbonizing the resin from the tree by
a method similar to that used in the Mediterranean region.
This method for producing carbon black was discovered in many
regions of the earth. During the Mayan civilization of Mexico,
black pigment was made by carbonizing resin from the chacak
tree. Carbon from cedar resin in the Mediterranean region was
added to cedar oil to make ink for writing on papyrus.

These notable achievements of antiquity in developing
processes for obtaining oleoresinous products from conifers
and utilizing them for making varnishes, lacquers, paints,
and inks remained the state of arts for many centuries. Except
for the eclectic writings of Theophratus (372-288 B.C.), Pliny
the Elder (23-79 A.D.), and Dioscorides (40-90 A.D.) the
literature of these processes suffered historical amnesia
well into the Renaissance.

THE MIDDLE AGES

An oil-varnish, prepared by dissolving molten resin in
hot linseed oil was introduced in 11th century Europe. During
the Renaissance, artists developed their own paints by using
different pigments with rosin and sandarac in linseed oil.

Leonardo da Vinci (1452-1519) used resin-coated pigments and Rembrandt (1606-69) used oleoresinous vehicles mixed with amber varnish. During the Renaissance, Venice turpentine (chiefly pinene) was to be found in wares of artists. In the 13th century, turpentine was produced in France from the oleoresin of their conifers for use as the vehicle for paints and varnishes that were being increasingly employed as a protective coating for buildings in Europe.

NAVAL STORES IN THE NEW WORLD
     From antiquity to well into the 19th century, the primary use of oleoresins from the 500 species of the five families of conifers was the tar and pitch for caulking wooden ships. The centers of production of tar and pitch were the Scandinavian and Baltic countries. England's need for these products for its navy and merchant ships prompted the establishment of a settlement at Jamestown, Virginia, in 1607. In 1608 the colony was producing tar and pitch for England. The term Naval Stores was introduced in England for these products from its first American colony for caulking their sailing vessels, waterproofing tarpaulins, and treating rope. English sailors who worked with the tar and pitch became known as Jack Tars. North Carolina, which became the primary naval stores producer in 1665, was called the tarheel colony. Naval stores production of crude gum and the tar-burning process for producing tar and pitch was the first American industry. Up to the 20th century, the methods for obtaining oleoresin from the tree and for producing pitch, tar, spirits of turpentine, and rosin changed very little from those used in antiquity.

NAVAL STORES FOR COATINGS
     By mid-19th century, iron boats began to displace wooden boats. This prompted naval stores producers to separate turpentine from oleoresin to meet the growing demand for it as a solvent for paint and varnish. In the 20th century, the demand for rosin in paint and varnish was so great that its production reached 8.7 million pounds in 1900 and over 10 million pounds in 1910, the highest in the history of gum naval stores. Then wood rosin dominated the supply and since the 1970s tall oil became the major source for turpentine and rosin.

COATINGS INDUSTRY FORMATION
     Although coatings technology was first described by Watin in his 1773 book, which was reprinted 14 times up to 1900, the first varnish factory was established in England in 1790. Shortly after, factories were established in several European nations and in the U.S. It was not until the 20th century, however, that the technology left the art stage. Copal and amber continued to be the principal varnish resins and turpentine the thinner.
     A major advance in coatings technology was made in 1880 by Henry Alden Sherwin who with Edward Williams introduced the world's first ready-mixed paint. The Sherwin-Williams Co. was formed in Cleveland, Ohio, to market the product. Professional and amateur painters thereafter could abandon the onerous task of combining white lead, linseed oil, turpentine, and colored pigments.

CHEMISTRY OF COATINGS
     Until chemistry became a science in the 19th century, the history of terpenoids in coatings was not one of steady

transformations and improvements but rather a series of uneven
lurches and trade-offs. The composition of oleoresin was a
mystery except for the art of separating it into tar, pitch,
and turpentine. Practically nothing was known about terpenes,
resin acids, and their derivatives. Terpene chemistry could
not blossom before organic chemistry became a discipline of
chemistry. The first elementary analysis of turpentine was
made in 1818 showing an empirical formula of C5H8, which
became known as the isoprene unit. From 1852 to 1863, Marcellin
Berthelot (1827-1907) characterized camphene and pinene as
constituents of turpentine. The word Terpene was introduced by
Kekule in his 1866 textbook. Wallach began his extensive
studies on the characterization of terpenes in 1884. Until
high-plate distillation columns and sophisticated analytical
spectroscopy and chromatography were available terpene chemists
were confronted with difficulties in separating the closely
boiling terpenes that occur in turpentine and other essential
oils and resolving their constitutional problems as many
terpenes undergo structural rearrangement under relatively
mild conditions. Resin acid chemistry was even more difficult
and complex than that of the terpenes in turpentine. Successful resolution of these compositional, constitutional, and
structural problems occured in the 1940s.
     Before W.W.2, the knowledge of terpenes and resins in
protective coatings essentially was in the art stage and
coating formulas were closely guarded secrets. Rosin and other
natural resins, e.g., amber, copal, kauri, and congo, were
used in the manufacture of varnishes through the 18th and
19th centuries by cooking them plus a drier with linseed oil.

ROSIN ESTERS IN COATINGS
     Since Maly first esterified rosin in 1865, practically
every conceivable ester of rosin has been prepared, including
the glycerol ester, called ester gum, which was introduced in
1900 for lacquers and printing inks. The methyl ester of
hydrogenated rosin is used as a plasticizer in lacquers.
Pentaerythritol esters, unmodified and modified with maleic
anhydride and phenol-formaldehyde, are used in varnishes. In
1937, the coatings industry used 20 million pounds of phenolic
resins modified with ester gum; in 1987, 66 million pounds
were used. Phenolic resins modified with up to 80% ester gum
were called four-hour enamels. These enamels possess excellent
resistance to hot water and alkali. Ester gum replaced damar,
copal, and kauri gums in paints, lacquers, and enamels in
combination with tung oil.

PHENOLIC RESINS IN COATINGS
     After ester gum, the first synthetic resin of importance
to the coatings industry, phenolic resins were introduced in
the early 1900s as a substitute for shellac. Phenol-formaldehyde, which was first prepared in the 1850s, was introduced
in the 1920s for varnishes, paints, and lacquers by fusing it
with rosin to make it oil soluble.

TERPENOID ALKYD RESINS IN COATINGS
     A major contribution of naval stores products to the
coatings industry occurred in the 1920s and 1930s with the
production of alkyd resins by the reaction of polyhydric
alcohols, viz., glycerol and pentaerythritol with rosin-maleic
and terpene-maleic Diels-Alder condensation products. The
name alkyd was coined by Kienle in 1927.

Rosin-maleic glyceride yields a lighter colored and
non-yellowing varnish than the phenolic resin varnishes, and
is more compatible with nitrocellulose in lacquers. Reaction
of the rosin-maleic adduct with a dihydric alcohol followed
by heating with a monohydric alcohol ester of an unsaturated
resin yields an alkyd with a very low acid number for use
with vinyl polymers in coating applications. Although
terpene-maleic adducts are more expensive than phthalic anhydride, its glyceride alkyd has the advantage of greater
solubility in cheaper solvents and better compatibility with
superior adhesive and flowing properties. With the advent of
nitrocellulose lacquers, e.g., Duco, in the early 1920s for
finishing automobiles, then later for furniture, camphor and
ester gum were used as plasticizers allowing the lacquer to
be applied by dipping or spraying. Rosin-maleic adduct
esterified with a polyol yields a hard resin which has been
used with nitrocellulose lacquer for finishing furniture.

ROSIN SALTS IN COATINGS
    Rosin salts of polyvalent metals, e.g., Ca, Zn, Pb, and
Mn, have been used as driers for paint and varnish and in
printing inks. Six million pounds of rosin salts were produced
in 1980, mostly from tall oil rosin, The NH4 and Na resinates
have been used in emulsion paints.

EPILOGUE
    Turpentine has long been used as a thinner or solvent
for paints and varnishes. Its solvent and wetting properties
are superior to those of straight petroleum solvents. Nevertheless since W.W. 2 turpentine has been replaced steadily by
petroleum solvents and eliminated by the growing popularity
of water-based paints. Gum navel stores production has
decreased markedly since 1930 when wood naval stores took over.
Then in the 1950s, tall oil, a by-product of Kraft paper
manufacture became the dominant source for turpentine and
rosin. Since the mid-1970s, the total naval stores market in
coatings has been declining. Only ester gum, other synthetic
resins of rosin and terpenes, and a number of rosin and
terpene derivatives have experienced a relatively steady
market in coatings.
    The history of naval stores in coatings illustrates how
long a technology can exist and progress without the backing
of scientific knowledge, then how far it can advance in
partnership with chemical research. The history also
illustrates how well-established products have been replaced
by new products resulting from research and development.
    Both the history of naval stores and of coatings and
their interrelationship over the centuries make a fascinating
story. Browsing in these two related histories is like making
a pilgrimage in a time machine through many lands. I ended my
trip with great respect and hubris for those who made the
technology and science of naval stores and coatings.

SUPPLEMENTARY REFERENCES

1. Dioscorides, "Historia Naturalis", 77 A.D. (translated
    by C. Sprengel, ed., Leipzig, 1829-30. See ref. 13).
2. J. Drew, "Naval Stores, the Adaptable Resource".
    Chemistry 51, 17-19 (1978)
3. E.A. Gerry, Naval Stores Handbook, U.S. Dept. of Agr.,
    Forest Service (1935).
4. Kirk-Othmer "Encyclopedia of Chemical Technology"-sections
    on Terpenes and Terpenoids and Rosin and Rosin Derivatives,
    Wiley & Sons, New York, NY.
5. M. Kranzberg et al, "Technology in Western Civilization",
    Vol. I, Oxford Univ. Press, New York, NY (1967).
6. R.L. Maly, J. Prakt. Chem. 96; 145 (1865).
7. J.J. Mattiello, ed., "Protective and Decorative Coatings,
    Paints, Varnishes, Lacquers, and Inks", 5 Volumes,
    Wiley & Sons, New York, NY (1941-5).
8. Pliny, "Historia Naturalis" (77 A.D.) (translated by J.
    Bostock and H.T. Riley, Bohn's Classical Library,
    London (1856-93)-see ref. 13).
9. J.L. Simonsen et al, "The Terpenes", 5 volumes, Cambridge
    Univ. Press, New York, NY (1947-52).
10. H. Skolnik, "The Terpenes Around Us", Hercules Chemist,
    No. 36, 13-17 (1959).
11. H. Skolnik et al, "The Literature of Wood Naval Stores"
    in "Literature of Chemical Technology", 349-61, Advances
    in Chemistry Series, N. 78, Am. Chem. Soc., Washington,
    DC (1968).
12. H. Skolnik, "Production, Processing, and Utilization of
    Naval Stores", in Handbook of Processing and Utilization
    in Agriculture", Vol. 2: Part 2, 467-506, CRC Press, FL,
    (1983).
13. M. Stillman, "The Story of Alchemy and Early Chemistry",
    Dover Publications, New York, NY (1960).
14. Theophrastus, "Enquiry into Planets" (3rd century B.C.),
    (translated by A. Hart, London (1916)-see ref. 13).
15. D.F. Zinkel, "Chemicals from Trees", Chem. Tech. 5,
    235-41 (1975).

HISTORY OF CELLULOSIC COATINGS

CHARLES H. FISHER
Chemistry Department
Roanoke College, Salem, Virginia   24153

ABSTRACT

The primitive humans who made paints some 25,000 years
ago could not use cellulosic materials for the simple reason
suitable cellulosic materials were not available. Many
thousands of years passed before Schonbein made a suitable
cellulosic material, the nitrate, by a practical method in
1846; it was soon discovered cellulose nitrate is useful as
a coating, explosive, and plastic. In the latter half of
the 19th century and in the 20th century, many esters and
ethers of cellulose were prepared. Several of these continue to be used for various purposes, including protective
and decorative coatings.

HISTORY

Primitive humans, hunters and cave dwellers, are credited with making
paints about 25,000 years ago [1-3]. Cellulosic materials, however, were
not used in coatings until the middle of the 19th century for the simple
reason suitable forms and derivatives of cellulose were not available.
Cellulose itself is poorly suited for use in coating formulations because
it decomposes at temperatures (200-270°C) [4] far below its proposed
melting point of 450°C [5] and is insoluble in the organic solvents that
are commonly used in applying coatings.

Cellulose, however, is an excellent starting material for preparing
cellulose derivatives having suitable melting points and solubilities in
the common organic solvents. The attractive characteristics of cellulose
include its availability in large quantities at low cost. Cellulose, the
most abundant organic material on earth, comprises about one-third of all
biomass for an annual production of about 50 billion tons [6].

Cellulose is a linear macromolecular polysaccharide consisting of
anhydroglucose units joined together by beta acetal linkages. The number
of anhydroglucose units or degree of polymerization (D.P.) in naturally
occurring cellulose may be as high as 36,000. Cellulose occurs as an
approximately pure polymer in cotton, but it is mixed with lignin, hemi-
celluloses and minerals in wood [6,10].

Research to develop new or improved cellulose derivatives may be
divided into two major types:

(1) The cellulose fibers, usually wood pulp or cotton linters, are
converted into soluble derivatives. Solutions of these derivatives can be
used either to prepare coatings or, by extrusion, to prepare fibers and
films.

(2) The cellulose fibers are converted into cellulose derivatives
without loss of fibrous form [7-9].

According to Battista [11], cellulose and many other polymers can be
converted into polymer microcrystals. The industrial end uses being

Published 1990 by Elsevier Science Publishing Co., Inc.
Organic Coatings: Their Origin and Development
R.B. Seymour and H.F. Mark, Editors

22

evaluated for the microcrystals include water-based coatings for paper, prime coatings for glass and other materials, and various other coatings.

Some of the important advances in cellulose science and technology are listed below:

1832. Cellulose was nitrated by French naturalist, Henri Braconnot (1781-1855), professor and director of the botanical gardens at Nancy [1,12].

1838. Frenchman Anselme Payen (1795-1871) isolated cellulose from plants. His classic work on cellulose isolation and characterization was described in his articles entitled "Study of the Composition of the Natural Tissues of Plants And of Lignin" and "Concerning a Means for Isolating the Elemental Tissue of Wood" and published in 1838 in Comptes Rendu [13,14].

1838. Theophile-Jules Pelouze (1807-1867) prepared cellulose nitrate by nitrating paper [15].

1846. Christian Friedrich Schonbein (1799-1868), a professor of chemistry at the University of Basel in Switzerland, in his kitchen accidentally broke a flask containing nitric and sulfuric acids. Unable to find a mop, Schonbein wiped up the mess with his wife's cotton apron. Thus was formed cellulose nitrate [1,16]. He improved the nitration process by using a mixture of nitric and sulfuric acids. He also explored the potential of cellulose trinitrate ("gun cotton") as an explosive. In 1847, a factory making gun cotton exploded, killing twenty-one persons [17].

1848. Boston physicians, J. P. Maynard and S. L. Bigelow, introduced collodian (cellulose nitrate solution) for medical use, i.e., protective coatings for skin [18].

1851. Englishman Frederick Scott Archer (1813-1857) used cellulose nitrate to manufacture photographic film [17,18].

1854. Burgess in America subjected poplar chips to a one-stage pressure digestion with sodium hydroxide to get alkaline wood pulp [19].

1862. Alexander Parkes (1813-1890), a British inventor, displayed Parkesine (cellulose nitrate dissolved in wood naphtha) at the International Exhibition in London [15,17,18].

1863. The American firm Phelan and Collender offered a $10,000 prize to anyone who would assign it patent rights to a satisfactory process for manufacturing billiard balls without the use of solid ivory [18].

1865. Alexander Parks put plasticizers (cottonseed oil, castor oil, gums, stearine, and tar) into nitrocellulose to make "Parkesine," the forerunner of celluloid. Articles made from such mixtures were exhibited before the Royal Society of Arts. Parkes received a patent covering Parkesine [15,17,18].

1865. Cellulose acetate, the first organic ester of cellulose, was prepared by Frenchman Paul Schutzenberger (1829-1897) by heating cotton and acetic anhydride to about 180°C in a sealed tube until the cotton dissolved [20].

1867. American chemist Benjamin Chew Tilghman received a patent describing the preparation of sulfite wood pulp [19].

1869. John Wesley Hyatt (1837-1920), an American inventor, improved
Parkes' cellulose nitrate product and developed celluloid (cellulose
nitrate and camphor) as the first commercially successful plastic. Hyatt
was the only non-chemist to be honored with the Perkin Medal of the
American Section of the Society of Chemical Industry in 1914. Both
Celluloid and Parkes' Xylonite were widely used. Hyatt made many other
inventions and obtained more than 200 patents [15,17,21].

1879. Franchimont acetylated cotton at relatively low temperatures
with the aid of sulfuric acid catalyst [20].

1883. The English inventor, Joseph Wilson Swan (1828-1914), developed
a method for making cellulose nitrate fibers. These fibers, after being
charred, were used as filaments in early electric lamps. Swan was knighted
in 1904 [17,21].

1884. French chemist, Count Louis M. Chardonnet (1839-1924), who was
independently wealthy, produced fibers by forcing solutions of cellulose
nitrate through small openings and allowing the solvent to evaporate. In
1884 he obtained a patent on the process, as Swan had done previously in
England. Chardonnet hydrolyzed the nitrate fibers to cellulose fibers and
marketed these as "artificial silk." At the Paris Exposition of 1891,
"Chardonet silk" was a sensation [17,21].

1884. Wilson and Story coated cloth with cellulose nitrate in castor
oil and amyl acetate, thereby producing "imitation leather" [22].

1892. In England, Charles Frederick Cross (1855-1935) and Edward John
Bevan (1850-1921) dissolved cellulose in alkali and carbon disulfide to get
a thick solution called viscose. This viscous solution, by extrusion and
acidification, was converted into cellulose fibers or film (cellophane)
[1,21].

1894. Englishmen Charles Frederick Cross and Edward John Bevan
patented the first industrial process for making cellulose acetate. Industrial development was slow because (a) the cellulose acetate was soluble
only in toxic and expensive solvents such as chloroform, (b) it was difficult to dye, and (c) its production required expensive acetic anhydride
[1,15,18].

1903. Miles described partially hydrolyzed ("secondary") cellulose
acetate and distinguished it from the triester by its acetone solubility.
The solubility of secondary cellulose acetate in inexpensive solvents
contributed greatly to the development and commercialization of this
material [20,23].

1905. W. Suida made cellulose methyl ethers by treating cotton with
dimethyl sulfate and caustic soda [6].

1912. C. Dreyfus and Lilienfield investigated cellulose ethers [24].

1914-18. Airplane fabric wings were doped with cellulose nitrate
(pyroxylin) solutions. The cellulose nitrate was later replaced by less
flammable cellulose acetate [18].

1916. British Cellulose and Chemical Co. produced cellulose acetate
lacquers for aircraft wings to replace cellulose nitrate [24].

1919. Improved cellulose nitrate solutions, made possible by the use
of better solvents, caused an increase in pyroxylin lacquer production from
500,061 gallons in 1919 to 12,980,400 gallons in 1926 [18].

1923. The automotive industry adopted cellulose nitrate lacquers [12].

1928. The manufacture of partially acetylated cotton fibers (Cotopa 30) began in England. This was followed by the manufacture of fully acetylated cotton fibers (Cotopa 60) [9].

1935. First commercial production of ethyl cellulose [12].

1954. A general-purpose, low-viscosity coating grade cellulose acetate butyrate was introduced [20].

1956. Continuous nitration of cellulose [12].

1963. Innovations in cellulose nitrate lacquers had increased their acceptance so that they ranked second among all industrial finishes, trailing only alkyd finishes [12].

1969. A coating grade of cellulose acetate propionate was made available commercially [20].

1987. Commercially-available, ozonized cellulose acetate butyrate has advantages in several coating applications [25].

Wint [12] has provided the following chronology of water-soluble cellulose ethers (Table I):

TABLE I. Chronology of water-soluble cellulose ethers.[a]

Cellulose ether	Synthesis and isolation	U.S. commercialization
Methyl	1912	1939
Hydroxyethyl	1920	1937
Sodium carboxymethyl	1921	1943
Hydroxypropyl methyl	1927	1948
Hydroxypropyl	1960	1969

[a]Ref. 12.

CELLULOSE INORGANIC ESTERS

Cellulose has many hydroxyl groups (three per anhydroglucose unit) and hence can be used to prepare numerous derivatives, including esters and ethers [7-9]. Cellulose nitrate, first prepared by Frenchman Henri Braconnot in 1832, is today the most important and the only commercially available inorganic ester of cellulose [28]. Christian Schonbein was the first to make cellulose nitrate by the practical method of using a mixture of nitric and sulfuric acids [1]. Cellulose nitrate is soluble in many solvents, compatible with many plasticizers, and yields tough films [6,28]. Cellulose nitrate has the unique distinction of being the first synthetic—or semisynthetic—polymer to be used in plastics, explosives, and coatings.

In the early days of the airplane, fabric wings were coated with cellulose nitrate (pyroxylin) solutions. This application was very helpful to the development of aviation because this treatment waterproofed the fabric (which prevented fluctuations in weight due to absorption of

moisture) and tautened the fabric, thereby improving the aerodynamic characteristics of the wings [18]. During World War I, the Allies found it
necessary to replace cellulose nitrate coatings with less flammable
coatings for their military planes. Cellulose acetate coatings were
selected to replace cellulose nitrate [18].

In the early days of the automobile, nearly all automotive original
and refinish paints were based on cellulose nitrate. As an original paint
on the production line, cellulose nitrate was early replaced by acrylic and
alkyd systems. As a refinish paint, cellulose nitrate has persisted to the
present, albeit in diminished quantities [2].

The largest industrial use of cellulose nitrate is in lacquer coatings
for decorative and protective purposes. Automobile and wood furniture
coatings consume the greatest amount of the cellulose nitrate lacquers. A
substantial amount of the nitrate esters is used in coatings for leather
finishes and fabrics, and in nail polishes [28].

Cellulose sulfate and acyl sulfate have found limited use as
antistatic coatings for photographic films. Cellulose phosphate esters are
flame resistant and are of interest in the manufacture of textiles and
paper [28].

CELLULOSE ORGANIC ESTERS

Many organic esters have been prepared by acylating one or more of the
cellulose hydroxyl groups (three per anhydroglucose unit). The first
organic ester, the triacetate, was prepared in 1865 by Schutzenberger by
treating cellulose with acetic acid, acetic anhydride and sulfuric acid
[6]. The acetate began to replace flammable cellulose nitrate as a coating
for airplane wings and fuselage fabrics during World War I [28].

Cellulose acetate is today the most important organic ester of cellulose; it has broad applications, primarily in plastics and fibers.

Cellulose acetate is prepared with varying degrees of substitution,
ranging from the water-soluble monoacetate to the hydrophobic triacetate
[28]. The cellulose acetates most suitable for coatings have acetyl
contents ranging from about 38-40% and a hydroxyl range between 3.0 and
4.0% [20]. Cellulose acetate propionate has many applications, including
film products and lacquer formulations. Cellulose acetate butyrate is used
extensively in film-forming applications such as lacquers. Acetate
butyrate esters provide excellent barrier coating properties for molded
polystyrene and many other plastics. Cellulose acetate butyrates have been
formulated with acrylic polymers to provide weather-resistant automobile
coatings that have good color fastness, excellent pigment control, and good
flow properties [28].

Some of the properties of cellulose esters are given in Table II.

The following cellulose esters have been prepared but without
attaining commercial importance: Formate, propionate, butyrate, benzoate,
methacrylate, crotonate, acetate phthalate and most of the n-alkanoates up
to and including the 16-carbon palmitate [28,29,30].

The lower esters of cellulose (acetate to valerate) can be made by the
conventional method, i.e., using acid anhydrides with sulfuric acid
catalyst. The higher n-alkanoates, however, cannot normally be prepared by
this method because of excessive degradation of the cellulose and low
reactivities of the higher acid anhydrides. Acid chlorides, instead of

TABLE II. Properties of typical cellulose esters.[a]

	Nitrate	Acetate	Acetate propionate	Acetate butyrate
Melting temp., °C		230	190	140
Specific gravity	1.35–1.40	1.22–1.34	117–124	1.15–1.22
Tensile strength, MPa	48.3–55.2	13.1–62.1	13.8–53.8	17.9–47.6
Elongation, %	40–45	6–70	29–100	44–88
Refractive index	1.49–1.51	1.46–1.50	1.46–1.49	1.46–1.49

[a]Ref. 28.

acid anhydrides, have been used to make the higher esters (C_6–C_{16}) of cellulose [28,30].

The melting points (T_mK) and densities (d) of the cellulose triesters (acyl carbons, C_2–C_{16}) are given in Table III. Because Eq. 1 includes the limiting melting point [33] of 414K (using 10^6 to represent infinity), it seems reasonable to conclude Eq. 1 defines the melting points of all the higher cellulose triesters (C_{10}–C_∞). Eq. 2 correlates the melting points of the lower triesters (C_0–C_5) with the number of acyl carbons:

$$C/TmK = 0.003737 + 0.0024155\ C \qquad (1)$$

$$(TmK)^{1/2} = 26.882 - 1.3977\ C \qquad (2)$$

The densities of the cellulose triesters (for acyl carbons, C_5–C_{16}) are defined by Eq. 3:

$$C/d = -0.96622 + 1.0753\ C \qquad (3)$$

The calculated densities (Eq. 3) agree well with the literature densities (Table III).

CELLULOSE ETHERS

Cellulose reacts with sodium hydroxide to give alkoxides or alkali cellulose. Alkali cellulose reacts with various reagents to give cellulose ethers [6]: dimethyl sulfate gives methyl ethers; alkyl halides give alkyl ethers; sodium chloroacetate gives the carboxymethyl ether; epoxyalkanes give hydroxyalkyl ethers; and acrylonitrile (CH_2:CHCN) gives the cyanoethyl ether.

Cellulose ethers have a wide diversity of properties, ranging from organic-soluble, thermoplastic products to water-soluble materials. The commercial products can be divided into two groups, the water-soluble ethers having the largest markets:

Water-soluble cellulose ethers
Sodium carboxymethyl
Sodium carboxymethyl 2-hydroxyethyl

TABLE III.   Properties of Cellulose n-Triesters.[a]

Acyl C	Esters	TmK	Eq. 1	Eq. 2	Density, g/mL	Eq. 3	Moisture regain %[b]
0	(Cellulose)	723		722.6	1.52		10.8
1	Formate			649.6			
2	Acetate	579		580.4	1.28		2.0
3	Propionate	507[c]		515.1	1.23		0.5
4	Butyrate	456		453.7	1.17		0.2
5	Valerate	395		396.2	1.13	1.134	0.2
6	Caproate	367			1.10	1.094	0.1
7	Heptylate	361			1.07	1.067	0.1
8	Caprylate	359			1.05	1.048	0.1
10	Caprate	361	358.5		1.02	1.022	0.1
12	Laurate	364	366.7		1.00	1.005	0.1
14	Myristate	379[c]	372.8		0.99	0.9938	0.1
16	Palmitate	378	377.5		0.99	0.9853	0.1
18	Stearate		381.2			0.9788	
10^6		414	414.0				

[a] Data from References 28 and 30; the cellulose melting point is hypothetical [5].

[b] Moisture regain, % at 50% relative humidity.

[c] Omitted from calculations.

2-Hydroxyethyl
Methyl
2-Hydroxypropyl methyl
2-Hydroxyethyl methyl
2-Hydroxybutyl methyl
2-Hydroxyethyl ethyl
2-Hydroxypropyl

Organic-soluble cellulose ethers
Ethyl
Ethyl 2-Hydroxyethyl
2-Cyanoethyl

Cellulose ethers are used in many applications, including water-based paints and in lacquers. The five common ether substituents are carboxymethyl, methyl, ethyl, hydroxyethyl, and hydroxypropyl [31,32]. Carboxymethyl cellulose is applied to fabrics to inhibit soiling and to paper to obtain oil and grease resistance. Methyl cellulose and 2-hydroxypropyl cellulose have many uses, including latex paint, paint removers, and coatings of pharmaceutical tablets. Ethyl cellulose is used in lacquers, varnishes, and coatings for paper, glass, fluorescent tubes, and pharmaceutical tablets [31,32].

The allyl ethers of starch and cellulose have been investigated as components of coatings [34]. Cellulose derivatives have been made by graft polymerization [26], but apparently the products have not found markets in coatings.

28

REFERENCES

1. Seymour, R. B., J. Chem. Ed., 65, (4) 327-334 (1988).
2. Lambourne, R., Editor, Paint and Surface Coatings, John Wiley & Sons, Inc., New York, 1987.
3. Myers, R. R., in History of Polymer Science and Technology, R. B. Seymour, Editor, Marcel Dekker, Inc., New York, 1982.
4. Brandrup, J., and E. H. Immergut, Editors, Polymer Handbook, 2nd Edition, John Wiley & Sons, Inc., New York, 1975.
5. Back, E., S. Nordin, and J. Nyren, Textile Res, J., 44, (11), 915-917 (1974).
6. Seymour, R. B., and C. E. Carraher, Jr., Polymer Chemistry, 2nd Edition, Marcel Dekker, Inc., New York, 1988.
7. Andrews, B. A. K., and I. V. de Gruy, in Encyclopedia of Textiles, Fibers, and Nonwoven Fabrics, John Wiley & Sons, Inc., New York, 1984.
8. Vail, S. L., in Encyclopedia of Textiles, Fibers, and Nonwoven Fabrics, John Wiley & Sons, Inc., New York, 1984.
9. Fisher, C. H., in History of Polymer Science & Technology, R. B. Seymour, Editor, Marcel Dekker, Inc., 1982.
10. Harris, F. W., and R. B. Seymour, Structure-Solubility Relations in Polymers, Academic Press, New York, 1977.
11. Battista, O. A., in Applications of Polymers, R. B. Seymour, Editor, Plenum Press, New York, 1988.
12. Wint, R. F., and K. G. Shaw in Applied Polymer Science, 2nd Edition, R. W. Tess and G. W. Poehlein, Editors, American Chemical Society, Washington, D.C., 1985.
13. Phillips, M., in Great Chemists, E. Forbes, Editor, Interscience Publishers, Inc., New York, 1961.
14. Reid, J. D., and E. C. Dryden, Textile Colorist, 62, 43 (1940).
15. Morris, P. J. T., Polymer Pioneers, Center for History of Chemistry, Publication No. 5, Philadelphia, PA, 1986.
16. Mark, H. F., Am. Scientist, 72, 156-162 (1984).
17. Asimov, I., Biographical Encyclopedia of Science and Technology, 2nd Revised Edition, Doubleday & Co., Inc., Garden City, New York, 1982.
18. Wakeman, R. L., Chemistry of Commercial Plastics, Reinhold Publishing Corp., New York, 1947.
19. Stephenson, J. N., Editor, Preparation and Treatment of Wood Pulp, McGraw-Hill Book Co., Inc., New York, 1950.
20. Curtis, L. G., and J. G. Crowley, in Applied Polymer Science, R. W. Tess and G. W. Poehlein, Editors, American Chemical Society, Washington, D.C., 1985.
21. Ihde, A. J., The Development of Modern Chemistry, Dover Publications, Inc., New York, 1984.
22. Yarsley, V. E., et al, in Elastomers and Plastomers, R. Houwink, Editor, Elsevier Publishing Co., Inc., New York, 1949.
23. Considine, D. M., Editor, Van Nostrand's Scientific Encyclopedia, 5th Edition, Van Nostrand Reinhold Co., New York, 1976.
24. Kennedy, et al, Cellulose and Its Derivatives, Hallstead Press, New York, 1985.
25. Sand, I. D., Preprints, American Chemical Society PMSE Division, 57, 57-63 (Fall 1987).
26. Arthur, J. C., Jr., Cellulose Chemistry and Technology, American Chemical Society, Washington, D.C., 1977.
27. Seymour, R. B., and C. E. Carraher, Structure-Property Relationships in Polymers, Plenum Press, New York, 1984.
28. Bogan, R. T., et al, Kirk-Othmer Encyclopedia of Chemical Technology, 3rd Edition, 5, 118-143 (1979).
29. Hiatt, G. D., and W. J. Rebel, in Cellulose and Cellulose Derivatives, 2nd Ed., Vol. V, N. M. Bikales and Leon Segal, Editors, Wiley-Interscience, New York, 1971.
30. Malm, C. J., et al, Ind. Eng. Chem., 43, 688-691 (1951).

31. Savage, A. B., in Cellulose and Cellulose Derivatives, N. M. Bikales and Leon Segal, Editors, Wiley-Interscience, New York, 1971.
32. Greminger, Jr., G. K., in Kirk-Othmer Concise Encyclopedia of Chemical Technology, John Wiley & Sons, Inc., New York, 1985.
33. Fisher, C. H., J. Am. Oil Chem. Soc., 65, 1647-1651 (1988).
34. Nichols, P. L., Jr., and E. Yanovsky, J. Am. Chem. Soc., 66, 1625 (1944).

History of Wax Coatings

Raymond B. Seymour
University of Southern Mississippi
Hattiesburg, MS 39406

Abstract

Naturally occurring paraffin wax has been used as a waterproof coating for over six thousand years. The widely used petroleum wax, which was obtained commercially by chilling petroleum in 1867 is now used, in the U.S., at an annual rate of over 500 million pounds for waterproofing wood and paper.

Waterproofing Waxes

Today's coatings technologist is familiar with modern polymeric hydrocarbons, such as polyethylene, polypropylene, polybutylene, and copolymers of the monomers used to produce these homopolymers. However, because of the excellent physical properties of these high molecular weight alkanes, he may have forgotten that much of the technology for the application of melt coatings was developed by ancient artisans who used carnauba wax, beeswax, sperm whale wax and ozocerite for hot melt coatings.

Beeswax, which is obtained by heating honeycombs in boiling water, was used several thousand years ago as a hot melt caulk and coating and later as a binder in paints. Carnauba wax, which is still used for candles and polishes is obtained from the Brazilian carnauba palm tree. Sperm whale wax (spermacetti) is obtained by removing the melting oils from the head and blubber of the sperm whale.

Mineral waxes, unlike the animal and vegetable derived waxes cited above, include ozocerite, which is a mineral wax mined in Utah, Austria, and Galicia. This hydrocarbon wax as well as the bituminous montan wax have also been used as hot melt coatings. However, the most widely used naturally occurring wax is petroleum wax, which has been used for centuries as a water proof coating.

Paraffin wax was used by the Egyptian six thousand years ago as an alternate to beeswax for sealing coffins. The wood wax, which is derived from the Anglo-Saxon word weax meaning beeswax, is widely used for many types of melt coatings including petroleum wax. The first commercial paraffin wax was separated from chilled petroleum in 1867 in Cory, Pennsylvania.

The "ice box" process for separating the undesired wax from petroleum was replaced by a solvent extraction process in the 1930's. However, a chilling process has been used since the 1970's for the production of a refined wax.

Applications

Wax coated cartons were widely used as replacements for glass milk containers in the 1930's and 1940's but these containers have been replaced, to a large extent, by polyethylene milk bottles. Polyethylene, which is a higher molecular weight homologue of the paraffin series has a higher melting point and does not require paperboard for structural stability.

Published 1990 by Elsevier Science Publishing Co., Inc.
Organic Coatings: Their Origin and Development
R.B. Seymour and H.F. Mark, Editors

32

Paraffin wax, which is produced in the US at an annual rate of over 500 million pounds, is widely used as a coating for particle board, wall board, corrugated boxes, paper containers, and paper, such as butcher's wrap.

References

Marsel, C.J.; "Natural Waxes," Chem. Ind. 67 563 (1950).

Warth, A.H.; "The Chemistry and Petroleum of Waxes," Reinhold Publishing Co. New York 1947.

Zwicker, D.A.; "Petroleum Wax," The Lamp 70 (4) 8 (1988).

HISTORY OF COATINGS EDUCATION

RUDOLPH D. DEANIN* AND JOHN A. GORDON JR.**
*University of Lowell, Lowell, MA 01854; **Pacific Technical
Consultants, Torrance, CA 90505

ABSTRACT

Coatings education in the U.S. began in 1906 and has grown
with the industry until it now is offered by 20 colleges and
universities, with a total of 54 full-length courses leading to
3 BS, 6 MS, and 2 PhD programs, as well as at least 19
intensive short courses and 41 meetings, symposia, seminars,
and workshops.

INTRODUCTION

Painters made their own materials for many thousands of
years, handing down their formulations and techniques from
father to son and from master to apprentice. Commercial
manufacture began in 1867. In those early days, quality varied
widely and paint performance was often very poor. Users
naturally suspected that some unscrupulous manufacturers were
cheating on expensive high-quality ingredients and substituting
inferior low-cost ingredients in their place. In particular it
was generally believed that a good white paint needed a lot of
lead to make it high quality!

Edwin F. Ladd, Professor of Chemistry at North Dakota
Agricultural College for over a quarter of a century, starting
in 1890, led a campaign to convince the state legislature that
they should enforce honest labelling of paint ingredients and
quality. The legislature passed such a law in 1906, and
assigned Prof. Ladd to enforce it.

NORTH DAKOTA AGRICULTURAL COLLEGE → STATE UNIVERSITY

Starting in 1906, Prof. Ladd employed and trained students
to analyze paints chemically and evaluate their performance in
the laboratory and on test fences. Later, with Prof. Bolley
promoting the growth of flax as a major crop in North Dakota
and a major source of linseed oil for paints, the College
rapidly broadened its interest in paints and varnishes,
including them in industrial chemistry courses in the School of
Chemical Technology, under Prof. Leo L. Carrick 1922-1948, with
the assistance of Profs. Pearce and Dallas S. Dedrick. This
program graduated 5-8 paint technologists per year in the
1920's, growing to 18 in 1930. In the 1930's and 40's the
Department of Industrial and Physical Chemistry listed paint,
varnish, and lacquer as a strong feature of the School of
Chemical Technology and offered 3 years of course work in
protective coatings. Prof. Wouter Bosch 1948-1959 strengthened
the program further to a Department of Paints, Varnishes, and
Lacquers in the School of Chemical Technology, offering MS and
PhD degrees, and also developing intensive short courses for
people from industry.

Published 1990 by Elsevier Science Publishing Co., Inc.
Organic Coatings: Their Origin and Development
R.B. Seymour and H.F. Mark, Editors

While the Agricultural College became the State University, under Prof. A. E. Rheineck 1960-1971 the Department of Polymers and Coatings progressed from practical technology to fundamental theory as well, and offered BS, MS, and PhD degrees in Chemistry with a major in Polymers and Coatings. This development continued under Zeno W. Wicks Jr. 1972-1983 and Frank N. Jones since 1983 as chairmen, with the addition of S. Peter Pappas, Loren W. Hill, J. Edward Glass, and Marek W. Urban to the faculty. The current program offers 9 coatings courses, with polymers and coatings options in the BS Chemistry and Mechanical Engineering programs and MS and PhD in Chemistry with a major in Polymers and Coatings. Current enrollment is about 20-30 undergraduate students per year and 25 graduate students. Over the years these courses have been given to 2500 students. In addition the department continues to offer intensive short courses on campus and around the world.

BROOKLYN POLYTECHNIC INSTITUTE → POLYTECHNIC UNIVERSITY

Brooklyn has offered a course in coatings technology, for many years, at least since 1947, possibly as early as 1935 or sooner. Over the years, it has been taught successively by Joseph J. Mattiello, William H. Gardner, Henry F. Payne, Sidney Lauren, and Joseph W. Prane. Current enrollment averages about 30 students.

MISSOURI SCHOOL OF MINES → UNIVERSITY OF MISSOURI-ROLLA

Wouter Bosch came from Holland in the mid 1940's, and after teaching at Oklahoma A&M and North Dakota Agricultural College, came to the Missouri School of Mines 1959-1971. He set up coatings courses and intensive short courses here as he had at the other schools. The school became the University of Missouri at Rolla. When Bosch retired, Lewis P. Larson took over the short courses 1971-1977, followed by John A. Gordon Jr. 1977-1985.

The current program, led by James O. Stoffer in the Department of Chemistry, offers BS, MS, and PhD degrees in Chemistry with specialization in Polymers and Coatings Science. Enrollment in these programs is typically 20-25 undergraduates, 14 MS, and 25 PhD candidates, working under 9 faculty members in the Graduate Center for Materials Research. The Department also offers about 7 intensive short courses.

MISSISSIPPI SOUTHERN UNIVERSITY → UNIVERSITY OF SOUTHERN MISSISSIPPI

Francis Schofield of the Paint & Varnish Association suggested that the new Tung Oil Association select J. Scott (Shorty) Long as research director and locate at Mississippi Southern University in the 1960's. He brought in Shelby Thames as his assistant in the Department of Chemistry. Thames started the Department of Polymer Science in 1970, and brought in Gary Wildman and Billy George Bufkin; when Thames became Dean and later Executive Vice President of the University of Southern Mississippi in 1974, Wildman became chairman and then

dean, and Bufkin in turn became chairman of Polymer Science.
Later additions included Raymond B. Seymour, Lon Mathias, and
current chairman Gordon L. Nelson, as well as many others.

The current program includes 13 faculty, with 100
undergraduate and 45 graduate students working for BS, MS, and
PhD degrees in Polymer Science. In addition to college and
short courses, the department organizes the annual Water-Borne
and Higher-Solids Coatings Symposium in New Orleans.

KENT STATE UNIVERSITY

Raymond R. Myers and Carl J. Knauss came to Kent to direct
coatings rheology research for the Paint Research Institute and
other sponsors, producing a series of MS and PhD degrees in
Physical Chemistry. In 1972 they began offering 3 courses in
surface coatings, leading to BS, MS, and PhD degrees in
Chemistry with a Coatings Option. At the same time they also
began offering intensive short courses for the industry. They
were later joined by Richard J. Ruch to strengthen the program
further. At present they offer 3 college courses in coatings
which have been taken by 778 students. They also offer a
steady list of intensive short courses, of which at least 6 are
specifically in coatings.

UNIVERSITY OF LOWELL

At the urging of the New England Society for Coatings
Technology, Rudolph D. Deanin began an evening MS program in
Coatings & Adhesives in 1974 for people working in the local
industries. With the help of visiting lecturers from industry,
this program attracts an average of about 30 students and
produces about 7 MS degrees per year.

In 1989, again at the urging of the local Society, the
program was broadened to include an evening BS in Applied
Chemistry with an Option in Coatings & Adhesives, primarily for
the advancement of technicians employed in the local
industries. It is expected that this program will attract
about 20-30 undergraduate students per years.

EASTERN MICHIGAN UNIVERSITY

A BS program in Polymers and Coatings was started in the
College of Technology by John C. Graham in 1980, with help from
Paul D. Kuwik, Alvin E. Rudisill, and Later John A. Gordon Jr.
and Taki J. Anagnostou. This later expanded to two BS programs
(Polymers & Coatings and Coating Process Technology) with 100
undergraduate students, producing 15 BS per year. It also
expanded to 2 MS programs, with student research located in the
Coatings Research Institute (1985) and the Paint Research
Association (1987). The current curriculum lists 9 courses in
coatings.

36

OTHER SCHOOLS MENTIONED IN THE HISTORICAL RESEARCH

Fragmentary reports indicated that several other schools have had coatings courses and/or research over the years. These included Case Western Reserve University, Drexel University, Oklahoma A&M, and Rutgers University. Undoubtedly more thorough research would discover many more.

CURRENT STATUS OF COLLEGE AND UNIVERSITY PROGRAMS IN COATINGS

A survey of coatings education in 1987 (1) found 14 colleges and universities offering a total of 178 credits in coatings courses. Five of these schools offered 11 degree programs which named coatings in the degree; 6 of these were BS, 3 MS, and 2 PhD degrees.

The Federation of Societies for Coatings Technology annually collects the education efforts of all the local Societies, either through local colleges and universities or independently by the Societies themselves (2). The current survey found 20 schools and 4 Societies offering a total of 54 full-length courses and 19 intensive short courses. For degree programs, they found 3 BS, 6 MS, and 2 PhD programs.

The same FSCT survey also collected lists of meetings, symposia, seminars, and workshops sponsored by the local Societies. For the current year, they found that 18 local Societies had sponsored 41 of these activities, mainly meetings.

ACKNOWLEDGEMENTS

The authors are very grateful to all the people who contributed information and reminiscences for this review:

Paul Bruins
John C. Graham
Frank N. Jones
Carl J. Knauss
Thomas A. Kocis
Lewis P. Larson
Joseph W. Prane
James O. Stoffer
Marek W. Urban
Patricia D. Viola
John C. Weaver
Zeno W. Wicks Jr.

REFERENCES

1. R. D. Deanin, "Survey Shows Growth of Coatings Education," Am. Paint Coatings J. Convention Daily, 10/20/88, Pg. 35.
2. Fed. Soc. Coatings Tech., "1989 Guide to Coatings Courses, Symposia, and Seminars."

SIXTY YEARS OF AUTOMOTIVE COATINGS FROM LACQUERS TO OLIGOMERS

A. G. ARMOUR, D. T. WU, J. A. ANTONELLI, J. H. LOWELL
E.I. DuPont De Nemours & Co., P.O. Box 3886, Philadelphia, PA 19146

ABSTRACT

Looking back over the years one is impressed by the
continuous dedicated effort of automotive coatings manufacturers to respond to customer needs. DuPont's discovery
of low-viscosity nitrocellulose in the 1920's made possible
production-line finishing of automobiles with fast-drying
lacquers. Alkyd-based coatings soon followed as a
non-buffing alternative to the above. In the 1950's acrylic
lacquers and enamels emerged providing unprecedented
durability. In the 1970's EPA's demand for less atmospheric
pollution was met partially by acrylic dispersion finishes.
A much larger step was achieved by using thermosetting
polyester and acrylic oligomers, but at an appearance
sacrifice. Borrowing from European technology, the
so-called colorcoat/clearcoat system was perfected; it
provided both low solvent-emission and exceptional glamour.
Undercoat quality has also kept pace through the development
of electrocoat primers. Many coat- ings manufacturers
contributed to this history of success. The story is
presented here with emphasis on DuPont's involvement.

INTRODUCTION

Looking back over the years one is impressed by the continuous dedicated
effort of automotive coatings manufacturers to respond to customer needs.
Many companies contributed their know-how and scientific talents to the
development of the many finishing systems available to the automotive
industry today. The story is presented here with emphasis on the DuPont
Company's involvement.

The Pre-"Duco" Era

When the automobile replaced the carriage in the early 1900s, it brought
with it the finishing system of its horse-drawn counterpart. Both the
primer and the topcoat were based on oleoresinous vehicles. A total of as
many as 25 brush-coats of material (undercoat and topcoat) were required for
adequate build with extended drying periods and processing operations
between coats. The cars were finished individually, the entire operation
requiring as much as two to three weeks, depending on atmospheric
conditions; thus, production line manufacture was out of the question.
While the newly finished cars had a reasonably satisfactory appearance, the
finish dulled rapidly on exposure and in a few months was badly chalked; it
ultimately failed by checking.

The Discovery of Low-Viscosity Nitrocellulose

The discovery of low viscosity nitrocellulose (NC) by DuPont in 1920 led
to the development of "Duco" NC Lacquer. This development sparked a
burgeoning research effort in the industry yielding a trickle of new
products in the early years and growing exponentially to a virtual explosion

Published 1990 by Elsevier Science Publishing Co., Inc.
Organic Coatings: Their Origin and Development
R.B. Seymour and H.F. Mark, Editors

today (CHART #1). These products include compositions of acrylics, polyesters, vinyls, epoxies, and urethanes in various forms - solution, dispersion, high solids, waterborne, basecoat/clearcoat, powder, electrocoating, and a host of specialty materials.

Prior to the "Duco" development nitrocellulose as a lacquer component was little more than a laboratory curiosity; its principal commercial application was in Collodion and clear metal lacquers. The main problem with NC as a practical coating ingredient was that its molecular weight was so high that sprayable solutions could be achieved only at an impractically low solids level. The result was thin films and the necessity of multi-coat application. This situation changed abruptly in 1920 when chemists at DuPont's finishes plant at Parlin, N.J. made a fortuitous discovery. A drum of experimental high viscosity NC film base that had been treated with sodium acetate was inadvertently allowed to stand in the hot sun for several days. When tested, the base was found to have undergone a marked decrease in viscosity. It was determined that the viscosity lowering resulted from the action of sodium acetate on the NC at the elevated temperature involved. This led to the first patented process for making low viscosity nitrocellulose (1) (2). Other, more sophisticated, methods of lowering the viscosity soon followed.

The Emergence of "Duco" Nitrocellulose Lacquer

Three more years were required to put the finishing touches on the development of "Duco" NC Lacquer. The product could be sprayed at a practical solids level, force-dried, buffed to a high gloss and spot repaired in less than six (6) hours. Thus the bottle-neck was broken and the production line manufacture of automobiles became a reality. The product and finishing process were demonstrated on an Oakland touring car in September, 1923, and by mid-1924 "Duco" had been adopted by the following automobile manufacturers: Buick, Cadillac, Cleveland, Franklin, Lexington, Marmon and Moon. Other GM units soon followed suit as did Ford and Chrysler for a time (3). Other paint manufacturers, following DuPont's lead, eventually supplied similar products to the industry under a DuPont license. Among the NC lacquer suppliers were Rinshed-Mason, Forbs and Glidden.

Unmodified NC has poor adhesion to metal and is not durable when exposed to the elements. However, pigments adsorb some of the sun's harmful rays and protect the vehicle to a considerable extent. Adhesion and good film properties are achieved by modifying the NC with suitable resins and plasticizers. Natural gums such as dammar were used originally, but lacked durability and were eventually replaced by non-drying oil modified alkyd resins (Chart #2). These were used in conjunction with monomeric phthalate type plasticizers and castor or blown castor oil.

The Advent of Alkyd Enamels

Chrysler's interest (and eventually Ford's) turned in the direction of alkyd enamel topcoats, which were introduced by PPG, DuPont and other suppliers in the early 1930s. The original vehicle was mainly glyceryl phthalate polyester modified with drying oil fatty acids such as those from linseed and soya bean oils (Chart #3). Small amounts of nitrogen resin such as urea- or melamine-formaldehyde were often added to reduce baking time and improve hardness. DuPont originally supplied alkyd topcoats to Ford under the name "Dulux" enamel but subsequently sold the formulas and know-how to Ford to enable them to manufacture a major part of their own finishes. PPG continued to be a major supplier of alkyd topcoats to Chrysler.

Lacquers Vs Enamels - A Property Comparison

While the NC and alkyd systems were essentially equivalent in metal-protective characteristics and weatherability, they differed in balance of properties (Chart #4).

1. The ingredient and application costs of enamels were lower than those of lacquers because of lower solvent costs and higher spray solids.

2. Enamel and lacquer processing costs were about a stand-off. While lacquers exhibited poor flow and had to be compounded and buffed for satisfactory smoothness and gloss, enamels were sufficiently smooth-flowing to permit elimination of this step. However, the resulting cost advantage was largely offset by the added labor cost of providing the extra smoothness in the undercoat that a non-buffing topcoat required.

3. Another important cost and labor advantage of lacquer derived from its fast drying characteristics and the fact that the films were permanently soluble. Spot repairs made on the line could be force-dried in a matter of minutes and ready for buffing. The repair lacquer blended well with the soluble original coating and provided patches of excellent color uniformity. Enamel coatings, on the other hand, being insoluble at the repair stage, did not blend with the repair material and the result was off-color spots and rings. As a consequence, the practice of refinishing whole areas of the car was adopted. This necessitated recycling the cars through the baking ovens; discoloration of light pastel shades often resulted from the extra heating.

4. Metallic lacquers (those containing aluminum flake) had a more glamorous appearance than their enamel counterparts. General Motors was convinced that this appearance advantage helped sell automobiles.

The glamour of metallic finishes is associated with their so-called two-tone characteristics. By two-tone is meant the color differences observed when the finish is viewed at different angles. After being applied, the wet film is in turbulent motion as a result of solvent evaporation and non-uniform surface tension effects. Aluminum flake is carried to the surface and becomes partially oriented in a more-or-less flat position in the direction of flow. As the film dries the viscosity eventually becomes too high to allow further motion of the flake. Beyond this point, instead of flowing, the film shrinks. As a result additional flakes are drawn into a position more nearly parallel with the film surface. It follows that when viewed head-on the surface appears lighter and more silvery than when viewed at a more grazing angle. This is because at a smaller viewing angle light penetrates further into the film under the edges of the flake and exposes to the eye more of the dark prime pigment that makes up the color. NC lacquers show much more two-tone than alkyd enamels because of differences in rheology that occur as the film dries. The no-flow point for lacquers is in the neighborhood of 35-40% solids, so that 60-65% of the drying mechanism involves shrinkage. Alkyd enamels, on the other hand, do not cease flowing until the solids level has reached 80% or more. Hence there is only about 20% of shrinkage available to help increase the concentration of flake lying approximately parallel to the surface.

Over the ensuing years both NC lacquers and alkyd enamels continued to be upgraded in quality and served the automotive industry well for over a quarter of a century. However, in spite of the acceptable performance of these two systems, it was recognized that there were remaining weaknesses in a number of areas. In 1948 the DuPont Company undertook a research program aimed at developing a new automotive finish that would combine all the good features of "Duco" and "Dulux" and eliminate the remaining deficiencies (Chart #4). Radically improved durability was a major goal. A specification was drawn up embodying all the major attributes of a "perfect" finish (Chart #5):

Outstanding durability	>	"Duco" or "Dulux"
Superior blister resistance	>	"Duco" or "Dulux"
Wider color range	>	"Duco" or "Dulux"
Glamorous appearance	=	"Duco"
Non-buffing finish	=	"Dulux"
Low solvent cost	=	"Dulux"
High spray solids	=	"Dulux"
Good stain resistance	=	"Dulux"
Rapid dry	=	"Duco"
Easy spot-repair	=	"Duco"
Over-bake discoloration resistance	>	"Dulux"

A thorough survey of the available types of polymers and resins strongly indicated that film-formers capable of combining all of these characteristics in a single finish were not likely to be found. The research effort was therefore split, with one objective aimed at finding the best possible lacquer coating and a second directed toward the development of a superior enamel. From this work came "Lucite" acrylic lacquer and "Dulux 100" alkyd enamel.

Work on the lacquer development was begun first. Poly(methylmethacrylate) (PMMA) was selected for initial study because it was known to strongly resist degradation by UV light and hydrolysis. The choice was supported by theoretical considerations: since the methyl methacrylate polymer molecule has no tertiary hydrogens along its chain, DuPont chemists reasoned that it would be less susceptible to oxidative attack. Unfortunately, in the early evaluation, chemical stability appeared to be about the only redeeming feature of the polymer. The available grades of PMMA were high molecular weight types and the first "Lucite" vehicles, consisting of PMMA plasticized with monomeric phthalate plasticizer, could not be sprayed above 5% solids without webbing, ie, issuing from the spray gun in filaments instead of fine droplets. Other problems involved cracking of the plasticized film on exposure, crazing of the original finish during spot repair and poor adhesion to standard undercoats.

The webbing problem was solved by a polymerization technique that reduced the average molecular weight of the polymer and narrowed the molecular weight distribution. Cracking of the film on exposure was found to be due to migration of monomeric phthalate plasticizer from the topcoat into the undercoat. This embrittled the topcoat film, softened and swelled the undercoat and set up stresses that exceeded the tensile strength of the film. The problem was eliminated by increasing the undercoat bake and replacing a portion of the monomeric plasticizer with a polymeric one. This substantially reduced the migration tendency.

The mechanism of crazing was found to be somewhat similar to that of cracking. Shrinkage stresses are set up in the original film during dry. Also, PMMA films dry with a case-hardened surface that is not readily solvated. Film layers beneath the surface, however, dissolve more readily.

Solvent penetration of the surface through imperfections and by diffusion
preferentially soften and swell the under layers. The resulting stresses
set up in the case-hardened surface are relieved by crazing. If the film is
above its glass-transition temperature when repaired, it will not craze.
Modification of the vehicle through the addition of softer copolymers, a
readily soluble cellulose ester (cellulose acetate-butyrate) and a polymeric
plasticizer lowered the glass transition temperature of the film enough to
reduce the crazing tendency without rendering it impractically thermoplastic
(Chart #6). At the same time when the finish was baked at a temperature of
110^0C or higher it reflowed to an acceptably smooth, glossy film that did
not require buffing. The problem of poor adhesion to standard undercoats
was solved through the use of an intermediate coating that adhered well to
the primer and to which the acrylic lacquer in turn adhered. Later,
undercoats were developed to which "Lucite" would adhere directly.

   "Lucite" acrylic lacquer (4) was released to the automotive industry in
1957-58 after nine years of intensive research. While not all of the goal
properties were achieved, the major objective of exceptional durability was
accomplished. While many "Duco" colors dulled and chalked sufficiently to
require polishing after six to eight months in service, the same colors in
"Lucite" needed no polishing for a matter of two years. Since "Lucite" had
rheological properties similar to those of "Duco" in the drying film it
possessed the same outstanding two-tone properties; this led to a much
broader range of glamour colors than could be tolerated in the less durable
"Duco" vehicle. The buffing requirement for acceptable smoothness and gloss
was eliminated, the rapid dry and easy spot repair characteristics of "Duco"
were retained and the finish was considerably more resistant to staining by
surface contaminants (oil, tar, etc.) than its predecessor. Unfortunately
the goals of higher spray solids and lower ingredient cost were not
realized.

The Development of "Dulux-100" Enamel

   Failure to achieve the above objectives underscored the need for an
improved enamel for those automobile manufacturers who preferred this type
of coating. Concurrently with the acrylic lacquer program, work was
underway in the enamel area. In addition to providing gloss retention
comparable to that of "Lucite" acrylic lacquer, it was hoped that film
blistering and checking, which were weaknesses of conventional alkyd
enamels, could be improved to the same degree. Although these deficiencies
were not great, it was generally observed that in areas and periods of heavy
rainfall alkyd enamels were somewhat less blister-resistant than NC
lacquers. While the pigmentation of the undercoat (presence of water-
soluble salts) often abetted this weakness, it was well-known that the
auto-oxidative curing mechanism of drying-oil modified alkyds contributes
water-sensitive by-products which promote blistering and occasional film
failure. The obvious approach, therefore, to the solution of this problem
was to select a vehicle that did not cure via the auto-oxidative mechanism.

   Actually, the basic vehicle for an improved alkyd enamel was already in
existence, but had never been given serious consideration because of cost.
However, the competitive incentive provided by acrylic lacquer made it
desirable to reexamine the cost/property balance. This composition was a
blend of non-drying oil modified alkyd and melamine-formaldehyde resins - a
vehicle that had found extensive use in appliance enamels (Chart #7). The
enamel based on this vehicle required a higher bake than conventional alkyd
enamels (125^0C vs 110^0C), but when properly cured provided the sought-for
improvements in gloss retention, particularly in light pastel shades and
metallics and thus provided a wider range of durable colors. Improved
resistance to moisture-blistering and checking were also achieved. An

important benefit from the automobile manufacturer's viewpoint was the resistance of the finish to overbake discoloration. In order to finish a car in two or three colors, it is necessary to process the body through the baking oven after each color is applied. Conventional alkyds in pastel shades often showed a discernible yellowing when baked repeatedly, while the non-drying oil variety did not. The DuPont version of the improved alkyd enamel was released to the automotive industry in 1956 under the label "Dulux 100". Other major paint suppliers marketed similar products.

The Development of Thermosetting Acrylic Enamels

The "Dulux 100" type enamel was in use only a comparatively short time. The success of acrylic lacquer led enamel manufacturers to develop cross-linkable acrylic polymers. Complex hetropolymers were made involving styrene as well as acrylic and methacrylic monomers in the proper proportions to yield the desired balance of cost and film properties. Included was a hydroxyl-containing monomer such as hydroxyethyl acrylate. A small amount of acrylic or methacrylic acid was also included as a crosslinking catalyst and pigment dispersant (Chart #8). The final vehicle was a blend of this heteropolymer and butylated MF resin which reacted to crosslink the hydroxylated acrylic.

The resulting enamels had all the excellent weathering characteristics of acrylic lacquers. They were also harder and more damage resistant. However, they did not have the thermoplasticity of acrylic lacquers, could not be as easily spot repaired and could not be reflowed in the oven to improve film smoothness. Eventually modifications were made to incorporate reflow capabilities in these compositions. This was done by introducing a moderate amount of cellulose acetate-butyrate (CAB) into the formula and replacing the butylated MF with a methylated type (Chart #9). This enamel could be given a short bake at about 80°C to eliminate most of the solvent without initiating any crosslinking. The CAB functioned to produce a lacquer type dry, so that the film could be given a light sanding to reduce surface roughness. It could then be reflowed at about 150°C. at which temperature the methylated MF would subsequently react to crosslink the film.

The Impact of EPA Regulations

In 1967, Rule 66 (5), regulating the use of potentially harmful solvents in industrial coatings, was invoked in Los Angeles County, California. Using this rule as a guide, the Environmental Protection Agency (EPA) restricted the use of photochemically reactive hydrocarbons and oxidants that react with nitric oxide in the presence of UV radiation to produce an irritating smog. Subsequently the agency decided that all organic solvents were photochemically reactive and published guidelines drastically restricting the amount of solvent discharged into the atmosphere from industrial finishing operations. To meet the requirements, calculations showed that automotive topcoats would have to be sprayed at nearly 60% volume solids. This was a hopeless goal for solvent-borne lacquers and would require about a 75% increase in the spray solids of conventional enamels.

The Development of Acrylic Dispersion Lacquers

As a step toward higher spray solids, DuPont undertook a study of acrylic dispersion polymers. Imperial Chemical Industries in England had developed a process for polymerizing acrylic monomers in a non-aqueous non-solvent medium in the form of very finely dispersed particles (6)(7). These dispersions were analogous to the acrylic latex systems used in architectural paints, but employed a hydrocarbon dispersion medium instead of water. Lacquers made from plasticized PMMA dispersions were very fluid and theoretically could be sprayed at a sufficiently high solids level to comply with EPA regulations. In practice, however, the sag resistance was so poor as to render spray application unmanageable. The dispersions had to be modified to include a partially soluble graft copolymer which increased the spray viscosity to a practical level (Chart #10). In the process, however, a substantial part of the solids advantage had to be sacrificed.

The modified dispersion lacquers coalesced readily in the baking oven to smooth, glossy films having essentially the same properties as their solution lacquer counterparts. They were gradually adopted by General Motors during the 1970s and solution lacquers were simultaneously phased out. While this change resulted in only a modest reduction in solvent released to the atmosphere it served notice to the EPA that their restrictions were being taken seriously. Concurrently, enamel manufacturers were using reactive acrylic dispersions to make higher solids thermosetting enamels with film properties similar to those of their solution counterparts.

Water-Borne Finishes

Several avenues were considered for meeting the EPA regulations. Compositions based on water-soluble or water-dispersible resins (8) underwent extensive production line testing in the 1970s, and General Motors subsequently adopted Inmont's water-based topcoats in two of its plants (9). The main obstacle to the across-the-board adoption of these finishes was that spray application is strongly affected by changing relative humidity. Since water is the preponderant dispersing liquid or solvent, high humidity retards evaporation and encourages sagging and metallic mottling. Low humidity, on the other hand, increases volatility, dry spray and poor leveling. Since significant changes in humidity can occur in the space of a few hours, it is very difficult to maintain a uniform appearance. The best solution to the problem would be to air-condition the entire plant, but automobile manufacturers generally regard this as economically prohibitive. Other problems involved poor transfer efficiency (high overspray loss), gassing in the container because of water/aluminum flake interaction and water sensitivity of the dried film.

Powder Coatings

Powder coatings (10)(11) were briefly considered as a means of reducing atmospheric pollution. In this process, dry, finely divided paint particles are sprayed in a strong electrostatic field, in which the automobile body is one electrode and the spray gun, the other. The powder adheres to the body by electrostatic attraction, the thickness of the coating self-limited by the magnitude of the potential employed. The coating is given a high-temperature bake which coalesces it to a continuous film. In practice undesirably rough films were obtained which required a great deal of processing for satisfactory appearance. In addition, metallics lacked the typical flash and two-tone exhibited by conventional topcoats.

High Solids Coatings

The most successful approach has been in the direction of high-solids enamels based on less radical departures from conventional systems (Chart #11). Thermosetting acrylic and polyester oligomers have both been used (9)(12). In both cases the very low molecular weight polymers (5,000-10,000 for acrylics and 1,000-2,000 for polyesters) contain a higher-than-normal concentration of polar functional groups (carboxyl, hydroxyl, epoxy, etc.) to provide the crosslinking capabilities needed to develop satisfactory film properties and durability. These compositions are borderline for smoothness on vertical surfaces and lack metallic glamour relative to their low solids counterparts. Hybrids of acrylics with polyesters or polyurethanes have also been tested (9). Complementary functionality, where applicable, is furnished by low molecular weight melamine-formaldehyde resins.

Colorcoat/Clearcoat Systems

Colorcoat/clearcoat systems have recently come to the fore. As the name suggests, this type of coating consists of a heavily pigmented colorcoat followed by a relatively thick coat of clear enamel. The colorcoat/ clearcoat systems originally used in Europe in the early 1970s were based on alkyd vehicles. While possibly sufficiently durable for the European climate, they were not suitable for the more severe conditions encountered in the southern United States. Celanese was the first to develop an all-acrylic system. However, Ford, in the mid-1970s, became the first U.S. automobile manufacturer to use a colorcoat/clearcoat combination commer- cially (9).

The colorcoat/clearcoat system is currently being used as follows: the colorcoat (Chart #12a), preferably the polyester type, formulated on a high- solids basis and at a higher-than-normal pigment level, is first applied at a relatively low film thickness (0.5 - 1.0 mil), but sufficiently thick to provide complete hiding. Clear high solids acrylic enamel (Chart #12b) is then applied wet-on-wet at 1.5 - 2.0 mils. The combination is baked 30 min. at 120°C. The brilliant gloss achieved through this technique is very attractive and recaptures the glamour lost in conventional high solids enamels. There had been some concern that the clear topcoat might lack the necessary outdoor durability. Conventional UV screening agents, because they are gradually volatilized from the film on exposure are not satisfac- tory; however, a combination of sun-screeners developed by Ciba-Geigy that is less volatile, seems to be doing an adequate job. A recent DuPont development in which the sun-screen agent is chemically attached to the polymer chain is expected to provide an even more permanent effect (13). The gloss retention of these systems on exposure is outstanding.

One of the problems encountered in the application of the colorcoat/ clearcoat system and high solids coatings in general was the excessive viscosity drop in the baking oven (Chart #13). To counter the sagging tendency, thixotrope systems were developed by DuPont that are less affected by heat than conventional types (14)(15)(16). An additional benefit in the use of these agents is a certain amount of film reinforcement.

Undercoat Systems

The foregoing has been primarily a discussion of topcoats. Over the years there also have been revolutionary changes in undercoat compositions and application methods. In the early days the sheet-metal parts (fenders and hoods) were primed on a separate line from the bodies. The sheet-metal primers, generally based on blends of oleoresinous and alkyd vehicles, were applied by flow- or dip-coating and given a 200°C bake; they were then ready for topcoating after spot-sanding to remove imperfections. Bodies were spray-coated first with a primer of moderate pigment concentration. After baking and sanding, a surfacer, highly pigmented for easy processing, was applied, baked and thoroughly sanded to the smoothest practical surface for topcoating. An alternate method, employed less extensively, was to use a single hybrid primer-surfacer to coat the metal.

The rust-protective performance of the body undercoats, generally based on modified alkyd vehicles, was not uniformly good on all areas of the car. Recessed and partially enclosed sections of the body such as rocker panel interiors were difficult to coat completely; also when the body entered the oven, the still-wet undercoat, fluidized by the heat, tended to flow away from the edges of the metal. As a result many rusted-out rocker panels were encountered in service. An effort to improve this situation involved dip-priming the lower half of the body. This ensured that all the recessed and boxed-in areas were uniformly coated; however, it did not prevent the wet primer from sagging in the oven.

Electrocoat Primers

Electrodeposition, a radically different method of priming automobiles, was pioneered by the Ford Motor Company in the 1960s. The development of the electrocoating concept from the laboratory stage to full production is summarized by Burnside, Bogart and Brewer (17)(18). In electrocoating, the primer vehicle is an aqueous dispersion of polymeric materials carrying either positive or negative ionic groups, thus providing for either cathodic or anodic deposition. The anodic type is typically an amino- or alkalie-solublized polycarboxilic resin and the cathodic type, an acid salt of an amine-treated resin such as an epoxy. The car body is coated by immersing it in a tank containing the liquid primer and subjecting it to a direct current charge of 100-450 or more volts. Thus the body becomes either the anode or the cathode of an electrolytic system. The tank wall functions as the oppositely charged electrode. The applied voltage causes the dispersed particles to migrate to the car body sweeping the pigment particles along with them. As they are deposited, the transfer of electrons provides an electrically neutral deposit. During the process, electroendosmosis comes into play, squeezing water out of the deposited coating and leaving it in a relatively firm state. With this process, improved coverage is achieved in recessed areas, on sharp edges, etc. as well as on flat surfaces. Also, when the film is baked to coalesce and cure it, there is less sagging of the primer away from irregularities and metal edges.

The first commercial electrocoat primer was the anodic type. While this was a substantial improvement in rust resistance over previous primers, it was determined that the cathodic type was even better. In 1976 PPG released the first cathodic primer and this has become the standard of the automotive industry. Initially it was customary to apply a supplemental spray coat of an epoxy ester primer-surfacer over the electrocoat primer for optimum thickness and appearance. However, the composition was subsequently improved to provide a thicker deposited film which renders the extra step optional.

CONCLUSION

In summary, the combination of (a) a water-borne cathodic primer, with or without the supplemental spray coat of epoxy ester primer-surfacer, (b) a colorcoat based on a melamine-formaldehyde crosslinked polyester oligomer, and (c) a similarly crosslinked acrylic oligomer clearcoat, is currently the front runner among several competitive finishing systems (Chart #14). Water-based colorcoats and a variety of clearcoats with novel compositions are being considered. Among these are durable epoxies, powder clears, fluorocarbon polymers and crosslinking agents such as polysilanes and polyisocyanates.

Chart 15 summarizes the field performance properties of a number of the more important automotive topcoats that have seen extensive service at one time or another over the past 60 years. The patience and skill of automotive industry personnel during the commercialization of these products is hereby gratefully acknowledged.

CHARTS

CHART #1

AUTOMOTIVE & REFINISH
TOPCOAT COMMERCIALIZATION YEARS

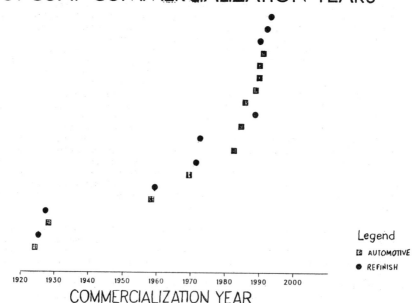

Legend

▣ AUTOMOTIVE
● REFINISH

1920   1930   1940   1950   1960   1970   1980   1990   2000

COMMERCIALIZATION YEAR

CHART #2

TYPICAL NITROCELLULOSE LACQUER VEHICLE

*	NITROCELLULOSE (3/8" BLEND)	50.0 %
	COCONUT OIL - MODIFIED ALKYD	35.0 %
	MONOMERIC PHTHALATE PLASTICIZER	7.5 %
	CASTOR OR BLOWN CASTOR OIL	7.5 %
		100.0 %

* N_2 CONTENT - 11.8-12.2%

* VISCOSITY IN STANDARD 12.2% SOLUTION - 100 CP

* OIL LENGTH 50% (APPROX.)

CHART #3

TYPICAL ALKYD ENAMEL VEHICLE

* DRYING OIL-MODIFIED GLYCERYL PHTHALATE RESIN 95 %

  MELAMINE-FORMALDEHYDE RESIN   <u>5 %</u>

                                    100 %

* OIL LENGTH - 65-70%

  THE OIL ACIDS MOST OFTEN USED:

      SOYA   LINSEED

      TUNG   DEHYDRATED CASTOR

CHART #4

COMPARISON OF "DUCO" AND "DULUX"

DESIRABLE CHARACTERISTICS

"Duco"	"Dulux"
RAPID DRY	NON-BUFFING
EASY SPOT REPAIR	- GOOD FLOW
GLAMOROUS APPEARANCE	- HIGH GLOSS
	LOW INGREDIENT & APPLICATION COST
	- CHEAP SOLVENTS
	- HIGH SPRAY SOLIDS
	GOOD STAIN RESISTANCE
	- (OILS, TARS, ETC.)

UNDESIRABLE CHARACTERISTICS

"Duco"	"Dulux"
ONLY MODERATE DURABILITY	ONLY MODERATELY DURABLE
- (CHALK & DULLS)	- (CHALKS & DULLS)
HIGH INGREDIENT & APPLICATION COSTS	NOT EASILY SPOT REPAIRED
- EXPENSIVE SOLVENTS	REQUIRES HIGH BAKE
- LOW SPRAY SOLIDS	LACKS GLAMOUR
POOR STAIN RESISTANCE	
MUST BE BUFFED	

CHART #5

SPECIFICATION FOR A "PERFECT" FINISH

OUTSTANDING DURABILITY	>	"DUCO" OR "DULUX"
SUPERIOR BLISTER RESISTANCE	>	"DUCO" OR "DULUX"
WIDER COLOR RANGE	>	"DUCO" OR "DULUX"
GLAMOROUS APPEARANCE	=	"DUCO"
NON-BUFFING FINISH	=	"DULUX"
LOW SOLVENT COST	=	"DULUX"
HIGH SPRAY SOLIDS	=	"DULUX"
GOOD STAIN RESISTANCE	=	"DULUX"
RAPID DRY	=	"DUCO"
EASY SPOT REPAIR	=	"DUCO"
OVER-BAKE DISCOLORATION RESISTANCE	>	"DULUX"

CHART #6

TYPICAL ACRYLIC LACQUER VEHICLE

* POLY(METHYLMETHACRYLATE)//METHYL METH-
ACRYLATE/BUTYL ACRYLATE COPOLYMER   BLEND   60%

CELLULOSE ACETATE-BUTYRATE (1/2" GRADE)   20%
MONOMERIC PHTHALATE PLASTICIZER   10%
POLYMERIC PHTHALATE PLASTICIZER   10%
                                                  100%

* MOL. WT. - 75,000-85,000

CHART #7

TYPICAL "DULUX-100" TYPE ALKYD VEHICLE

* COCONUT OIL-MODIFIED ALKYD RESIN   75%
MELAMINE-FORMALDEHYDE RESIN   25%
                                      100%

* OIL LENGTH - 55-60%

CHART #8

TYPICAL THERMOSETTING ACRYLIC ENAMEL VEHICLE

* THERMOSETTING ACRYLIC RESIN   70%

   BUTYLATED MELAMINE-FORMALDEHYDE RESIN   30%

                                                    100%

* THIS HETEROPOLYMER IS BASED ON A BLEND OF MONOMERS CHOSEN TO ACHIEVE THE DESIRED PROPERTY BALANCE AND MAY USE SEVERAL OF THE FOLLOWING: STYRENE AND METHYL METHACRYLATE (FOR HARDNESS), BUTYL OR 2-ETHYLHEXYL ACRYLATE (FOR FLEXIBILITY) AND HYDROXYETHYL ACRYLATE OR METHACRYLATE (FOR CROSSLINKING ACTIVITY). A SMALL AMOUNT OF ACRYLIC OR METHACRYLIC ACID IS ALSO INTRODUCED AS A CROSSLINKING CATALYST AND A PIGMENT WETTING AGENT.
  - ACID NUMBER 16-20
  - RELATIVE VISCOSITY 1.06-1.09

CHART #9

TYPICAL THERMOSETTING ACRYLIC
ENAMEL VEHICLE (REFLOW TYPE)

* THERMOSETTING ACRYLIC RESIN   62.5%

   METHYLATED MELAMINE-FORMALDHYDE RESIN   22.5%

** CELLULOSE ACETATE-BUTYRATE (1/2" GRADE)   15.0%

                                     100.0%

   * HETEROPOLYMER SIMILAR TO THAT SHOWN ON CHART #8

   ** 50% BUTYRYL CONTENT

CHART #10

TYPICAL DISPERSION TYPE ACRYLIC LACQUER VEHICLE

* METHYL METHACRYLATE/2-ETHYLHEXYL ACRYLATE
  GRAFT COPOLYMER DISPERSION SOLIDS   80%

   POLYMERIC PHTHALATE PLASTICIZER   15%

   MONOMERIC PHTHALATE PLASTICIZER   5%

                                     100%

   * MOL. WT. - 70,000-80,000

50

CHART #11

TYPICAL HIGH SOLIDS ACRYLIC ENAMEL VEHICLE

*   THERMOSETTING ACRYLIC OLIGOMER   70-50%

** MELAMINE-FORMALDEHYDE CROSSLINKER   30-50%

                                             100%

    *   MONOMER COMPOSITION:
       STYRENE/ALKYL ACRYLATES/HYDROXYETHYL ACRYLATE MOL. WT. OF
       OLIGOMER 5,000-10,000

    ** FULLY ALKYLATED LOW MOLECULAR WEIGHT MELAMINE-FORMALDEHYDE
       REACTION PRODUCT

CHART #12

TYPICAL COLORCOAT/CLEARCOAT VEHICLES

A. COLORCOAT VEHICLE

    *   THERMOSETTING POLYESTER OLIGOMER   30%

    ** MELAMINE-FORMALDEHYDE CROSSLINKER   30%

       RHEOLOGY CONTROL AGENT   30%

       PIGMENT DISPERSION RESIN   10%

                                          100%

    *   MOL. WT. - 1000 (APPROX.)

    ** FULLY ALKYLATED LOW MOLECULAR WEIGHT MELAMINE-
       FORMALDEHYDE REACTION PRODUCT

B. CLEARCOAT VEHICLE

    *   THERMOSETTING ACRYLIC OLIGOMER   70-60%

    ** MELAMINE-FORMALDEHYDE CROSSLINKER   30-40%

                                         100%

    * MONOMER COMPOSITION:
       STYRENE/ALKYL ACRYLATES/HYDROXYETHYL ACRYLATE

       * MOL. WT. OF OLIGOMER - 5,000-10,000

       ** FULLY ALKYLATED MELAMINE-FORMALDEHYDE REACTION
       PRODUCT SAME AS IN COLORCOAT

CHART #13

l
o
g
v
i
s
c
flash
bake
gel
time

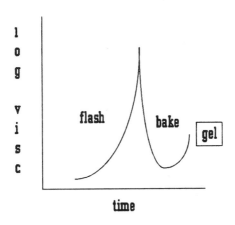

CHART #14

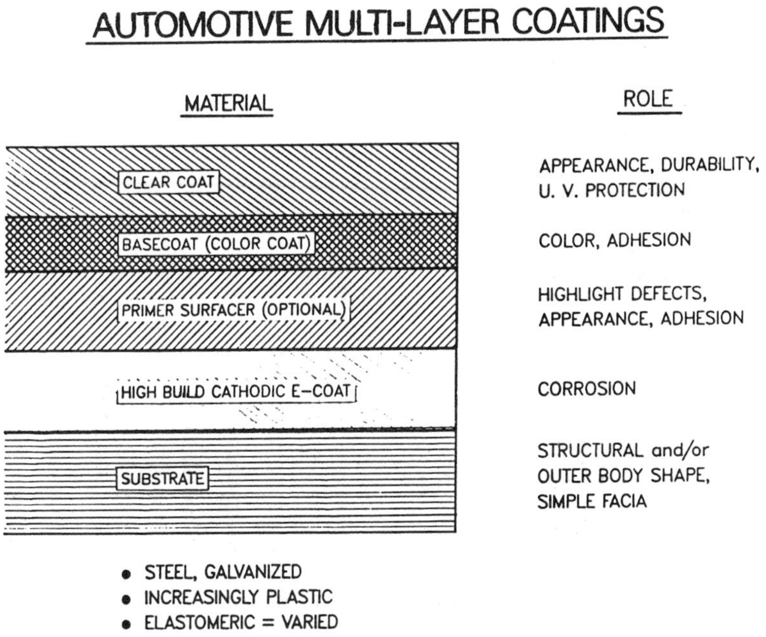

AUTOMOTIVE MULTI-LAYER COATINGS
MATERIAL
ROLE
CLEAR COAT
APPEARANCE, DURABILITY, U. V. PROTECTION
BASECOAT (COLOR COAT)
COLOR, ADHESION
PRIMER SURFACER (OPTIONAL)
HIGHLIGHT DEFECTS, APPEARANCE, ADHESION
HIGH BUILD CATHODIC E-COAT
CORROSION
SUBSTRATE
STRUCTURAL and/or OUTER BODY SHAPE, SIMPLE FACIA
• STEEL, GALVANIZED
• INCREASINGLY PLASTIC
• ELASTOMERIC = VARIED

CHART #15

PERFORMANCE PROPERTY COMPARISON

	"Duco" Lacquer	Lucite Acrylic Lacquer	"Dulux" Alkyd Enamel	"Dulux 100" Alkyd Enamel	"Acrylic" Disp. Lacquer	Convent. Acrylic Enamel	Hi-Solids Acrylic Enamel	Colorcoat Clearcoat System
GLOSS RETENTION (% 60° Gloss)								
3 mo. Del.	51%	94%	87%	85%	Equal to "Lucite" Acrylic Lacquer	--	--	--
6 mo. Fla.	19	90	24	82%		--	--	--
12 mo. Fla.	5	65	5	34%		80%	80%	90%
STAINING BY:								
Road Tar	1	8	6	4	8	9	9	10
Hypoid Grease	2	9	6	5	9	9	9	10
BLISTER RESIST.	8.5	10	7.5	9.5	10	10	10	10
COLD CRACK RESIST.	8	10	10	10	10	10	10	10
CHECK RESIST.	9	10	7	10	10	10	10	10
GASOLINE RESIST.	10	8	10	10	8	10	10	10
YELLOWING RESIST.	8	10	8	10	10	10	10	10
CHIP RESIST.	7	7	7	7	7	7	7	7
EASE OF FIELD REPAIR	10	9	8	7	9	7	7	6

10 = EXCELLENT

REFERENCES

1. U.S. Patent 1,629,999 (May 24, 1927), Edmund M. Flaherty (to E. I. DuPont de Nemours & Co., Inc.)
2. U.S. Patent 1,710,453 (April 23, 1929), Maurice V. Hitt (to E. I. DuPont de Nemours & Co., Inc.)
3. DuPont Refinisher News, January - February, 1974, No. 184, pp. 3-4
4. U.S. Patent 2,934,510 (April 26, 1965), L. W. Crissey and J. H. Lowell (to E. I. DuPont de Nemours & Co., Inc.)
5. C. R. Martens in C. R. Martens, ed., Technology of Paints, Varnishes and Lacquers, Robert E. Krieger Publishing Co., Inc., New York, reprinted with corrections, 1974, Chapter 1, pp. 9-10
6. K. J. E. Barrett and M. W Thompson, Dispersion Polymerization in Organic Media, John Wiley & Sons, Inc., New York, 1975, Chapter 5, pp. 207-212
7. W. J. Elms and R. D. Spitz, J. Paint Technology, 46 (594), 40 (1974)
8. C. Martens, Emulsion and Water-Soluble Paints and Coatings, Reinhold Publishing Corp., New York, 1964
9. Chemical Week 135 (1), 30, (July 4, 1984)
10. M. T. Gillies, Power Coatings - Recent Developments, Noyes Data Corp., Park Ridge, N.J., 1981
11. S. B. Levinson, J. Paint Technology, 44 (570) (Pt.2), 36, (1972)
12. J. E. Huber, ed., Kline Guide to the Paint Industry, 5th ed., Charles H. Kline & Co., Fairfield, N.J., 1978, pp. 46 - 47
13. U.S. Patent 4,495,325 (January 22, 1985), Michael Debergalis and Robert P. O'Fee (to E. I. DuPont de Nemours & Co., Inc.)
14. U.S. Patent 4,591,533 (May 27, 1986), Joseph A. Antonelli and Isador Hazan (to E. I. DuPont de Nemours & Co., Inc.)
15. U.S. Patent 4,238,387 (December 9, 1980), Joseph A. Antonelli (to E. I. DuPont de Nemours & Co., Inc.)
16. U.S. Patent 4,455,331 (June 19, 1984), Robert J. Barsotti (to E. I. DuPont de Nemours & Co., Inc.)
17. R. L. Yates, Electroplating, 2nd ed., Robert Draper LTD, Teddington England, 1970, pp. 178 - 207
18. G. L. Burnside, G. G. Strossberg, and G. F. Brewer, American Electroplaters' Society, Baltimore, May 1, 1965, and Milwaukee, May 5, 1965

HISTORY OF COATINGS FOR AIRCRAFTS

Jan W. Gooch

Georgia Tech Research Institute
Atlanta, Georgia 30332

ABSTRACT

The use of coatings on aircrafts began in 1783 with the
successful flight of the first Montgolfier hot-air balloon,
disregarding mythological flights. The carriage of the
aircraft was modestly decorated with contemporary artist's
paint for aesthetic purposes. Successive aircrafts were
conservatively marked or colored since identification nor
decoration were desirable. However, weight was critical and
coatings were not structural materials and could be eliminated
as they often were. Occasional clear varnishes were utilized
to seal wooden structures. During the World War I era,
aircraft development was accelerated, and "dopants" were
utilized to stream-line and water-proof the fabric skin.
Rarely were colors or decals utilized until military aircrafts
had to be distinguished as friend or foe. Camouflage coatings
made their debut during the first world war, emphasizing that
the visual appearance of aircraft was important after the
shooting started. As more metals were incorporated into
aircrafts, coatings were utilized for prevention of corrosion,
and by World War II standardized coatings were on most military
aircrafts for adjustment of visibility and/or corrosion
prevention. Commercial and military aircraft utilized the
Alclad treated aluminum exterior finish except where color was
important. Today coatings are employed on aircraft interior
and exterior surfaces for aesthetic appearance and protection
of metal surfaces. Organic coatings cannot fly faster than
Mach 3 for long periods of time since temperatures generated
from air friction cause degradation. Most contemporary
commercial and military aerospace crafts are coated with
standardized light-weight and tenacious polysulfide and epoxy
primers topcoated with polyurethane. Flying colors and
corrosion prevention continue to be the primary motivations
for using coatings on aircrafts.

Published 1990 by Elsevier Science Publishing Co., Inc.
Organic Coatings: Their Origin and Development
R.B. Seymour and H.F. Mark, Editors
                                                         55

INTRODUCTION

It was the same everywhere, the notion that a human being might move through the sky at will aboard a machine that was many times heavier than the colorless, odorless, apparently insubstantial fluid that supported it. "Flight was generally looked upon as an impossibility," Orville Wright[1] recalled in 1908 at Fort Myer, Virginia, "and scarcely anyone believed in it until he actually saw it with his own eyes." After the first powered flight in 1903, the mythical flight from Crete of Icarus and his father Daedalus must have occurred to at least some one that witnessed the event. Icarus discovered at high altitude and near the sun that beeswax was not a wise choice of material for cementing and coating feathers. The quest for the selection of perfect coating materials continues.

Coatings have flown on most aircraft as nonstructural materials, which means that the aircraft could fly without them. During the World War I era, an aircraft could have been shot if not painted with the correct color or pattern. This fact made an aircraft's appearance of critical importance and, therefore, coatings were utilized to adjust the visual image of an aircraft. Prevention of corrosion took a longer period to appreciate, but incorporation of corroding metals into aircraft designs also required protective coatings. The initial wooden structures benefited from varnishes and shellac to "seal" the surfaces from moisture, and they were also useful as light-weight materials for stream-lining fabric surfaces which covered wings and fuselage parts. Weight was always and continues to be a critical consideration in designing an aircraft.

The evolutionary development of dopants, paint and coatings influenced use of the colored stuff used on aircraft. Before the nineteenth century, naturally-drying oils (e.g., linseed) and rosins from sources in nature were all they were available for making "dope, dopants or paint". In the nineteenth century and thereafter, improved synthetic resins were prepared and catalytically-cured to form more tenacious films more properly referred to as coatings. Performance of coatings continue to improve for aerospace vehicles, and the industry of coatings for aircrafts has become so enormous that special environmental laws have been passed to regulate volatile organic compounds.

The first term for a material used for marking, water-proofing or coloring an aircraft was called a dopant (D., doop, adj.) which commonly meant, "any of varnish-like products for waterproofing and strengthening the fabric of airplane wings." The term paint (M.E., peinte, n.) worked its way into the aviators vocabulary by the 1920s and commonly meant, "a substance composed of a solid coloring matter suspended in a liquid medium," and the more modern term coating (M.E., cote, n.) was used after synthetic systems were developed such as polurethane and epoxy materials resins in high solids topcoat and primer coatings.

1783 - 1894 BALLOONS AND GLIDERS

On June 4, 1783 in France, Joseph and Etienne Montgolfier invented the hot-air balloon[2]. This famous balloon ascended into the stratosphere carrying the world's first and nonpaying passengers, barnyard animals. The balloon's carriage was modestly decorated with national colors and with contemporary paint formulated from drying oils and ground pigments. The use of coatings was purely aesthetic, but set a precedent for the contemporary commercial blimps today. Later in the 1780s, Jacque Charles invented the hydrogen-filled balloon. The Montgolfier brothers didn't know that they were designing a machine for twentieth century aerial coverage of football games we enjoy on television today. It was obvious even then that a large painted sign on a high flying balloon could be seen from a long distance.

The first balloons were constructed of paper without an undercarriage and there was no impetus to coat them. As the sport of ballooning became popular in central Europe, the navigators felt the need to decorate their crafts for identification and self-gratification. A balloon has a lot of surface area and were new targets for artistic patterns. The art of painting balloons is practiced today.

In 1894, the Chanute-Herring Glider[3] cruised for short distances. A dopant was used to seal the Japanese Silk skin from water and stream-line the structure and prevent the bare fabric skin from flopping in the wind. The dopant[4] consisted of unpigmented cellulose nitrate (balsa wood treated with nitric acid). Still, the dopant added weight and was used sparingly. Other available materials were India rubber, shellac in mineral spirits, and marine glue in naphtha. Lilentha[5] made more than 2000 flights with his "hand gliders" which ended on August 9, 1896 with a crash. His gliders were constructed of wood and fabrics that was probably doped, but with materials of unknown orgin.

1903-1914 WRIGHT AIRPLANES

In 1901, Wilbur Wright learned to fly gliders which was valuable training for a historical event to come. On December 17, 1903, Orville and Wilbur Wright successfully tested their first powered aircraft near Kitty Hawk and Devil Hills, North Carolina[6]. The skin of this aircraft was silk and was not doped or coated since weight was critical. Even the wood frame was uncoated or protected in anyway. However, coatings did not entirely miss this historic flight since some small parts of the engine were coated with a dark paint as observed from inspection of the aircraft in the National Air and Space Museum and original photographs. The Wright Flyer made its debut in 1909 at Fort Myers in Virginia, and set off an avalanche of aviation developments. The Wright EX[7] was the first aircraft used for advertising a product, "Vin Fiz Grape Drink." The lettering was green (ink) and the aircraft skin was natural. With this success, the aerial advertising industry was born and formed a natural marriage with the coatings industry.

1914-1918 MILITARY AIRCRAFT

The aircraft used for military purposes required markings for identification to prevent shooting down the wrong aircraft. The markings were actually "inks" of very low percent solids which were printed on fabric. As the markings became more important, they also became larger. Doping the fabric skin for various camouflage and color continued. The famous "Red Baron" airplane of the Imperial German Army Air Corp. was a bright red color and was flown by Baron Manfred von Richthofen. The arrogant display of a brightly colored red airplane in a war zone was a fatal mistake since he was detected and shot down. Camouflage coatings were available at that time and would have served the Baron much better.

The French army utilized camouflage dope[8] containing aluminum "fleck" for reflecting sunlight and reducing the contrast against the bright sky. These coatings were usually formulated from drying oils and resins. Near the end of the war, most aircraft were colored by some coating method to reduce visibility to the enemy while retaining a distinguishable marking which could be seen at close range. The role of coating materials on an airplane at this time was not defined. Should the airplane be seen or hidden to the observer?

1927 THE SPIRIT OF SAINT LOUIS

The Spirit of Saint Louis was flown by twenty-five year old Charles Augustus Lindbergh from New York to Paris in 1927. Lindbergh accepted the challenge made by Raymond Orteig and won the prize of $25,000. Although he was the seventy-ninth person to cross the Atlantic Ocean by air, his was the first solo, nonstop leap between the North American and European Mainlands. Some of the metal parts of the Wright J-5 Whirlwind engine cowling and skin of the fuselage were coated with an aluminum powder-pigmented cellulose nitrate[9] The basic silvery appearance of the aircraft is due to the aluminum coating. The designator number "N-X-211" on the underside of the wing and rudder was black formulated from a carbon-filled alkyd resin. The printed name on the aircraft was painted using the same material.

1926 - 1938 ALUMINUM VERSUS COATED SURFACES

The economic reasons for using dopants during this period deserve to be mentioned. Dope painted onto fabric surfaces tightens, shrinks and hardens the skin, providing strength not found in unpainted fabric. Dope also waterproofs and prevents absorption of moisture and prevention of rot. Lacquers and oil based enamels were the two most common paints for metal surfaces. Aluminum alloy skin possesses inherent surface strength and does not absorb moisture, although aluminum does oxidize. Intergranular corrosion is a problem and is responsible for fatigue.

The Air Corps' first experience with an unpainted all metal aircraft involved the Ford XC-3 Trimotor[10] airplane. The

corrugated aluminum alloy sheet covering this aircraft was
coated with an electrochemcical film (annodizing) to inhibit
oxidation. In 1932, after 1300 hours of flight time and four
years of service, a depot inspection revealed minute grayish-
black spots on the wing surface. Further testing revealed that
intergrannular corrosion was present and the structural
integrity of the wing was destroyed. In 1934, another similar
airplane (C-9) was deployed for about the same time and was
inspected, and no corrosion problems were discovered. Both
airplanes were built from the same grade of aluminum, but the
C-9 had been coated with 1.5% pure aluminum. This sheet
treatment known as Alclad was able to resist scratching and
offer superior protection for minor weight increases. Alclad
became widely used and enabled construction of aircraft without
the use of painted surfaces. The economic benefits were a
savings of weight and more importantly, overhaul time. Weight
savings in 1934 were 25 pounds for a pursuit ship and 80 pounds
for a bomber, but depot time to remove old paint was 175 to 400
hours, and refinishing averaged a cost of about $250 per
airframe. In 1935, the Air Corps approved plans to take
advantage of unpainted metal tactical aircraft, only to be
delayed by an economic consideration: supplies of Light Blue
and Yellow paint had to be used before a new color scheme could
be adopted. Unpainted tactical aircraft would not be accepted
from manufacturers until 1937 and Technical Orders for aircraft
in service were not changed until March 1938.

By the same plan above, fabric-covered aircraft acting in
a tactical role, or fabric covered portions of metal aircraft,
were to be doped aluminum to achieve uniformity with metal
aircraft.

1921-1945 CAMOUFLAGE COATINGS

The "Report on Camouflage of Day Airplanes'1[1]" published
by the Air Services' Engineering Division in January 1921,
established the guidelines of camouflage schemes and coatings
that would influence our air force for 20 years. Schemes to
decrease an aircraft's visibility when viewed from above, were
called "terrestrial camouflage" or "ground camouflage," and
the report suggested that these consisted of three-color
patterns tailored to fit local terrain colors. Permanent
camouflage dopes were prescribed, with additional colored dopes
to be added as conditions changed. One of the major
innovations in the study of concealment was "celestial
camouflage" to decrease the visibility of aircraft viewed from
below. In tests, clear-doped aircraft with a light yellow cast
became visible at an altitude of 17,000 feet. When
camouflaged, this altitude was lowered to less than 10,000
feet. It was found that when so camouflaged, visible aircraft
appeared to be flying at a much higher altitude, an advantage
in confusing anti-aircraft fire. "Shadow shading" was the
subject of some experimentation, with lighter colors being used
to reduce shadow areas between wings and beneath the tail. The
first major recommendation of this 1921 report was that one or
both national insignia should be eliminated from the wings of
any camouflaged aircraft. Funding was in short supply and the
project was still-born. In ·1930, the Materiel Division became
interested in a temporary means of covering the yellow flying

surfaces of aircraft during the annual field maneuvers. A commercial water based paint was ordered in Olive Drab color, and tested on an XCO-8 observation plane. The result was a relative durable coating which added less than 10 pounds of weight and cost only $1.25 per airframe. Temporary olive drab was stocked and recommended for use in situations where high visibility colors were undesirable.

The next logical step was to apply camouflage on an aircraft as recommended in the Engineering Division's 1921 report, but using water-based temporary paints. This was accomplished in 1932, and a new era of aircraft color schemes followed shortly. Removal of the paint required washing in cold water and light scouring with a rag, although there were exceptions to this rule. Certain supplies of Purple and Black (carbon black) paints clung tenaciously to any surface they contacted, but special cleansers were available for their removal.

The use of permanent camouflage paints in 1940 brought about the decline of temporary schemes. Olive Drab and Neutral Gray camouflage, while predominant, was not the exclusive color scheme used on USAAF combat aircraft. The Air Corp Board recommended the use of Medium Green 42 with Olive Drab over predominantly green terrain; Sand 26 could be similarly employed over desert areas. One of the problems faced by every combatant air force has always been the rapid identification of friendly vs. enemy aircraft during the action of a battle. Accurate aircraft recognition during the World War II was not an exception, with inadequate training, poor visibility, or confusion of the moment being contributing factors. Often the case was "shoot first and ask questions later."

First issued in 1919, Quartermaster Corps Specification 3-1 was the standard for Army paints until 1943. As originally published, the specification included a color card with 24 paint samples; the later camouflage colors were sent to the QM Corps for approval, and carried consecutive numbers, but were never added to the color card. Air Corps Bulletins were printed and circulated with these samples. To complicate matters, a number of other specifications were employed by the Air Service, Air Corps, and Air Force and it is likely that supplies from different specifications could be found in the same stockroom on any given date. Throughout the '20s and '30s, the Air Corps and the Navy had maintained separate and unrelated standards for colors. Although both services used red, white and blue insignia and painted their wings a high visibility yellow, the colors were distinctly different even to the casual observer. By 1939, agreement was reached on a single "Army-Navy Aircraft" standard for all peacetime colors; paints remaining in stock continued to be used until depleted. It was not until August 1942 that camouflage colors were brought under similar control.

1943 EARLY STEALTH AIRCRAFT

One of the first "stealth" aircrafts consisted of a Mosquito airframe of the British Special Duties Squadron and powered by twin Rolls-Royce Engines. This aircraft was

primarily constructed of plywood and coated with a non-glossy
black finish. The plywood construction foiled radar of that
day, and the carbon black and nonspecular appearance reduced
visibility at night. This aircraft was called a "Moon
Airplane"[12] since it flew only at night and at high altitudes,
above 12,000 feet. Its mission was to fly between England and
Europe to deliver special agents, one of whom was Dr. Niels
Bohr. Dr. Bohr was secretly flown from Sweden to Scotland in
1943 to assist the Allies in the construction of the atomic
bomb and deprive the Nazis of his work in nuclear physics. On
this particular mission, the plane was forced to fly below
12,000 feet due to a defect in Dr. Bohr's oxygen supply, but
the plane was not discovered until it reached Scotland, the
flat-black coating and the low radar cross-section made the
aircraft invisible to German interceptors. However, with
todays technology, the infrared emissivity of that coating
would have been detectable against the very low emissivity of
the sky.

1950-1960 EXPERIMENTAL SUPERSONIC AIRCRAFT

     The X - series of Bell Corporation experimental aircraft
build for the U. S. Air Force was rocket powered and usually
coated. The sound barrier was shattered in 1947 by the Bell X-
1 (Glamorous Glennis) aircraft[13] and was piloted by Charles E.
"Chuck" Yeager. The X-1 was coated with an orange pigmented
enamel coating. The X-2 aircraft was coated with a similar
material, but with a white color.

     The X-15 aircraft did not possess an organic coating due to
the Mach 6.8 air speed. The following equation[14] shows the
relationship between skin temperature and velocity.

$$T = 75 \ (M)^2$$

where T = Temperature, $^{\circ}$F, of the skin
      M = Mach number or number of times the speed of sound

At Mach 2.53, the skin temperature reaches 482°F(250°C) which
is the decomposition temperature of most organic coatings.
This relationship has prohibited the use of organic coatings on
space
craft which experience high temperatures.

1970 - 1989 CONTEMPORARY COATINGS FOR COMMERCIAL AND MILITARY
            AIRCRAFTS

     The research philosophy of the fomulator changed from
improving or developing coatings for new applications to
developing new coatings for low Volatile Organic Compounds
(VOC) applications·1[5] The rediscovery of corrosion resistant
primers and exterior polyurethane topcoats, utilizing new
technologies, comprise exempt solvents, water-borne and higher
solids coatings. Before discussing new formulations, it is
necessary to define VOC per California Rule 1124[16] (July 1989).

Volatile Organic Compounds

Volatile organic compounds (VOC) as defined in California Rule 1124 - Aerospace Assembly and Component Coating Operations is any volatile compound of carbon, excluding the following:

> Methane,
> Carbon dioxide,
> Carbon monoxide,
> Carbonic Acid,
> Metallic Carbides or carbonates
> Ammonium carbonate,
> 1,1,1-Trichloromethane,
> Methylene chloride,
> Trifluoromethane (CFC-23),
> Trichlorotrifluoroethane (CFC-113)
> Dichlorodifluoromethane (CFC-12)
> Trichlorofluoromethane (CFC-11)
> Chlorodifluoromethane (CFC-22)
> Dichlorotetrafluoroethane CFC-14)
> Chloropentafluoroethane (CFC-115)

Grams of VOC per liter of "coating," less water and less exempt compounds, is the weight of VOC per combined volume of VOC and coating solids.

The maximum VOC values for aerospace coatings per Rule 1124 less water and exempt compounds are listed below:

Coating	VOC,g/l
Primer	350
Topcoat	600
Phosphate Ester - Resistant Primer	650
After 1 January 1988	
Pretreatment Coating	780
Adhesive Bonding Primer	850
Flight-Testing Coating	840
Fuel-Tank Coating	720
Electric or Radiation-Effect Coating	800
Maskant for Chemical Processing	1200
Solid Film Lubricant	880
Temporary Protective Coating	250
Space Vehicle Coatings	
Electrostatic Discharge Protective Coating	800
Other Space-Vehicle Coatings	1000

Coatings Specifications

Specifications for military and commercial aircraft are similar, but possess differences with regard to testing procedures. The major commercial and military specifications for topcoats and primers are thoroughly discussed in documents referred in the Reference section[17-23].

Exempt Solvent Technology

The use of exempt solvents was one of the first answers to low VOC coatings as reported by Fujihara.[24] Halogenated solvents were substituted for other organic solvents in established topcoats and primers, and this type of formulating uncovered a number of inadequacies. Instability during storage and application properties have proven to be the major obstacles in formulating with exempt solvents. The problem with storage instability is caused by the poor solvent characteristics of halogenated solvents as methylene chloride and 1,1,1-trichloroethane. Although methylene chloride is more polar than 1,1,1-trichloroethane and therefore a better solvent, it is not used extensively because of the associated health hazards. In many cases 1,1,1-trichloroethane is difficult to use and has warranted the development of different resin systems for established applications. The same properties of the exempt solvent that caused the deficiencies in storage stability also contributed to application problems.

The poor solvent characteristics and the rapid vaporization rate of halogenated solvents can be correlated to dry spray, poor wetting and uneven leveling properties that often are associated with exempt solvent primers. The spraying characteristics of exempt solvent coatings can be improved with polyacrylate leveling agents or silicone flow additives, but careful selection and blending of the organic solvents has proven to be the most effective solution to the application problems.

At this time, some corrosion resistant primers based on exempt solvents already have found use both in commercial and military applications. The MIL-P-23377 primers are formulated using 1,1,1-trichloroethane and they basically use the same resin system as their high VOC counterparts. These epoxy polyamide primers meet all requirements of the MIL-P-23377 with a VOC of less than 350 g/l.

The formulation of Skydrol hydraulic fluid (especially phosphate ester type) resistant primers based on trichlorethane for commercial aircraft has been more difficult due to the solubility and instability problems of the established resins in 1,1,1,-trichloroethane. These systems have required improved resins.

Exempt solvent-based topcoats have found limited use in the aircraft industry. Some aircraft manufacturers in the eastern United States currently use 420 g/l polyurethane topcoats. The coatings are reported to meet the requirements of the Air Force MIL-C-83286B specification.

Water-Reducible Technology

The second low VOC category is water-borne coatings, which utilize solvents and resins that are compatible with water. The water is used to dilute the coatings to a spray-application viscosity. At present, only the military has authorized the use of water-reducible primers for critical areas of aerospace applications with the MIL-P-85582 specification. Current test results of waterborne primers have not yet proved this type of primer to be suitable for commercial airline application.

High Solids Technology

Experience with high solids coatings indicates two major difficulties in meeting the aircraft industry's performance requirements. The first comprises balancing dry times with pot life. Generally, high solids two-component coatings have a shorter pot life than low solids coatings at equal dry times. The second problem is that the viscosity in high solids coatings compound the difficulties in maintaining flexibility without sacrificing fluid resistance. The logarithm of viscosity varies inversely with volume of solvent, depending on the compatibility of the solvent.

High solids primers currently are under development. Along with the problems normally associated with high solids coatings, the primers also must contend with excessive film build-up. A dried film thickness of less than 1.0 mil required for primers has been difficult to achieve due to the increased non-volatile content and higher viscosities of high solids primers.

The development of high solids polyurethane topcoats has progressed more quickly than their epoxy counterparts. The military aircraft industry has written and approved specifications for high solids polyurethane topcoats, U. S. Air Force MIL-C-83286B, and U. S. Navy MIL-C-85285 (specifically for high solids polyurethanes).

The rapid development of high-solids polyurethanes can be linked directly to the advent of new resins. The use of low viscosity acrylic resins can provide the dry times typical of the conventional polyurethanes, but without the flexibility. The need for flexible polyurethane topcoats has led to the development of low viscosity branched polyester resins used in conjunction with various isocyanate coreactants. The problem associated with the use of these less viscous polyester resins for high solids coatings has been balancing dry times and pot life.

Electrostatic spraying of high solids coatings on airframes improves transfer efficiency, and at least one manufacturer is exploring this application of high-solids compliant coatings. However, the resistivity (minimum of 0.15 megaohms) of the liquid coating is critical and not always suitable for electrostatic spraying. In addition, recoating an aircraft requires "grounding" or creating an electrical charge on the surface of the aircraft which is objectionable with regard to interference with sensitive instruments within the aircraft.

A one-coat system for Navy aircraft applications was reported by Hegedus·2[5] The novel one-coat system is a self-priming coating and has the potential of replacing the conventional primer-topcoat systems. Laboratory testing indicates acceptable salt spray chamber results. Field testing of the F-14 Tomcat is planned.

1958-1989 SPACECRAFT COATINGS

The National Aeronautics and Space Administration (NASA) had its orgins in 1958 and officially referred to as NASA during President Kennedy's Administration. Research facilities for coatings include the Langley Research Center in Hampton, Virginia, the Johnson Space Center in Houston, Texas and the Marshall Space Center in Huntville, Alabama. Among special coatings projects have been the Lunar Obitor, Apollo, Viking Lander and Shuttle Orbitor. The Space Station Freedom is presently in the research stages. Present research concerning this project focuses on the durability of paints and coatings to withstand an extended period of time in space. The return of a Long Duration Experiment Facility satellite that has been conducting coatings and materials experiments for the last five years will give valuable insight into coating's performance in space.

A major emphasis of current research concerns the Space Station Freedom and highly stable coatings[26] than can perform acceptably at altitudes of 165 to 300 nautical miles from the earth, low-earth orbit. The radiation including ultraviolet is a factor, but not the primary problem in this environment. A vacuum of about 10^{-4} mm Hg and thermal cycles of about $-100^{o}F$ ($-73.3^{o}C$) to about $100^{o}F$ ($37.8^{o}C$) exists. Chemically aggressive atomic oxygen is found at these altitudes which has a degrading effect on almost all polymer systems and other materials. Typical aircraft coatings will not perform in this environment as shown from actual testing[27]. Silicone-based and fluorinated polymers are better suited for this environment. A mixture of magnesium fluoride and polytetrafluoroethylene has given good resistance to atomic oxygen flux. Chromic acid-anodized and sulfuric acid-anodized are being investigated to provide the 30 year lifetime which will experience 175,000 thermal cycles. Second-surface coatings consists of a transparent polymeric film applied over a reflecting, opaque metal as silver, which will reflect radiation and maintain thermal control of the surface. Besides the surface of the basic structure, coatings are required for instruments and reflecting solar dishes.

Further out into space, there exists another environment of the solar-wind and electron/proton effects. The increased vacuum has little effect, and ultraviolet radiation intensity would be similar. Thermal effects would be similar without as many thermal cycles as in low-earth orbit.

REFERENCES

1. Boyne, Walter J.; "The Smithsonian Book of Flight"; Orion Books: New York, New York, 1987, pp. 126-138.

2. Ibid. pp. 33-39

3. Ibid. pp. 46-48

4. Letter from A. Herring to O. Chanute, Dece. 12, 1894

5. Ibid. ref. 1, pp. 38-39

66

6. Ibid. ref. 1, pp. 44-45, 50-54

7. Ibid. ref. 1, pp. 68

8. Bell, D.; "Air Force Colors, Vol. 1"; Squadron/Signal
    Publications: Carrollton, Texas, 1979, pp. 6-95.

9. Wilson, L., Archives Division, National Air and Space
    Museum, Smithsonian Institution, August 2, 1989,
    personal communication

10. Ibid. ref. 8, p. 28

11. Ibid. ref. 8, pp. 54-59

12. Stevenson, W.; " A Man Called Intrepid"; Ballantine Books:
    New York, New York, 1976, pp. 478

13. Ibid. ref. 1, pp. 203

14. Hall, J.; Hebard, R., "Structures, Theory, and Materials
    Method", Aeronautical Engineering Review, (December 1953),

15. Gooch, J.W., "Aircraft Coatings Meet Challenge of VOC
    Compliance", Modern Paint & Coatings, 1989, 79(9), p.36-
    38.

16. Rules and Regulations, South Coast Air Quality
    Management District, Rule 1124 - Aerospace Assembly
    and Component Coating Operations, Amended 3 April 1987,
    9150 Flair Drive, El Monte, California 91731.

17. NAS No. 1545, Primer Coating: Low Volatile Organic
    Content Compound, Chemical and Solvent Resistant,
    National Aerospace Standard, Aerospace Industries
    Association of America, Inc., 1725 De Sales Street,
    N.W., Washington, D.C. 20036

18. NAS No. 1545, Coating: Low Volatile Organic Content
    Compound, for Topcoat Applications.

19. MIL-P-233770 (Int. Amendment 5) Military Specification
    Primer Coatings: Epoxy-Polyamide, Chemical and
    Solvent Resistant.

20. MIL-P-87112, Military Specification Primer Coatings:
    Polysulfide (#PR1432GP).

21. MIL-C-83286B (Amendment 2) Military Specification
    Coating, Urethane, Aliphatic Isocyanate, for Aerospace
    Applications.

22. MIL-P-85582, Military Specification Primer Coatings:
    Epoxy, VOC Compliant, Chemical and Solvent Resistant.

23. MIL-C-85285, Military Specification for Aerospace
    Equipment, Urethane Coating, Aliphatic.

24. Fujihara, G., "New Technology Is Used For Old
    Applications In Aerospace Applications", Modern
    Paint and Coatings, September 1988, pp..

25. Hegedus, C. R., Development of a Primer/Topcoat and
Flexible Primer for Aluminum, Phase Report, Project
No. R534A52, Naval Air Development Center,
Warminister, Pennsylvania 18974-5000.

26. Resha, K., "The Long Term Demands of Space Age Coatings",
Modern Paint & Coatings, 1989, 79(9), p.44-45.

AKNOWLEGEMENTS

A special thanks to Dana Bell and Howard Wolko of the
National Air and Space Museum, Washington, D.C. for their
valuable information; Eugene Bishop and David Ellicks of
Warner-Robins Air Logistics Center, Robins Air Force Base for
their cooperation and knowledge of Air Force coatings
materials; the Air Force Wright Aeronautical Laboratory,
Wright-Patterson Air Force Base; the Air Force Museum, Dayton,
Ohio; and the Deutsches Museum, Munich, Federal Republic of
Germany.

HISTORY OF APPLIANCE COATINGS

Thomas J. Miranda
Whirlpool Corporation
Benton Harbor, MI 49022

INTRODUCTION

The appliance industry traces its beginning to the early part of the twentieth century when hand operated washing machines, iceboxes, and wood, kerosene, coal or gas fueled ranges were common. The turning point was the development of the universal electric motor, which led to the development of the vacuum cleaner by the Birtman Electric Company of Chicago [1]. By 1909, vacuum cleaners and washers were being powered by fractional horsepower motors and by 1920, the first million washers and vacuum cleaners had been sold. With rural electrification, the major appliance industry took hold, such that from 10,000 refrigerators in 1920, the industry now sells over 7 million today.

The first Maytag washer was built in 1907 with a wooden tub and a hand cranked dolly to move clothes through the water. Two years later, a wringer was added, then a pulley which permitted an electric motor (1910) or gasoline engine (1914) to drive the unit. In 1919 the famous square cast aluminum tub was added; then in 1936, a porcelain tub. The automatic washer was introduced by Maytag in 1949 [2].

Whirlpool Corporation began in 1911 with the Upton Machine Company which produced electric powered washers. In 1916, the first orders for Sears washers were received which marked the beginning of a long relationship between the two companies. In 1947, Whirlpool produced the first automatic washer for Sears and for Whirlpool in 1948. Today, Whirlpool is the world's largest manufacturer of major appliances.

The projected sales for major appliances in 1989 is some 53 million units. Major appliances include compactors, dishwashers, disposers, dryers, freezers, microwave ovens, ranges, refrigerators, washers and water heaters. In addition, some seven million air conditioners are expected in the same year. This large volume represents a considerable amount of coatings, metal preparation chemicals and metal [3].

Appliances are classified as white goods, those mentioned above and brown goods which include television, VCR'S, air conditioners, and furnaces and traffic appliances which include hand mixers, toasters, irons, fans, blenders and other portable units.

Coatings used on appliances range from organic coatings to inorganic coatings which include porcelain on washer baskets, range tops, cavities and burner boxes and in some cases heat exchangers. Coatings must provide both an aesthetic function, as well as, protection against corrosion.

In this chapter, we examine the history of appliance coatings from early times, the changes which occurred as a result of new polymer developments and those brought about by the energy crisis and government edicted rules pertaining to environmental issues [7]. Future trends in appliance coatings will also be cited.

EARLY COATINGS

For a review of appliance coatings the reader should consult the works of Shur [4], Miranda [1,5,6] and Brendley and Bakule [8] In the years prior to World War I, appliance coatings were derived from natural oils and resins or varnishes prepared from them. These coatings required long drying times at ambient temperatures or long baking times.

Published 1990 by Elsevier Science Publishing Co., Inc.
Organic Coatings: Their Origin and Development
R.B. Seymour and H.F. Mark, Editors

69

The most durable were the "japans", black coatings obtained by cooking bitumens with natural oils.

The history of appliance coatings traces a number of quantum leaps as new technologies emerged. For example, varnish technology improved with the use of phenolics and tung oil modified phenolics, which greatly speeded up the drying time. At the end of World War I, another leap forward occurred, when faced with an excess of gun cotton, acetone and butanol, nitrocellulose esters became available which afforded fast drying laquers, revolutionizing the industry. Other modified low viscosity nitrocelluloses followed, as well as, modified lacquers plasticized with dewaxed damar gums for top coats on refrigerators.

The next big advance was the development of the alkyd resin by Roy Kienle [9]. Early alkyds were slow drying with bake times up to 2 hours at 250°C. After this, solution vinyls and urea resins (1936), melamines in 1940, silicones in 1944 and epoxy resins in 1947 were used in appliance primers and top coats [4].

These coatings were then supplanted with the development of the thermosetting acrylic by Strain in 1939 [10]. Up to the time of Strain's work, only thermoplastic acrylic resins were available and though they produced excellent top coats, they did not have the performance required to overcome detergents and corrosion requirements of the appliance market. Strain prepared copolymers of acrylic esters and acrylamide, then formylated the amide to produce a hydroxymethyl group capable of further reaction:

$$CH_2 = CH + CH_2 = CH = --CH_2 - CH - CH_2 - CH ---$$
$$\quad\quad |\quad\quad\quad\quad |\quad\quad\quad\quad\quad |\quad\quad\quad |$$
$$\quad COOR\quad\quad CONH_2\quad\quad\quad COOR\quad CONH_2$$

$$--CH_2 - CH - CH_2 - CH --- + n\ CH_2\ O = --CH_2 - CH - CH_2 - CH --$$
$$\quad |\quad\quad\quad |\quad\quad\quad\quad\quad\quad\quad\quad\quad\quad\quad\quad |\quad\quad\quad |$$
$$\quad COOR\quad CONH_2\quad\quad\quad\quad\quad\quad\quad COOR\quad\quad CONH$$
$$\quad\quad\quad\quad\quad\quad\quad\quad\quad\quad\quad\quad\quad\quad\quad\quad\quad\quad\quad |$$
$$\quad\quad\quad\quad\quad\quad\quad\quad\quad\quad\quad\quad\quad\quad\quad\quad\quad\quad CH_2OH$$

The hydroxy function was capable of further reaction with melamines to form a crosslinked structure. The acrylic developed by Strain was unstable and it was not until the early sixties that Roger Christenson developed a stable thermosetting acrylic [11]. This led to another revolution in appliance finishes as thermosetting acrylics became the Hallmark of appliance finishes.

COATING PRACTICE

METAL TREATMENT

Appliances are manufactured in large highly automated plants in which miles of conveyers are employed. As a result, one of the largest manufacturers can produce over 15,000 washers or dryers in a single day. The coating process is generally tied to this conveyerized system and involves: cleaning and metal treatment, priming and baking and topcoat and baking.

Cleaning and metal treatment are the most critical steps in any meaningful coating process. The surfaces most widely used in making appliances are cold rolled steel, and more recently galvanized steel. These surfaces contain oils, dirt, oxides and other contaminants from the manufacturing process which must be removed. Alkaline cleaning with detergents is widely used at temperatures of 74°C to melt the fats and oils and to remove smut from the milling process. This is followed by a phosphate treatment of either iron or zinc to create a layer of zinc or iron phosphate, which provides corrosion protection and a surface for the coating to adhere. Generally, iron phosphate is deposited at 70-100 mg per sq. ft. while zinc phosphate is deposited at 180-210 mg per sq. ft. The amount of phosphate can be determined by acid stripping or more recently, zinc phosphate can be measured using infrared measurements directly [5]. Of more importance is the size and shape of the phosphate crystals which can be examined using Scanning Electron Microscopy.

PRIMING

Priming of appliances was usually conducted by dipping. This involved large tanks filled with flammable paint which was a fire hazard. This method gave way later to flowcoating in which the amount of coating was limited to a few drums which was recirculated. Unfortunately a vapor chamber is required which is an unsafe situation in a large plant. In fact, the General Motors Transmission plant in Wixam, Michigan was destroyed by a fire which began when a welder's torch ignited a flowcoating line. Following the application of primer, the appliance parts are baked in ovens which are gas, oil, electric or steam fired at temperatures ranging from 275° to 450°F.

Primers used for appliances include alkyds modified with melamine, urea or phenolics which were used into the early 1960's. These gave way to epoxy ether resins modified with melamine or phenolics [4]. Epoxy resins modified with fatty acids and cured with melamines would pass corrosion and detergent resistance required by appliance manufacturers [12]. Other primer developments will be discussed under electrocoating.

TOPCOATS

Topcoats are applied by hand spray, air or electrostatic, electrostatic bells or discs or in special cases by flowcoating or electrocoating. (For example, airconditioner cabinets are topcoated in a single step by electrocoating.) The topcoats are then baked usually at lower temperatures than primers i. e. 275 - 350°F.

RESIN TECHNOLOGY

Alkyds

Following the development of alkyds by Kienle, the alkyd based coatings enjoyed a large share of the appliance market. Improvements were made with efforts to minimize the problem of long term yellowing and to improve alkali and stain resistance [13]. The primary color for appliances was white, hence the name 'white goods' and this could be achieved with drying oils like coconut oil or pelargonic acid and melamine resins. To speed up the drying time, styrenated alkyds could be used with some loss in color stability. Using non drying oils, these alkyds required epoxy primers.

These oil based alkyds were then replaced with non drying polyester type alkyds [14]. This development occurred in 1958 and became more dominant after that, with the availability of isophthalic acid by Amoco. Oil free polyesters could be formulated from 2-ethyl hexoic acid, trimethylolethane, and propylene glycol. Other approaches include polyesters made from p-tertiary-butyl benzoic acid, pelargonic acid, phthalic anhydride, trimethylol ethane, glycerol, and maleic anhydride [15] which can be acrylated.

Acrylics.

While the alkyd-melamine resin system was an excellent coating, the long term yellowing on refrigerators and the lower corrosion and detergent resistance was not that desirable. Acrylic resins were widely used in automotive industry, but were thermoplastic and not suitable for appliances. The discovery by Strain of a thermosetting acrylic resin launched an intensive research effort by many investigators to make practical thermosetting acrylics and to develop patent positions which would allow firms to sell these without infringement.

The earliest success story came when Roger Christenson of PPG Industries succeeded in synthesizing a practical acrylic system. His work also involved an acrylamide based polymer backbone which was then reacted with formaldehyde to provide functional groups for further reaction. Unfortunately, the hydroxy function was not stable and required an additional step for stabilizing the hydroxy function. Once this was done, the thermosetting acrylic was here to stay. Christenson's method was complicated since a number of steps at different pH was required, to complete the reaction. Sekmakas, Desoto, Inc., developed a simple straight forward means for accomplishing the synthesis using an alkaline medium [15]. For a review of acrylic developments, the reader should consult the works of Gerhart [16], Brown [17], Mercurio [18], Solomon [19] and Paul [20].

Other approaches to developing thermosetting acrylics via functional groups include, acid by Gaylord, who copolymerized glycerol monoallylether/methyl methacrylate/butyl acrylate and 10% methacrylic acid. Crosslinking was effected by blending with urea or melamine resins [21]. Vasta prepared an acrylic terpolymer which he cures with epoxy resins [22]. Segall and Cameron prepared styrene, alkyl acrylate and acrylic acid copolymers which they cure with an epoxy resin catalyzed with a quaternary ammonium halide [23], while Applegath used liquid aliphatic epoxies to cure acidic copolymers [24].

Hydroxyl group modified thermosetting acrylics were prepared by Vasta, using an acidic copolymer which was then treated with 1,2 butylene oxide to form a hydroxyl group and subsequently cures with aminoplast resins [25].

Epoxy modified thermosetting acrylics were prepared by Simms using glycidyl acrylate or methacrylate. These resins are then cured with amines or dibasic acids [26]. Ravve and Khamis prepared copolymers of styrene, 2-ethylhexyl acrylate and glycidyl methacrylate which they cured with citric or phthalic acid [27].

A novel method for preparing a thermosetting acrylic involves the preparation of a copolymer of styrene, an acrylic ester and acrylic acid. The copolymer is then reacted with tris (hydroxymethyl) amino-methane to form an methylolated polymer which is cured with melamine resin [28]. Further heating converts the system to an oxazoline modified acrylic which can be self cured or cured with epoxy or amino plast resins. A summary of acrylic types is shown in Table 1.

Table 1

THERMOSETTING ACRYLICS

Type	Functional Monomer	Crosslinker
Acid	$CH_2 =CH\text{-}COOH$ (CH_3)	Epoxy resin

Acrylic (Methacrylic) Acid

Hydroxyl $CH_2=CH\text{-}CO$ Melamine
$\quad\quad\quad\quad\quad\quad | $
$\quad\quad\quad\quad\quad O\text{-}CH_2\text{-}CH_2OH$

Hydroxyethyl acrylate or Methacrylate

Epoxy $CH_2=CH\text{-}CO \quad\quad O$ Acid
$\quad\quad\quad\quad\quad | \quad\quad\quad / \; \backslash$
$\quad\quad\quad\quad O\text{-}CH_2\text{-}CH\text{-}\text{-}CH_2$

Glycidyl acrylate or Methacrylate

Methylol
$\quad\quad\quad\quad\quad CH_2=CH\text{-}CO$ Melamine
$\quad\quad\quad\quad\quad\quad | $ Epoxy
$\quad\quad\quad\quad\quad NH\text{-}CH_2\text{-}OH$

Methylolated acrylamide

Oxazoline $-CH_2\text{-}CH-$ Melamine
$\quad\quad\quad\quad\quad\quad | $
$\quad\quad\quad\quad\quad\quad C$
$\quad\quad\quad\quad\quad / \; \backslash\backslash$
$\quad\quad\quad\quad\quad O \quad\quad N$
$\quad\quad\quad\quad\quad | \quad\quad\quad |$
$\quad\quad\quad\quad CH_2\text{-} C \; (CH_2OH)_2$

REGULATORY CLIMATE

Thermosetting acrylics played a prominent role from their discovery until the late sixties, when a new concern overshadowed the coatings industry in the form of new government regulations. The early efforts in solvent emission reduction came with Rule 66 from the Los Angeles Pollution Abatement District. Citing paint solvents as contributors to ozone generation, this rule suggested that coatings contain less photoreactive solvents. Acrylics require strong solvents and being formulated at 34 % volume solids, produced significant emissions. Attempts to raise the solids level were not immediately successful since the equipment (electrostatic bells and disks) would not handle the higher viscosities. This caused a lot of concern in the appliance and other industries which now became more involved in environmental issues. Coupled with this was the Arab Oil Embargo, which laid havoc to an already stressful situation as the industry was now confronted with a triple threat; Energy, Materials and Government.

While all of this seemed gloomy, there was a bright side. As a result of these new concerns, there also was an opportunity to bring on new technologies which would never have surfaced without these outside influences. As a result major efforts were directed to water dilutable, high solids and powder coatings.

The major effort supported by the government and urged onto the appliance industry was to convert to exempt solvents. While this had good short term effects, it never did address the long term problem of solvent emissions since exempt solvents would still contribute the same amount of solvent to the atmosphere [29].

RESPONSE TO ENVIRONMENTAL/ENERGY CONSTRAINTS

One of the first successful responses to these concerns was the implementation of an acrylic coating on range doors replacing porcelain. The industry standard for range doors had been porcelain enamel which involves high firing temperatures, 1,500°F, and produced a brittle coating. Using the acrylic coating, significant energy savings were realized. This was an industry first which later became a standard in the Appliance Industry [7].

The next application for a porcelain replacement coating was on chest freezer liners. Two approaches were available; a two component urethane or powder coating. Powder coating was used in a highly automated process which eliminated a porcelain furnace which was fired with propane fuel.

WATER BORNE AND HIGH SOLIDS COATINGS

While the industry was scrambling to convert to exempt solvents, Whirlpool Corporation adopted a different approach and worked with their suppliers to employ new technology to meet environmental/energy goals. The first success was in the conversion of a solvent based flowcoater used for coating dryer drums at Findlay, Ohio. At the time, early seventies, no practical water borne flowcoating with appliance quality was available. It was here that the Glidden Company stepped in and committed a team, headed by Dr. Harry Kiefer, to develop such a coating which effectively solved the problem and some six months after the start of the program, dryer drums were coated with a water dilutable flowcoat. This represented another Industry First!

The water borne flowcoat system greatly improved emission reduction, but it was not applicable for topcoats since these are applied by electrostatic means. Top coats consisted of thermosetting acrylic applied at 34 volume percent solids on Ransburg disks revolving at 900 rpm. PPG introduced a higher solids polyester which could be applied using higher rpm disks so that a 52% solids coating was successfully applied. Although this did not satisfy the EPA objective of 62.5 % solids coatings, the improvement was significant in meeting the "bubble concept" permitting continued operation of the plant.

A key to the success of this effort was the joint efforts of the coating supplier, the equipment manufacturer and the user who worked in concert to address a problem of interest to all parties.

ELECTROCOATING

For many years, automotive and appliance primers were deposited by using a flow coat method. In this application, the liquid coating was pumped and flowed onto auto bodies and washer cabinets. The evaporating hydrocarbon solvents provided flow and leveling, but in recesses, such as rocker panels and window wells, the solvent also stripped the paint from the surface and caused early corrosion failures. As the paint drained from the metallic surfaces, the coating tended to bead up at the edges and gave a wedge effect i.e. thinner films at the top and thicker film at the bottom. In addition, a disastrous fire at a transmission plant in Wixom, Michigan called attention to the hazardous nature of the flow coat process. As a result, the coatings industry began a search for safer coatings and therefore, water dilutable coatings emerged as a viable means for achieving this goal[6,30].

Anodic Electrocoating.

The commercial development of anodic electrocoating must be credited to the pioneering work of Dr. George E. F. Brewer, a Staff Scientist at the Ford Motor Company, who was seeking a painting method which would overcome the limitations of the flow coat process [31]. Searching for a solution, Brewer turned to electrocoating using latex emulsions which were not very effective. He then asked the Glidden Company for assistance, where Alan Gilchrist developed maleinized oils, which proved the feasibility of the process. Because of more effective edge coverage and solvent resistance of the deposited coating, greatly improved corrosion resistance of automobiles was achieved. In the anodic process, the polymer carries a negative charge, which is obtained by neutralizing the acid function with ammonia, an amine or with an alkali metal hydroxide. The part to be coated is the anode. After the part is removed from the electrocoat tank, the deposited film is now less soluble in water and can be rinsed with water, then baked to form a crosslinked coating with excellent edge coverage. The rinse can be treated with an ultrafilter which removes water and returns the coating solids to the tank, making the process highly efficient in material usage.

Anodic electrocoating is widely used in metal coating for light fixtures, appliance primers and general metal finishing. Corrosion resistance is good and light colors are achievable.

[An interesting aside to this development is that when Dr. Brewer and his manager, Gilbert Burnside, approached Ford's management to obtain financial support for the project, the Executive listened politely to their proposal. They described how they would fill a 50,000 gallon tank with a water soluble paint, dip the entire car body into the paint, then apply 300-500 volts dc to the system and 'plate' a film which would now be insoluble. After dismissing the adventurous duo, he turned to his secretary and said:

"Not one dime for those two lunatics"!]

Today electrocoating is practiced around the world!

The first application of anodic electrocoating in the Appliance Industry occurred in 1968 at the Clyde Division of Whirlpool Corporation where automatic washer cabinets were coated replacing a flow coat operation. Here are examples of technology transfer between two major industries. In the first case, Anodic Electrocoating was developed and proven in automotive coating, while in the second, Cationic Electrocoating, was developed in the appliance industry and transferred to the automotive industry.

Cathodic Electrocoating

The development of cationic electrocoating is credited to the efforts of Joseph F. Bosso and Marco Wismer of PPG Industries [32]. The successful proof of this technology's viability was due to the commitment of a user, Whirlpool Corporation, which undertook the risk of testing the system on an operating production line. Here's how that happened.

Whirlpool coated their air conditioner compressors using an anodic system similar to that used in automotive coatings. However, this system did not have the high corrosion resistance needed to withstand tropical and Gulf Coast exposures and subsequently failed, particularly where the copper tubing entered the steel shell of the compressor. At this juncture, there is a trimetallic couple, copper, brass and iron, which favors electrolytic activity and hastened corrosion.

Whirlpool Research scientists began looking for an improved system and while reviewing the problem at a joint meeting with the PPG Research Staff, Dr. Wismer pointed out that they had developed an answer looking for a question. He pointed out that they had succeeded in developing a practical cationic electrocoat system, but could not prove it in the field. (This is because it is difficult to empty a large tank of existing paint and recharge it. For flowcoating, a simple flush and rinse of the pumping system is all that is required for a paint changeover.)

After evaluating the chemistry of the system, the process and the test results PPG had obtained, Whirlpool opted to determine if the process could be implemented in their plant. This involved reviewing the process with their engineers and setting up a changeover strategy with manufacturing engineering. The Evansville plant was on strike, so the meeting was held in a nearby office of a drive-inn theater with the author, G. Malcolm Slaney, (later Vice President of PPG), Erwin Kapalko of PPG and Whirlpool engineers who agreed to make a changeover on a Thanksgiving weekend on a 4,000 gallon tank. We planned to drain and store the existing paint, recharge the new material and change the electrode configuration by installing a carbon anode. The results were better than expected and the system ran for ten years until Whirlpool discontinued manufacturing compressors. The first successful cationic electrocoating tank is shown in Fig. 1.

[An interesting aside to this development was that chemists found that during the cationic electrodeposition process, hydrogen is generated at the cathode which would blow off the deposited film:

$$H_3O^+ \ + \ e^- \ = \ 1/2 \ H_2 \ + \ H_2O$$

As a matter of fact, published literature indicated that the process would never be viable and a consultant hired by PPG urged Bosso and Wismer to discontinue the effort since it would never work. So much for consultants!]

DEVELOPMENT OF CATIONIC ELECTROCOATING

Following the successful introduction of cationic electrocoating, other applications were investigated. The first large scale operation was employed to coat air conditioner cabinets. These cabinets were coated using a flowcoat acrylic primer over phosphated galvanized steel, followed by an acrylic topcoat. The process employed about 50 people to handle touch up and repair.

Whirlpool's Evansville division proceeded to install a 42,000 gallon tank to coat air conditioner cabinets which were formed from galvanized steel, phosphated and coated with a 0.5 mil cationic epoxy coating. This was a first for applying a single coat cationic E-coat as a topcoat which greatly improved corrosion resistance. The entire operation was supervised by a single operator so that the fifty production workers previously used could be assigned elsewhere.

The next application involved a dryer drum. The physical and chemical requirements of a dryer drum coating are stringent. In addition to appearance, the coating must have abrasion and chip resistance, be tolerable of high temperatures and of laundry aids, such as quaternary ammonium salts, which are used in fabric softeners. In one plant, an acrylic flow coat was changed to a cationic electrocoat. At the time, the throwing power (the ability of the coating to coat recesses) was such that only a 0.4-0.5 mil coating could be deposited. The cationic electrocoat performed in this application and was used for many years until it was replaced by a powder coat system in 1985.

The next application was as a primer on washing machine cabinets. Here again the improved corrosion resistance of the cationic electrocoat system was employed in a new installation which was successfully operated for ten years until the plant was closed in 1987. Cationic electrocoating is still being used on washer cabinets.

These applications completed the proof of feasibility of cationic electrocoating. Because of poor throwing power, cationic electrocoating was not favored by the automotive industry, but based on the success of the process in appliances, the automotive industry began an investigation and finally implemented the process for car bodies and small parts. After Whirlpool's successful introduction of cationic electrocoating in 1971, Amana was the next user about three years later.

COMPARISON OF ANODIC AND CATHODIC SYSTEMS

The differences between the two systems lies in the chemistry of the polymers used in synthesizing water dilutable coatings. Anodic electrocoating requires a polymer having a carboxylic function in the backbone which is subsequently neutralized to form an amine or alkali metal macrocarbanion:

$$CH_2=CH \quad + \quad CH_2=CH \quad + \quad CH_2=CH \quad = \quad --[CH_2-CH--CH_2-CH --CH_2-CH--]_n$$

C_6H_5	COOR	COOH	C_6H_5	COOR	COOH

Styrene	Acrylic ester	Acrylic acid	[I]

The polymer is prepared in a coupling solvent which is miscible in water, then the polymer [I] is neutralized with an amine to make a water dilutable vehicle in which pigment, additives and crosslinking additives are added.

$$[I] \quad + \quad R'NH_2 \quad = \quad --[CH_2-CH--CH_2-CH --CH_2-CH--]_n + R'NH(+)$$

$$C_6H_5 \quad COOR \quad COO(-)$$

Anionic Polymer

In a similar manner, a cationic electrocoat polymer must be prepared such that a positive charge exists on the macrocation. This is accomplished by reacting an epoxy resin with an amine as follows:

$$\underset{\text{Epoxy resin}}{-----CH----CH_2} \quad + \quad \underset{\text{Amine}}{R \ NH_2} \quad = \quad \underset{[II]}{------CH- \ CH \ -NR_2}$$

with O over the CH----CH₂ (epoxide ring) and OH over the right CH.

The amine modified resin is then reacted with acetic, lactic or other acid to form the cationic macroion:

$$\underset{[II]}{------ CH-CH_2-NR_2} \quad + \quad \underset{\text{Lactic acid}}{CH_3-CH-COOH} \quad = \quad \underset{\text{Cationic Polymer}}{------CH -CH_2-NR + CH_3-CH \ COO(-)}$$

with OH groups above and H(+) above.

A schematic of the process is shown in Figure 2 and a comparison of the two systems in Table II.

Table II

A Comparison of Anodic and Cathodic Electrocoating

ANODIC	CATHODIC
CHEMISTRY	
Amine or base soluble	Acid soluble
Tank ground	Carbon or stainless steel anode
Phosphate dissolution	Low phosphate loss
CORROSION RESISTANCE	
300 - 500 hours	1,000 hours or better
Trapped ions in film	Few ions in film
FILTRATION	
Good	Some problems
COST	
Lower	Higher
COLOR	
White	Difficult to obtain appliance white

Shortly after the anodic process became commercial, May showed that poor corrosion was due to the presence of soluble ions trapped in the film. This occurs during deposition where the phosphate coating is dissolved, while the film is being deposited, resulting in trapped ions in the film. These ions act as driving forces for diffusion of water into the film and leads to early corrosion failures [33]. It was assumed that cathodic electrocoating did not lead to early corrosion failures since no phosphate was dissolved. This misconception was exposed when Anderson showed that there is some dissolution of phosphate in the cathodic process, but not to the extent which occurs the anodic process [34].

Porcelain Replacement Coatings.

Armed with these successes, and wanting to take advantage of the corrosion resistance of cationic electrocoating primers, efforts were then made to develop porcelain replacement coatings. The driving force for this was the high energy required to operate porcelain furnaces, and the capital investment required for new furnaces. Targeted was the washer top and lid as potential application possibilities. The system would consist of a cationic electrocoat primer and a high solids polyester topcoat. Corrosion resistance of the primer would contribute to long term durability and the top coat would provide a porcelain like appearance. A major concern was detergent resistance required for laundry use. Specifications for detergent resistance called for 250 hours, but was raised to 500 hours as a safety factor. This was achieved and exceeded as tops and lids as well as tubs and baskets were coated and tested.

The results were that the technical feasibility for replacing porcelain in laundry applications was successfully demonstrated and the performance of the cationic primer proved its worth.

PLATING REPLACEMENT

Metal plating is used on chest freezer baskets to provide corrosion resistance and an aesthetic appearance. Plating however, is a process which requires cleanup of the waste water, which adds to the process costs. The possibility of replacing plating with an electrocoat system was evaluated. The system selected was an acrylic white which passed all the necessary requirements and was installed within a few months after completing these tests. The acrylic system does not have the corrosion resistance of the epoxy system, but has high gloss, excellent weather resistance and is non-yellowing. In fact, such a system has been used effectively on farm tractors to provide a single coat electrocoat.

OTHER APPLICATIONS

A recent installation of a cationic electrocoating system by the Heil Quaker Corporation utilizes a highly automated production line in which the electrocoat serves as a single coat over galvanized steel. This system is cost effective and improves the quality of the product. The coating is an acrylic which has excellent exterior durability and does not chalk like an epoxy coating [35].

POWDER COATING APPLICATIONS IN APPLIANCES

Powder coatings were developed in Germany in 1952 to coat metal objects using a fluid bed method. Here the parts to be coated were heated, then lowered into a fluid bed of powder suspended by an air blanket, at which time the powdered material fused to the metal. Upon withdrawing the coated part, the residual heat caused fusion and filming to occur. Coatings were generally thick, over 10 mils, and non-uniform if the metal thicknesses varied greatly. However, this was an excellent method for coating pipe, wire goods and certain large castings. Other methods of application followed using cloud chambers and electrostatic spray and discs. In this discussion we shall limit our discussion to the application of powder coatings to appliances.

EARLY HISTORY

Fluid bed powders have been used at Whirlpool for many years. One of the first applications was in dishwasher racks. Dishwasher racks are made from welded steel wire of varying thicknesses. The welded basket is cleaned and phosphated, then treated with a low solids methacrylate primer, followed by heating to 500°F and placed into a fluid bed of vinyl powder. The basket is withdrawn from the fluid bed and water quenched. Coating thickness is over ten mils. This process is still used today.

A second application for powder coatings involved the use of an electrostatically sprayed epoxy coating on wire grills for the central air conditioner condenser. This grill is subjected to continuous outdoor exposure and must have excellent corrosion and weather resistance. The operation lasted for several years until the grill metal was replaced with an engineering thermoplastic.

WHY POWDER COATING?

In the early seventies, the industry was faced with a growing concern for environmental control of paint solvents. Until the end of the sixties, coatings were low solids, about 34-40%, thus emitting significant amounts of solvents to the atmosphere. This caused both coating suppliers and users to reexamine their coating technology to determine what could be done to eliminate or greatly reduce solvent emissions [29,36]. On careful study of present and emerging coating technologies it became clear that three approaches to solving the emerging environmental crisis were available. These approaches were: High Solids, Water Dilutable (including Electro-deposition) and Powder Coatings. All three would contribute to solvent reduction, but powder would be the least polluting.

Powder was not a first option, since there was also an emerging energy crisis. Another problem with adopting powder was that it would require scrapping existing equipment, some of which was relatively new. Other options like high solids coatings, or two component urethanes, would provide the best of both worlds -- but at a price.

FREEZER LINER

One of the first opportunities to use powder was in replacing porcelain on a chest freezer line. The existing facility had a porcelain furnace which was fired with propane and costly to operate. After evaluating a number of alternatives, Whirlpool used an electrostatically applied epoxy powder and an acrylic powder to coat freezer liners. After some difficulties with cross contamination of the acrylic and epoxy powders, a conversion to all epoxy solved the problem. A marketing concern was the perceived value of porcelain against a 'painted' liner. This turned out to be no problem since it could be effectively demonstrated that the powder coating had excellent impact resistance over porcelain. In addition, a new trademark was coined -DYNAWHITE - which was eagerly accepted by the sales force.

POWDER REPLACES PLATING

Another manufacturing problem was resolved using powder coating involved removing heavy metals from waste water. This was particularly important in plating operations which were used to coat refrigerator racks. The excellent abrasion resistance of powder suggested that plating might be eliminated in this operation. A test program was set up to evaluate abrasion and humidity resistance. Test results showed that there was a minimum film thickness below which the coating failed humidity resistance. As a result, the coating had to be deposited above a minimum film thickness of 4.5 mils. This also provided the necessary thickness to survive abrasion requirements. Other problems, i.e. food staining and oil and fat exposure were passed successfully. As a result, plating was replaced on the wire goods for refrigerators and freezers, which reduced heavy metal discharges and the need to rely on plated metal for these products.

PORCELAIN REPLACEMENT

Porcelain is used in washer tops and lids, tubs and baskets, dishwasher tubs, range tops and oven cavities and heat exchangers for furnaces. Because of the high temperature requirements of furnace heat exchangers, oven cavities and range tops, it is unlikely that organic coatings, including powders, will replace porcelain. However, as in the case of freezer liners, it was felt that it was possible to replace porcelain in certain laundry applications with organic coatings like urethanes and powder coatings.

The first step was to develop a profile methodology to replace porcelain. This work was outlined on paper, in as much detail as possible, to include all the critical issues involved in bringing about this change. Some of the elements of this profile include supplier choice and acceptance, sales and marketing agreements, product engineering input and manufacturing division assistance to bring the program to a successful conclusion. Once this was accomplished, work was begun with a single supplier to develop the feasibility of the technology between research centers [37,38,39,40].

After some time the technical feasibility of using a cathodic electro-coat primer and a two component urethane topcoat and another system using a direct on epoxy powder over zinc phosphate was demonstrated. Subsequently, it was shown that a high solids polyester liquid topcoat would be effective and cost efficient over the urethane. Both of these approaches were used in production which demonstrated that organic coatings could indeed replace porcelain in tough laundry applications. More will be said about porcelain replacements below.

DRYER DRUM AND BULKHEAD

The dryer drum and bulkhead coating is subjected to a severe environment. Some of the major concerns include impact and abrasion resistance, softening agents, heat, moisture, stain and yellowing resistance and corrosion. Coatings for this application have included solution epoxy, cationic electrocoating and acrylic copolymers. Recently, the increased use of fabric softeners which contain quaternary ammonium compounds have caused softening of epoxy systems including powders. To resolve this, changes in crosslinkers and increased baking temperatures have been used.

As a result, dryer drums and bulkheads are now coated with an automated electrostatically applied epoxy powder. This coating provides the necessary impact and pencil hardness. One of the problems encountered has been dye staining. This is more noticable since previous dryer drum coatings have been brown or gray, while the new coating is white. One of the dyes which has been a problem is indigo, which is common in blue jeans. If the dye is not completely set, there can be bleed through which stains the finish. This has been extensively studied by powder coating suppliers and the research centers of epoxy resin suppliers with some encouraging results.

COIL COATING

Coil coating was developed by J. Hunter in 1935 to coat metal strips in a continuous manner. By 1943 there were eight coil coaters operating in the United States. Because of increasing pressure of environmental controls, coil coating offers manufacturers an opportunity to dispose of inhouse coating lines and move the pollution control process one step back in their processes [41].

The earliest applications include a microwave oven plant in Columbia, Maryland built by General Electric in 1972 which was designed specifically for coil coating. This was followed by a Westinghouse plant in Michigan using coil coating for chest freezers. In 1972, Whirlpool began using coil coating in a six cubic foot chest freezer which consisted of a precoated aluminum liner and a precoated steel shell. Other applications include precoated sheets which are inserted in the door frame of a dishwasher permitting the user to change the color of the door on demand. General Electric built a plant in Decatur, Alabama to manufacture 12 and 14 cu. ft. refrigerators from coil [42].

The driving force for precoated metal in appliances came from the National Coil Coaters Association's need to expand their capacity such that in the early eighties, they began a concerted effort to increase their share of the appliance market [43]. The refrigeration area provided a good opportunity in that the raw edges of shorn steel could be rolled under and is covered by insulating foam. Coil coated steel found application in liners, cabinets, decks and related parts. A problem with coil coated steel is in fastening. Unlike post coated metal, welding burns the coating and due to the insulating property of the coating, spot welding is difficult, if not impossible. Coating suppliers have supplied weldable primers to attempt to overcome the problem.

Here is where adhesive bonding plays a key role in the successful application of coil coated metal. Adhesives are being used in the manufacture of coil coated appliances. The refrigeration sector of the appliance industry will probably convert to coil coating by the early nineties since by mid 1989 over 81% of the refrigerators manufactured by the majors are by coil coated metal.

One of the salient marketing features of refrigerators is the patterned steel door which masks fingerprints and provides an aesthetic appeal. This is accomplished by pressing a pattern into the steel then coating either on a coil line or post painting. Recently, Roll Coater has developed a post embossing process which can eliminate the premium cost of embossed steel [44]. Problems associated with embossed steel have recently been reviewed [45]. An excellent monograph on coating defects has been published by Pierce and Schoff [46].

Coil coating has been used on Ranges with good success. Laundry products present more challenging problems. For example, a laundry tub or basket would be difficult to prepare from a coil coated sheet since many holes are cut into the surface leaving raw edges where corrosion can easily begin. Washer and dryer cabinets could be made from coil, but the raw edge effect must be addressed. Effort to overcome this problem include using coated metals such as hot dipped or electrogalvanized steel where the wiping action of a die can smear zinc over the raw edge protecting the exposed surface from corrosion.

Future developments include methods for applying coil coating such as extrusion developed by Alcan and powder coating of coil [47].

FUTURE

There are a number of critical issues facing the appliance industry in the future. These include the use of halogenated hydrocarbons, formaldehyde, a significant factor in the crosslinking of coatings and waste disposal. This suggests that alternatives to suspect chemicals are needed and attention must be given to the handling and treatment of wastes. In addition, metal treatment and its subsequent wastes demand alternatives to reduce the cost involved in space and energy consumption.

Formaldehyde is a major factor in the crosslinking of coatings. Because of the pressure of environmental groups on this useful chemical and its importance in coating cure, it behooves the industry to develop alternatives to formaldehyde while we have the luxury of time. In addition, the opportunity to lower the energy of cure is another challenge. Two recent papers by Clemens and Rector offer room temperature curable coatings using acetoacetyl chemistry [48,49].

Prepainted metal offers appliance manufacturers the option of eliminating solvent emission, metal treatment and the cost of paint lines. The trade off is in expensive scrap, and flexibility of change in coatings. The refrigeration sector will probably be converted to precoated metal by the early nineties. Laundry products will not fare as well since there are more raw edges exposed in these products which can lead to corrosion failures. Design changes must be made to accomodate precoated metal in laundry products.

Powder coating will be challenged by coil coating in that thinner coats can be obtained with coil, although for high performance coatings, powders will be excellent selections.

Energy considerations will be a concern in the next two decades and must be addressed by the coatings and manufacturing engineering sectors of the industry. One aspect of energy utilization is radiation curing, particularly ultraviolet curing, which could be used in coil coating, decorative finishing or for general curing operations eliminating ovens and conserving floor space [50,51].

83

REFERENCES

1. T. J. Miranda, "Appliance Coatings" in Applied Polymer Science, Second Edition, Edited by Roy W. Tess and Gary W. Poehlein, ACS Symposium Series No. 285, p. 883.
2. J. Schrantz, Industrial Finishing, 65, No. 1, 14, 1989.
3. APPLIANCE, 46, No. 1, 69 (1989).
4. E. G. Shur, , "Treatise on Coatings"; Myers, R. R.; Long, J. S. Editors; Marcel Dekker: New York, 1975; Vol. 4 Chap 2.
5. T. J. Miranda, "Recent Advances in Coatings for Household Appliances"; J. Coatings Technology 55, No 696, 81, 1983.
6. T. J. Miranda, "Electrocoating in the Appliance Industry"; J. Coatings Technology, 60, No. 760, 47, 1988.
7. T. J. Miranda, "Reading the Signals of Society: Technology Push or Market Pull"; J. Coatings Technology, 57, No. 721, 22, 1985.
8. Brendley, W. H., and Bakule, R. D., "Chemistry and Technology of Acrylic Resins for Coatings"; in Applied Polymer Science, Second Edition, Edited by Roy W. Tess and Gary W. Poehlein, ACS Symposium Series No. 285, p. 1031.
9. Kienle, R. H., Ind. Eng. Chem., 41. 726 (1949).
10. Strain, D. E., U. S. Patent 2,173,005 September 12, 1939 to Dupont.
11. Christenson, R. M., U. S. Patent 3,037,963, June 5, 1962 to PPG Industries.
12. Dow Chemical Company, Epoxy Ester Appliance Primer Formula 35-0-21.
13. Payne, H. F. Organic Coating Technology, Vol. II Wiley, New York, 1961.
14. Allied Chemical Corp. Brit. Pat. 957,367 (1964).
15. Sekmakas, K., U.S. 3,163,615, December 29, 1964. To DeSoto, Inc.
16. Gerhart, H. L., Official Digest, 33, No. 680 (1961).
17. Brown, W. H. and T. J. Miranda, "Chemistry of Acrylic Solution Polymers," Official Digest 36, No 475, 92 (1964).
18. Mercurio, A. Official Digest 36, No 475, 135 (1964).
19. Solomon, D. H. Chemistry of Organic Film Formers, John Wiley & Sons, 1967.
20. Paul, S., Surface Coatings, Science and Technology, John Wiley & Sons, 1985.
21. Gaylord, N. G., U. S. Patent 2,853,463 (1958) to Interchem.
22. Vasta, J. A., U. S. Patent 3,065,195 (1962) to Dupont.
23. Segall, G. H. and J. L. Cameron, Can. Patent 534,261 to Canadian Industries Ltd.
24. Applegath, D. D., Ind. Eng. Chem., 53 (1961).
25. Vasta, J. A., Belg. Patent 634,310 (1963).
26. Simms, J. A., ACS Div. Paints, Plastics, Pigments, 19 No. 2, (1959).
27. Ravve, A. and J. T. Khamis, U. S. Patent 3,306,883 (1967) to Continental Can Co.
28. T. J. Miranda, "Oxazoline Modified Thermosetting Acrylics", J. Paint Technol. 39, 40 (1967).
29. T. J. Miranda, "Water-Based Finishes in Response to Environmental Constraints", Water Soluble Polymers, Edited by N. M. Bikales, Plenum Publishing Corp. New York, 1972.
30. T. J. Miranda, "Chemistry of Water Soluble Polymers" OFFICIAL DIGEST. 37, No. 469, 62, 1965.
31. G. E. F. Brewer, J. Paint Technology, 45, No. 587, 36 (1973).
32. J. Bosso and M. Wismer, CHEM ENGR. 78, No. 13, 11, 1971.
33. C. May, J. Paint Technology, 42, No. 552, 43, (1971).
34. D. G. Anderson, E. J. Murphy and J. Tucci, J. Coatings Technology, 50 No. 646. 38 (1978).
35. Appliance Manufacturer, August 1988, p 74.
36. T. J. Miranda, "Powder Coating Potential in the Appliance Industry", SME Paper FC-847, (1971).
37. T. J. Miranda, "Coatings in Transition", J. Coatings Technology. 49 No. 628, 66 (1977).
39. K. M. Biller, "Coating Research Benefits Appliance Manufacturers, Industrial Finishing, June 1984 p 60.
40. P. Gribble, Metal Finishing, Feb. 1987 p 77.
41. J. E. Gaske, Coil Coating Fed. Soc. Coat. Tech. Monograph. D. R. Brezinski, T. J. Miranda, Editors. Feb. 1987.
42. J. Schrantz, Industrial Finishing, 61 No. 5, 26 (1985).

43.   R. H. Braswell, Proceedings of the National Coil Coaters Association, Nov. 11, 1974, pp. 16-24.
44.   N. C. Remich, Jr., **Appliance Manufacturer**, August 1988 pp. 45.
45.   T. J. Miranda, "Appliance Coatings: Defects and Their Prevention" **J. Coatings Technology**, 60 No 765, 113 (1988).
46.   P. E. Pierce and C. K. Schoff, **Coating Film Defects**, Fed. Soc. Coat. Tech. Monograph. D. R. Brezinski, T. J. Miranda, Editors. Jan.1988.
47.   F. L. Church, **Modern Metals**, Feb. 1987 pp. 54.
48.   R. J. Clemens and F. D. Rector, **J. Coatings Technology**, 61 No. 770. 83 (1989).
49.   F. D. Rector, W. W. Blount and D. R. Leonard, **J. Coatings Technology**, 61 No. 771. 31 (1989).
50.   J. R. Costanza, A. P. Silveri and J.A. Vona, **Radiation Cured Coatings**, Fed. Soc. Coat. Tech. Monograph. D. R. Brezinski, T. J. Miranda, Editors. June 1986.
51.   J. V. Koleske and T. M. Austin, **J. Coatings Technology**, 58, No. 472. 47 (1986).

Figure 1
First Cationic Electrocoat Application

Figure 2
Schematic of a Cationic Electrocoat System

CATHODIC ELECTROCOATING

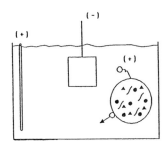

○   PAINT PARTICLE

←○   COUPLING SOLVENT

●   PIGMENT

▲   ADDITIVE

/   BINDER

CATHODE REACTION:

$$2H_2O \xrightarrow{+2E} H_2 + 2OH^-$$

$$\sim\!\!\sim\!\!\sim\!\!\sim N- \; + \; R\text{-}COOH \longrightarrow$$

$$\sim\!\!\sim\!\!\sim\!\!\sim N^+- \; \underset{R}{\overset{|}{C}OO^-}$$

COATINGS FOR MEMBRANES--PAST, PRESENT, AND FUTURE

A. K. FRITZSCHE

Romicon, Inc., a Rohm and Haas Company, 100 Cummings Park, Woburn, MA 01801

ABSTRACT

Coatings are an integral part of many commercial membrane systems. Coatings have been used in gas separating membranes to plug pores and defects to render permeation through the underlying membrane predominate. Charged coatings serve to increase retentate rejection and retard fouling in ultrafiltration membranes used for the recovery of cationic paints from the electrocoat process used in the appliance and automotive industries. Coatings also have permitted development of improved reverse osmosis membranes for desalination. Coatings have been applied by deposition from solution, by interfacial polymerization, by plasma deposition, and by grafting. This paper discusses coatings for membranes from a historical perspective with examples in which coated membranes have gained commercial success. Finally, recent advances in molecular engineering of ultrathin polymeric films and environmentally responsive coatings will be presented which suggest that the next generations of composite membranes will utilize "smart" coatings whose separation characteristics can be regulated.

INTRODUCTION

Any discussion about synthetic membranes must credit the contributions of Loeb and Sourirajan. Loeb and Sourirajan discovered a process to make integrally skinned cellulose acetate membranes by phase inversion (1,2). These membranes possess a dense skin of cellulose acetate subtended by a finely porous substrate of the same material which supports the skin. This discovery has been called the "dominant discovery in synthetic membranology to date" (3). Not only did this discovery yield commercially viable reverse osmosis membranes for desalination, but it was a progenitor for many technical advances in ultrafiltration, microfiltration, and gas separations, which ultimately resulted in commercial processes.

Now separations with synthetic membranes have become increasingly important in the chemical industry, in food and wastewater processing, and in medical applications as well as desalination. These advances have generated new industries in areas as diverse as controlled-release of pharmaceutical formulations to the inert blanketing of oil tankers and improved storage of perishables. The development and growth of this technology and many of the resultant applications have been extensively reviewed and a fraction of these are included in the reference section of this paper (3-10). For those seeking a comprehensive work suitable as a reference source, a beginning text, and an advanced text, the book by Kesting is highly recommended (11).

Coatings have functioned as an integral part in the development and subsequent commercial success of membrane technology. Coatings have been used to plug pores and defects in gas separating membranes to render permeation through the underlying membrane predominate, to increase the retentate rejection and retard fouling in ultrafiltration membranes, and to serve as the separating layer in composite membranes. Such coatings

have been applied by deposition from solution, by interfacial polymerization, by plasma deposition, and by grafting. It is the object of this paper
to discuss each of these approaches from a historical perspective and give
examples in which commercial success has been achieved. Finally, advances
in the molecular engineering of ultrathin polymeric films and environmentally responsive coatings suggest that the next generation of composite
membranes will be fabricated from "smart" coatings whose separation characteristics can be regulated.

THIN-FILM COMPOSITE MEMBRANES FOR REVERSE OSMOSIS

The cellulose acetate membrane developed by Loeb and Sourirajan
readily served to desalinate brackish water. However, problems were encountered during use to desalinate seawater. In this application, the
cellulose acetate membrane exhibited insufficient selectivity, marginal
chemical stability, and decreasing permeability with time. Two approaches
were taken to develop improved membranes for the desalination of seawater:
improved integrally skinned membranes and composite membranes.

Efforts to develop improved integrally-skinned membranes resulted in
the cellulose triacetate hollow fiber membranes by Dow (12-15) and the
aromatic polyamide hollow fiber membranes of DuPont (16). However, it must
be asserted that preparation of new asymmetric membranes is no trivial task.
Initially, a polymer must be identified possessing the required permeability, selectivity, and chemical stability. This polymer must then be fabricated into a flat sheet or hollow fiber membrane with a suitable asymmetric structure. Attainment of the appropriate structure requires determination of the dope constituents and their relative compositions as well as
the appropriate processing variables, i.e. coagulation bath temperature and
composition, processing speed, etc. Finally, the resultant membrane must
have sufficient mechanical properties to permit it to be handled and assem-
bled into a separator and survive the fluctuations in temperature and pressure encountered under operating conditions.

Upon consideration of the difficulties in preparing new asymmetric
membranes, the appeal of thin-film composites is evident. A coating with
the desired performance characteristics is placed on a well-characterized
porous substrate. The substrate may be used in a variety of applications
with only an alteration of the coating. The fabrication process itself is
limited to the controlled deposition of the coating unto the substrate to
yield a thin, pinhole-free selective layer. The first manifestation of
this class of membranes dates back to the mid-1960s with the work of
Cadotte and Francis (17-20) and Riley and colleages (21). In these studies,
very thin films of cellulose esters and other polymers were first cast on
a water surface from a very dilute solution. After the solvent evaporated,
the thin film was lifted off onto a porous substrate (22). Subsequently,
thin-film composite reverse osmosis membranes were developed by forming a
thin film of cellulose triacetate from dilute solution directly upon a
surface of a porous supporting membrane pretreated with poly (acrylic acid)
(23). However, the prerequisite techniques for making these composite
membranes were not amenable for large scale production.

The real breakthrough in the formation of thin layer composite
membranes resulted from the application to membranes of well-known inter-
facial-reaction chemistries (24). During interfacial polymerization, condensation polymerization occurs at the interface between two immiscible
solutions. The well-known nylon rope trick is an example of this art. A
diamine dissolved in water reacts with a diacid chloride dissolved in
hexane to form a polyamide film at the interface between the upper hexane
layer and the lower aqueous layer. The film initially forms very rapidly

because of the high reactivity of the reactants, but then film growth is limited because the film is relatively impermeable to both of the reactants.

The development of the interfacially polymerized membrane was pioneered independently by Cadotte (25-27) and Riley (28,29), and the process is illustrated in Figure 1. Both were able to perform the interfacial polymerization reaction at one surface of a finely porous membrane, which then becomes the support for the thin film. The original NS-100 membrane, invented at North Star Research Institute by Cadotte (26), was a thin film composite membrane comprised of a microporous polysulfone base layer, a topcoating of polyethylenimine, and an interfacially formed skin of polyurea or polyamide produced by contact of toluene diisocyanate or isophthaloyl chloride with polyethylenimine coating. The barrier layer is only 20 nm thick, and the remaining unreacted polyethylenimine is insolubilized by heat-curing at 110°C (30).

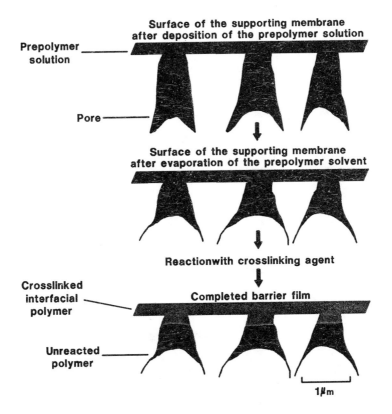

Figure 1. Schematic diagram illustrating the formation of a composite membrane by interfacial polymerization of polyethyleneimine with diisocyanate. H. Strathman, "Synthetic Membranes and their Preparation", in Synthetic Membranes: Science, Engineering, and Applications, Eds. P.M. Bungay, H. K. Lonsdale, and M.N. dePinho, D. Reidel Publishing Company, 1986.

90

The RC-100 and PA-300 membranes of UOP are prepared by interfacial polymerization of epiamine (an epichlorohydrin ethylenediamine adduct) with toluene diisocyanate or with isophthaloyl chloride, respectively (31). The former reaction yields a polyetherurea coating; the latter a polyether-amide coating. Interfacial polymerization is also used by Film Tec to pro-duce its FT-30 thin layer composite membrane by reaction of metaphenylene-diamine with trimesoylchloride to produce a polyamide (30). Cabasso and Tamvakis demonstrated that interfacial polymerization could also be used to prepare thin layer composite hollow fiber membranes (32).

CHALKED MEMBRANES FOR GAS SEPARATIONS

The basic requirement in fabrication of practical membranes is obtain-ing a minimum resistance to flow through the membrane coupled with maximum selectivity. If the effective thickness of the membrane is not minimized, then the optimum flow rates cannot be achieved, and the membrane's commer-cial potential is diminished. However, if the membrane is prepared so as to minimize the effective thickness of the separating layer, pores appear in the separating layer with a concomitant sacrifice in selectivity. Pores only 0.5-1.0 nm in diameter permit passage of most permanent gases across the membrane without actual permeation through the polymer and need to constitute less than 0.01% of the surface area of the effective separating layer for selectivity to be lost (33).

The discovery by Henis and Tripodi, as a result of suggestions by D.R. Paul and R.L. Leonard, of a proprietary coating procedure to effec-tively plug these pores with a nonselective highly permeable coating repre-sented a major breakthrough culminating in the commercialization of the PRISM® gas separating membrane systems for hydrogen, helium, and carbon dioxide recovery (34). The coating, i.e. silicone rubber, presents far less resistance to flow than the hole-free portion of the membrane but much more resistance to flow than the pores. It essentially blocks flow through the holes without changing the flows appreciably through the underlying effec-tive separating layer of the membrane. In other words, the use of a thin coating of a highly permeable, nonselective polymer, through which the gases move rapidly, enables the pores and defects in the membrane's surface to be plugged with insignificant loss of flow rate through the perfect portion of the membrane as is illustrated in Figure 2.

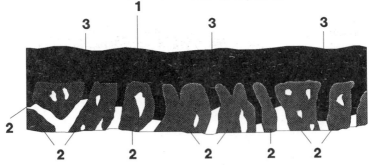

Figure 2. Schematic diagram illustrating a multicomponent membrane for gas separations in which 1 is the highly permeable, nonselective coating, 2 is the effective separating layer of the membrane, and 3 shows pores and defects penetrating the effective separating layer. J.M.S. Henis and M.K. Tripodi, U.S. Patent 4,230,463, October 28, 1980.

These hollow fiber membranes are operated in systems in which the feed
gas is on the outside of the membranes and the permeate is collected from
the hollow fiber bores. In this mode of operation, these coated hollow
fiber membranes have successfully functioned at 2,150 psig and at differential pressures of 1650 psi (35).

Another significant advance in hollow fiber membrane technology was the
discovery of a new class of membranes with graded-density skins prepared by
phase inversion from Lewis acid:base complex solvent systems (36). Such
membranes produce three to five times the nitrogen at equivalent purity to
that generated by an identical system composed of membranes from the same
polymer and of the same dimensions prepared by conventional phase-inversion
spinning processes. These PRISM® Alpha systems have encountered widespread
acceptance for offshore platform and shipboard inerting, chemical blanketing,
metal heat treating, and the preservation of perishables. For these hollow
fiber membranes to be rendered suitable for gas separations, they also must
be coated using the proprietary coating procedure of Henis and Tripodi (34).
Like the hollow fiber membranes used in the PRISM® separator systems, the
coating is applied to the outer surface of the hollow fiber membrane by
deposition of the silicone rubber from a dilute isopentane solution with a
vacuum applied to the hollow fiber bores. However, unlike the PRISM® separator systems, the feed stream of elevated pressure enters the hollow fiber
bores. Oxygen and water transport across the membrane. It is the dry
nitrogen emerging from the opposite end of the separator cartridge that is
the desired product. In this mode of operation, the coating demonstrates
maintenance of integrity at differential pressures in excess of 125 psi.

The advantages of chalked coatings, such as used in the PRISM® Alpha
gas separating system, lie in their ease of application and their applica-
bility to asymmetric membranes prepared from a variety of polymers such as
crosslinked polyphenyleneoxide (37,38). Coatings were also made by
A. Zampini which polymerized after deposition enhancing their stabilities
to aggressive feed streams. Furthermore, when coatings are applied to
asymmetric hollow fiber membranes which can be crosslinked by ionizing
radiation, such as asymmetric hollow fiber membranes prepared from styrene/
acrylonitrile copolymers, not only are both the coating and underlying
membrane crosslinked but the two are coupled by chemical bonds (39). Consequently, a coated asymmetric hollow fiber membrane results which demonstrates both improved resistance and improved coating adhesion when exposed
to deleterious environments.

This coating technology not only permitted the commercialization of
membranes for gas separations but has maintained Permea as the dominant
contender in the membrane gas separations arena throughout this decade.
Therefore, other companies desiring penetration into the gas separation
market have concentrated on development of thin film composite membranes.
Innovative Membrane Systems, a subsidiary of Union Carbide, has recently
introduced its NitroGen ® systems to separate nitrogen from compressed air.
It is suspected that the composite membrane is a separating layer of ethyl-
cellulose on a porous support of polysulfone. Ube manufactures composite
hollow fiber membranes in which the support is a porous polyimide on which
a polyamic acid layer is deposited. The polyamic acid is subsequently
converted to a polyimide layer by thermal treatment (40-43).

The resistance of a porous substrate often limits the performance of
composite membranes. The surface porosity of the substrate is only a fraction of the total surface area of the membrane, and the maximum pore size
is limited by the requirement that the pore be bridged by the coating. Gas
permeation through the coating is most efficient through the regions of the
coating in juxtaposition to subtending pores. In these areas, the distance

which the gas must permeate through the coating is equivalent to the thickness of the coating. However, if no subtending pore exists below the point
of entry of the gas molecule into the coating, it must diffuse a greater
distance through the coating to the nearest pore. Consequently, the
apparent thickness of the coating as indicated by gas flux measurements is
greater than the actual thickness of the coating. A novel approach to
circumvent this deficiency is the application of a highly permeable, nonselective coating of an aminoorganofunctional polysiloxane crosslinked with
a diisocyanate on the surface of a highly porous substrate to form a gutter
layer (44), as illustrated in Figure 3. A thin permselective layer is then
deposited on top of this highly permeable layer to effect the separation.
The "gutter layer" serves to channel the permeating species to the nearest
pore. Cabasso and Lundy are currently attempting to improve the performance characteristics of these three-layer composite membranes by casting
anisotropic permselective submicron separating layers upon the intermediate
channeling "gutter layer" (45). Although the concept is intriguing and is
the source of much challenging research, it is the opinion of this author
that its commercial potential is limited because of its complexity requiring numerous process steps: (1) the formation of the porous support, (2)
deposition of the gutter layer, and then (3) the formation of the anistro-
pic selective layer, which may also then require (4) a chalked coating.
Intuitively, it seems that such an approach would be more amenable to flat-
sheet membranes for use in spiral wound configurations than with hollow
fiber membrane separators.

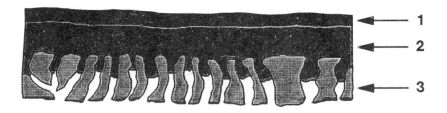

Figure 3. Schematic diagram illustrating the "gutter layer" composite
membrane of Cabasso and Lundy in which 1 is the thin permselective layer,
2 is the highly permeable, nonselective layer, and 3 is the porous
substrate support.

COATINGS BY PLASMA DEPOSITION FOR GAS SEPARATIONS AND REVERSE OSMOSIS

Plasma polymers, i.e. polymeric material formed by plasma or glow discharge polymerization, can be deposited in ultrathin layers. Their use to form both protective coatings and composite membranes has been the focus of extensive investigation for the last two decades (46). Primarily, these composite membranes were evaluated for reverse osmosis and gas separations but recently ion exchange membranes were prepared by this approach (47). This process has an appealing advantage in that flawless layers of less than 50 nm can be easily obtained while the preparation of ultrathin films from a conventional polymer becomes progressively more difficult as the thickness of the film decreases (48).

Plasma polymerization is also a unique ultrathin film technology which yields polymers having completely different properties from those of the more conventional polymers (49). Plasma polymers have no discernible repeating units and often are formed in highly crosslinked and highly branched networks consisting of very short segments, even if a well-defined monomer is used as the starting material. Consequently, the chemical and physical properties of a plasma polymer are not solely determined by the chemical nature of the monomer such as in the case of most conventional polymers.

Buck and Davar first used glow discharge polymerization of monomer vapor to prepare reverse osmosis membranes by depositing polymer onto the surface of a Millipore filter tightly fixed on the surface of the electrode (50). However, the size and shape of a membrane which can be prepared by glow discharge deposition onto the electrode are limited by the size and shape of the electrode itself. Therefore, Yasuda and Lamaze attempted to prepare reverse osmosis membranes utilizing electrodeless radiofrequency discharge to initiate the plasma (51). With electrodeless glow discharge, polymer deposition occurs on any surface exposed to the glow discharge. Yasuda and Lamaze observed that these plasma polymerized membranes exhibited (1) very stable performance independent of salt concentration and applied pressure, (2) salt rejection and water flux increases with time in the initial stage of reverse osmosis, and (3) very high salt rejection (over 99%) with high water flux at 1500 psi. Such membranes perform equally well under conditions of sea water conversion as well as brackish water treatment. Subsequent investigations revealed that the primary factors influencing the quality of the membranes are the choice of substrate material, the deposition time, and the power supplied to the discharge (52); that only nitrogen-containing compounds such as amines, aromatic amines, and heteroaromatic amines are easily converted to reverse osmosis membranes (53); and that the reverse osmosis characteristics of composite membranes are dependent on the combination of substrate and monomer (54). Among the best reverse osmosis composite membranes prepared by plasma deposition were those formed from allylamine and allylamine/nitrogen mixtures (52,55,56).

However, the advent and success of composite membranes prepared by interfacial polymerization and inherent technical difficulties of the plasma polymerization process have prevented reverse osmosis membranes prepared by this technology from meeting their initial promise. The difficulties arise from the difficulty in obtaining a suitable porous substrate and the plasma polymerization process itself. It is difficult to obtain porous substrates without defects, such as large pores or flaws. If these defects are too large, they cannot be covered or bridged by the thin layer of plasma polymer. These requirements in pore size are more stringent than those for conventional coating methods. Also, the plasma deposition process is a vacuum deposition process which would be both expensive to install and maintain in a commercial industrial process. An appreciation of the difficulty in translating this process to commercialization can be gained by the work of Lawton to improve the adhesion of tirecord by oxygen plasma ablation (57). Lawton had to construct a system in which the fiber could be continuously removed from the bobbin,treated with the oxygen plasma, and then subsequently rewound onto another bobbin--all while under a vacuum. In contrast, interfacial polymerization can be performed at atmospheric pressure, and the need for a vacuum system is eliminated.

The interest in composite membranes for gas separations formed by plasma deposition of ultrathin coatings originates with the research of Stancell and Spencer (58). These investigators used benzonitrile and cyanogen bromide as agents to deposit thin films on silicone/polycarbonate block copolymer and polyphenyleneoxide films, and they subsequently measured the hydrogen and nitrogen permeability of the resultant composites.

Research in composite membranes with plasma polymerized coatings has persisted for oxygen/nitrogen separations. Various fluorocarbons and silicone compounds have been plasma polymerized and the performance characteristics and substrate adhesion measured (59-65). Similarly, thin layers of tertiary carbon and tertiary silicone containing compounds have been impregnated into a surface of a heat resistant porous polymeric membrane and then subsequently plasma polymerized (66). As can be seen by examining the references, Japan is now in the forefront of activities to prepare suitable composite membranes by plasma polymerization deposition processes, and this effort is an indication of that country's commitment to capturing preeminence in membrane separation technology. It is also my opinion that these efforts are indicative of their attempts to circumvent the composite coating technology of Henis and Tripodi which have heretofore frustrated many attempts to produce commercially viable gas separation membrane systems.

COATINGS FOR ULTRAFILTRATION MEMBRANES

Many important applications for industrial ultrafiltration membrane systems exist in areas of pollution control and the recovery of valuable raw materials and by-products (67). One such application is the recovery of paint in the electrodeposition painting process widely used in the automotive and appliance industries (68,69). The part, as the cathode, is immersed in a dip tank containing the cationic polymeric electrocoating. When the treated part is subsequently removed from the dip tank, it is accompanied by significant amounts of entrained paint, which is then removed by rinsing with water. Separation of the paint from the rinse water recovers both for reuse reducing process costs and eliminating pollution.

This separation is most effectively accomplished by ultrafiltration membranes. The dominant supplier of ultrafiltration membranes for this application, Romicon, Incorporated, a subsidiary of Rohm and Haas, utilizes an ultrafiltration membrane which is internally treated with a cationically charged coating, preferably poly (vinylimidazoline) in the bisulfite form (70,71), and is illustrated in the schematic diagram in Figure 4. This coating is deposited on the internal surface of the hollow fiber ultrafiltration membrane which is in direct contact with the process solution. The coating of the cationic polyelectrolyte rejects the cationically charged paint macromolecules due to the concentration of high charge in the proximity of the membrane surface. While the coating is relatively uniform over the entire membrane surface, it does not necessarily need be a continuous film-like layer to be effective unlike membranes deposited on a porous support to form composite membranes. Rejection of the cationically charged paint macromolecules by the cationically charged coating not only reduces fouling of the membrane by the paint but facilitates cleaning of the membrane by backflushing. Consequently, not only does the coating improve the performance of the ultrafiltration membrane but it simultaneously extends the operational lifetime.

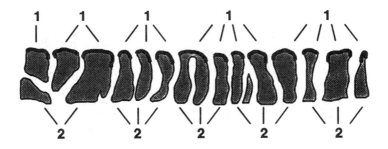

Figure 4. Schematic diagram illustrating a coated ultrafiltration membrane for cationic paint recovery in which 1 is the cationically charged coating and 2 is the ultrafiltration membrane.

COATINGS OF THE FUTURE

Membranology is a rapidly evolving science now yielding a plethora of commercial products. The fundamentals of the phase inversion process to form asymmetric membranes are being defined, and an understanding of the relationship between molecular structures of membranes and their permeabilities is being developed. These advances coupled with equally significant developments in coating technology will yield successive generations of new membrane products. Advances in the molecular engineering of ultrathin polymeric films will permit the formation of multilayers of especially chosen polymers for precise control of thickness, composition, density, and molecular orientation on the membrane (72). Depending on the molecules in each monolayer, unusual physical properties may be imparted to the final macromolecular assembly including high differential permeation rates for one species in contrast to another (73). Already coatings for composite gas separation membranes have been prepared using polymerizable Langmuir-Blodgett multilayers (74), as shown in the schematic diagram in Figure 5. The Langmuir-Blodgett deposition technique also offers an additional advantage beyond its capability for engineering multilayers. Choice of the final monolayer can determine whether the coating surface will be hydrophilic or hydrophobic as illustrated in Figure 5.

Coatings derived from liquid crystals may eventually be applied for gas separations, molecular filtration, active transport of metal cations, and thermocontrol of ion permeation. Not only can the structure of liquid crystals be regulated by temperature, but the orientation of nematic liquid crystals can be regulated by application of electric or magnetic fields. Their potential for various separations has already been demonstrated by forming composite membranes with the liquid crystalline material interdispersed in the polymer matrix (75-77).

Membranology is on the threshold of the "smart" membrane in which the coating is sensitive to the environment. Already grafting hydrophilic monomers onto the surfaces of porous substrates has yielded composites, which are sensitive to pH, solvent composition, and ionic species (78-80), as shown in Figure 6. Janos Fendler discussed the potential of membrane-mimetic polymers in membrane technology which will serve as the bridge between biological and polymeric membranes (81). I believe these membrane-mimetic polymers will find their first commercial applications as coatings. With advances in membrane coating technology, membranes should approach their ultimate potential.

96

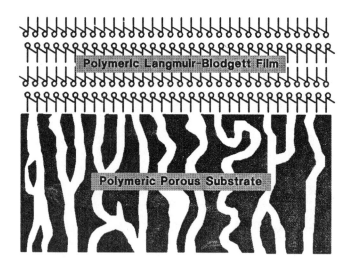

Figure 5. Schematic diagram of polymer membrane covered by an ultrathin LB layer. S.T. Kowel et al, "Future Applications of Ordered Polymeric Thin Films", in Molecular Engineering of Ultrathin Polymeric Films, Eds, P. Stroeve and E. Frances, Elseveir Applied Science, 377-403, 1987.

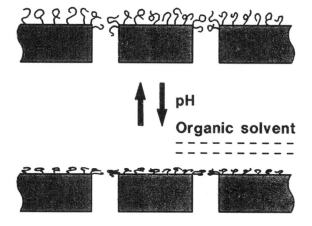

Figure 6. Schematic representation of the mechanism of sensitivity to the environment of a "smart" membrane. H. Iwata and T. Matsuda, J. Membrane Sci., 38, 185-199 (1985).

97

REFERENCES

1. S. Loeb and S. Sourirajan, Adv. Chem, Ser., 38, 117 (1962).
2. S. Loeb and S. Sourirajan, U.S. Patent 3,133,132, May 12, 1964.
3. H.K. Lonsdale, J. Membrane Sci., 10, 81-181 (1982).
4. J. D. Ferry, Chem. Rev., 18, 373-455 (1936).
5. T. Teorell, Ber. Bunsengellschaft fur Phys. Chem., 71, 814 (1967).
6. N.K.Lakshminarayanaiah, Transport Phenomena in Membranes, Academic Press, New York, 1969.
7. A.S. Michaels, Pure Appl. Chem., 46, 193-204 (1976).
8. V. Stannett, J. Membrane Sci., 3, 97-115 (1978).
9. H. Strathman, "Synthetic Membranes and their Preparation", in Synthetic Membranes: Science, Engineering, and Applications, Eds. P.M. Bungay, H.K. Lonsdale, and M.N. de Pinho, D. Reidel Publishing Company, 1-37, 1986.
10. A.K. Fritzsche, Polymer News, 13, 266-274 (1988).
11. R.E. Kesting, "Synthetic Polymeric Membranes, A Structural Perspective", 2nd Edn., John Wiley and Sons, New York, 1985.
12. H.I. Mahon, U.S. Patent 3,228,876, January 11, 1966.
13. H.I. Mahon, U.S. Patent 3,228,877, January 11, 1966.
14. E.A. McLain, U.S. Patent 3,422,008, January 14, 1969.
15. E.A. McLain and H.I. Mahon, U.S. Patent 3,423,491, January 21,1969.
16. J.W. Richter and H.H, Hoehn, U.S. Patent 3,567,632, March 2, 1971.
17. P.S. Frances, Fabrication and Evaluation of New Ultrathin Reverse Osmosis Membranes, Research and Development Report No. 177, U.S. Department of the Interior, Washington, D.C., February 1966.
18. L.T. Rozelle, J.E. Cadotte, and D.J. McClure, "Ultrathin Cellulose Acetate Membranes for Water Desalination", in: Membranes from Cellulose and Cellulose Derivatives, Ed. A.F. Turbak, John Wiley and Sons, New York, 61-71, 1970.
19. J.E. Cadotte, L.T. Rozelle, R.J. Petersen, and P.S. Francis, "Water Transport across Ultrathin Membranes of Mixed Cellulose Ester and Ether Derivatives", in: Membranes from Cellulose and Cellulose Derivatives, Ed. A.F. Turbank, John Wiley and Sons, New York, 73-83, 1970.
20. J.E. Cadotte and P.S. Francis, U.S. Patent 3,580,841, May 25, 1971.
21. R.L. Riley, G.R. Hightower, H.K. Lonsdale, J.F. Loos, C.R. Lyons, K.J. Mysels, and D.E. Wont, Development of Reverse Osmosis Modules for Seawater Desalination, Final Report, prepared by General Atomic Co. for the Office of Saline Water, NTIS (PB214973), Springfield, Virginia, 1972.
22. R.L. Riley, H.K. Lonsdale, C.R. Lyons, and U. Merten, J. Appl. Polymer Sci., 11, 2143-2158 (1967).
23. R. L. Riley, H.K. Lonsdale, and C.R. Lyons, J. Appl. Polymer Sci., 15, 1267-1276 (1971).
24. P.W. Morgan, Condensation Polymers: By Interfacial and Solution Methods, Interscience Publishers, New York, 1965.
25. L.T. Rozelle, J. E. Cadotte, K.E. Cobain, and C.V. Kopp, "Nonpolysaccharide Membranes for Reverse Osmosis: NS-100 Membranes", in Reverse Osmosis and Synthetic Membranes, Ed. S. Sourirajan, National Research Council of Canada, Ottawa, 1977.
26. J.E. Cadotte, U.S. Patent 4,039,440, August 2, 1977.
27. J.E. Cadotte, R.S. King, R.J. Majerle, and R.J. Petersen, J. Macromol. Sci. - Chem., A15, 727-755 (1981).
28. R. L. Riley, R.L. Fox, C.R. Lyons, C.E. Milstead, M.W. Seroy, and M. Tagami, Desalination, 19, 113-126 (1976).
29. R. L. Riley, C.E. Milstead, A.L. Lloyd, M.W. Seroy, and M. Tagami, Desalination, 23, 331-355 (1977).
30. R. J. Petersen, "Membranes for Desalination", in Synthetic Membranes, Ed. M.B. Chenowith, Harwood Academic Publishers for the Michigan Molecular Institure Press, New York, 129-154, 1986.

31. H.K. Lonsdale, J. Membrane Sci., 33, 121-136 (1987).
32. I. Cabasso and A.P. Tamvakis, J. Appl. Polym. Sci., 23, 1509-1525 (1979).
33. J.M. S. Henis and M.K. Tripodi, J. Membrane Sci., 8, 233-246 (1981).
34. J.M. S. Henis and M.K. Tripodi, U.S. Patent 4,230,463, October 28, 1980.
35. D. L. MacLean, C.E. Prince, and Y.C. Chae, Chem. Eng. Prog., 98-104 (March 1980).
36. R.E. Kesting, A.K. Fritzsche, M.K. Murphy, A.C. Handermann, C.A. Cruse, and R.F. Malon, U.S. and Foreign Patents applied for.
37. R.F. Malon and A. Zampini, U.S. Patent 4,468,502, August 28, 1984.
38. R.F. Malon and A. Zampini, U.S. Patent 4,472,175, September 18,1984.
39. A.K. Fritzsche, J. Appl. Polym. Sci., 32, 3541-3550 (1986).
40. H. Makino, Y. Kusuki, H. Yoshida, and A. Nakamura, U.S. Patent 4,378,324, March 29, 1983.
41. H. Makino, Y. Kusuki, T. Harada, and H. Shimazaki, U.S. Patent 4,378,400, March 29, 1983.
42. H. Makino, Y. Kusuki, T. Harada, and H. Shimazaki, U.S. Patent 4,440,643, April 3, 1984..
43. H. Makino, Y. Kusuki, T. Harada, H. Shimazaki, and T. Isidu, U.S. Patent 4,528,004, July 9, 1985.
44. I. Cabasso and K.A. Lundy, U.S. Patent 4,602,922, July 24, 1986.
45. I. Cabasso and K.A. Lundy, "Multilayer Composite Membranes in Gas Separation: Novel Concepts and Superior Membranes", paper presented at Second Annual National Meeting of the North American Membrane Society, Syracuse University, Syracure, New York, June 1-3, 1988.
46. H. Yasuda, "Plasma Polymerization", Academic Press Inc., New York, 1985.
47. J. Sakata and M. Wada, J. Appl. Polym. Sci., 35, 875-884 (1988).
48. H. Yasuda, J. Membrane Sci., 18, 273-284 (1984).
49. H. Yasuda, J. Polymer Sci., Marcomol. Rev., 16, 199-293 (1981).
50. K.R. Buck and V.K. Davar, Brit. Polym. J., 2, 238-239 (1970).
51. H. Yasuda and C.E. Lamaze, J. Appl. Polym. Sci., 17, 201-222 (1973).
52. A. T. Bell, T. Wydeven, and C.C. Johnson, J. Appl. Polym. Sci., 19, 1911-1930 (1975).
53. H. Yasuda, H.C. Marsh, and J. Tsai, J. Appl. Polym. Sci., 19, 2157-2166 (1975).
54. H. Yasuda, H.C. Marsh, E. S. Brandt, and C.N. Reilly, J. Appl. Polym. Sci., 20, 543-555 (1976).
55. D. Peric, A.T. Bell, and M. Shen, J. Appl. Polym. Sci., 21, 2661-2673 (1977).
56. P.V. Hinman, A.T. Bell, and M. Shen, J. Appl. Polym. Sci., 23, 3651-3656 (1979).
57. E.L. Lawton, J. Appl. Polym. Sci., 18, 1557-1574 (1974).
58. A.F. Stancell and A.T. Spencer, J. Appl. Polymer Sci., 16, 1505-1514 (1972).
59. M. Kawakami, Y. Yamashita, M. Iwamoto, and S. Kagawa, J. Membrane Sci., 19, 249-258 (1984).
60. M. Yamamoto, J. Sakata, and M. Hirai, J. Appl. Polym. Sci., 29, 2981-2987 (1984).
61. J. Sakata, M. Yamamoto, and M. Hirai, J. Appl. Polym. Sci., 31, 1998-2006 (1986).
62. N. Inagaki and J. Ohkubo, J. Membrane Sci., 27, 63-75 (1986).
63. N. Inagaki and H. Katsuoka, J. Membrane Sci., 34, 297-305 (1987).
64. J. Sakata, M. Hirai, and M. Yamamoto, J. Appl. Polym. Sci., 34, 2701-2711 (1987).
65. N. Inagaki, N. Kobayashi, and M. Matsushima, J. Membrane Sci., 38, 85-95 (1988).
66. K. Okita, U.S. Patent 4,533,369, August 6, 1985.
67. B.R. Breslau and R.A. Cross, "An Introduction to Membrane Separation

Technology" in: An Introduction to Separation Science, 2nd Edition, John Wiley, New York, 1982.

68. F. Forbes, Prod. Finish., 23, 24-29 (1970).

69. J. Zahka and L. Mir, "Ultrafiltration of Cathodic Electrodeposition Paints--Seven Years of Field Experience", Association for Finishing Processes of the Society of Manufacturing Engineers, Technical Paper FC79-684, 1979.

70. J. Latty, U.S. Patent 4,125,462, November 14,1978.

71. M.J. Hurwitz, E. Park, and H. Aschkenasy, U.S. Patent 3,406,139, October 15, 1968.

72. "Molecular Engineering of Ultrathin Polymeric Films", Eds. P. Stroeve and E. Frances, Elsevier Applied Science, New York, 1987.

73. H. Cacovic, J.P. Schwengers, J. Springer, A. Laschewsky, and H. Ringsdorf, J. Membrane Sci., 26, 63-77 (1986).

74. O. Albrecht, A. Laschewsky, and H. Ringsdorf, J. Membrane Sci., 22, 187-197 (1985).

75. T. Kajiyama, Y. Nagata, S. Washizu, and M. Takayanagi, J. Membrane Sci., 11, 39-52 (1982).

76. T. Kajiyama, S. Washizu, and M. Takayanagi, J. Appl. Polym. Sci., 29, 3955-3964 (1984).

77. T. Kajiyama, H. Kikuchi, and S. Shinkai, J. Membrane Sci., 36, 243-255 (1988).

78. Y. Osada, K. Honda, and M. Ohta, J. Membrane Sci., 27, 327-338, (1986).

79. H. Iwata and T. Matsuda, J. Membrane Sci., 38, 185-199 (1988).

80. T. Hirotsu and S. Nakajima, J. Appl. Polym. Sci., 36, 177-189 (1988).

81. J.H. Fendler, J. Membrane Sci., 30, 323-346 (1987).

HISTORY OF PRESSURE SENSITIVE ADHESIVE COATINGS

D.Satas
Satas & Associates

INTRODUCTION

Pressure sensitive adhesive tapes, labels and various other coated products are used for many diverse applications, such as hospital and first aid tapes, packaging tapes, office tapes, labels for many different product uses, sun light control films, and corrosion protective tapes for underground oil and gas lines. There is hardly a product group which has not benefited from pressure sensitive adhesives. Pressure sensitive adhesives are used because of their convenience, even though there are usually alternate methods to accomplish the same task. A hospital dressing does not have to be secured by a tape, it can be simply wrapped and tied. Labels can be attached by glue; instead of packaging tapes, one can use a string.

Pressure sensitive adhesive products, because they are products of convenience, are used much more frequently in industrially developed countries than in less developed areas of the world. Per capita consumption of pressure sensitive adhesive products might be a good index of the industrial development level.

A pressure sensitive adhesive is defined as an adhesive which forms a bond to a variety of surfaces on the application of light pressure..In England the term "self-adhesive" is preferred over that of pressure sensitive. In German tacky adhesive (Haftklebstoff) is used in addition to the translation from the American "pressure sensitive adhesive". Such adhesives are often used for temporary holding purposes. Therefore, it is expected that the adhesive bond will be easily broken at will without contaminating the surface. These adhesives are used as thin coatings (25-50 μm thick, although some products may have a much heavier adhesive coating).

MEDICAL TAPES

The start of the pressure sensitive adhesive industry is found in the medical application of adhesive tapes and plasters. The use of adhesive masses and plasters can be traced from the beginnings of recorded medical history.[1] A Chinese medical text published in 2055 BC mentions a medical plaster consisting of a blend of rosin, plant oil, milk fat, and animal glue coated on fabric.[2] European early plasters consisted of resins and beeswax. These were not pressure sensitive adhesives as we know them today. It is not clear at which point these masses began to resemble modern products.

We consider that the main ingredient of pressure sensitive adhesives is an elastomer. Early medical plasters, while tacky, lacked the elastomeric and cohesive properties associated with pressure sensitive adhesives. During the mid-nineteenth century, natural rubber was introduced to such medical plasters and these products already resembled pressure sensitive adhesives.

Published 1990 by Elsevier Science Publishing Co., Inc.
Organic Coatings: Their Origin and Development
R.B. Seymour and H.F. Mark, Editors

The invention of such adhesives is generally attributed to
Dr.Horace Day. Such medical plaster adhesives consisted of
India rubber, spirits of turpentine, turpentine extract of
cayenne pepper, litharge, pine gum, and other ingredients. It
already contained two main ingredients of pressure sensitive
adhesives: elastomer (India rubber) and tackifying resin (pine
gum) dissolved in a solvent for coatability. A U.S. patent was
issued in 1845 to Shecut and Day on an adhesive of this type.[3]

        The first large scale hospital tape business was started in
1874 by Robert Wood Johnson and George Seaburg in East Orange,
New Jersey. In 1886 Johnson founded Johnson & Johnson Co. which
is still a major manufacturer of hospital tapes.[4] European
developments paralelled those in the United States. A German
patent was issued in 1882 to a druggist named Paul C.Beiersdorf
for a medicated plaster based on gutta percha. The same year is
held to be the establishment date of Beiersdorf AG, one of the
oldest continuous pressure sensitive adhesive businesses with
headquarters in Hamburg. Beiersdorf in cooperation with a
physician- Paul Gerson Unna- developed a large number of medical
plasters. One of their researchers, physician Troplowitz
developed a zinc oxide containing adhesive. Zinc oxide
neutralized acidic resin and decreased skin irritation. Such a
product appeared on the market in 1901. These plaster adhesives,
however, had to be warmed up by body heat in order to develop
the required adhesion.[5] Today Beiersdorf AG is one of the
largest companies specializing in pressure sensitive adhesive
products in the world.

        In 1899, a natural rubber based zinc oxide containing
adhesive was introduced by Johnson & Johnson, which was quite
close in composition to modern adhesives. Apparently there is a
slight dispute who was first with a zinc oxide containing
adhesive: Johnson & Johnson or Beiersdorf. In Japan, the first
natural rubber based medical plaster was prepared in 1911 by
Takeuchi Chemical Laboratory.[2] Natural rubber initially was
the only elastomer used for pressure sensitive adhesives and
still remains an important one, despite the fact that many other
elastomers have been introduced into the pressure sensitive
field.

INDUSTRIAL TAPES

        Pressure sensitive adhesive tapes were used mainly for
first aid and hospital products for a long time. Developments,
such as Citoplast by Beiersdorf (1897)- a tape for bicycle tire
repair and for use as an athletic tape. Further development of
pressure sensitive tapes for industrial applications took a long
time, although medical tapes must have been used occasionally
for various other purposes. The industrial applications of
pressure sensitive tapes started to appear during the period
1920-1930. The first electrical tapes were made during this
period. They were not really pressure sensitive tapes, but so
called friction tapes, because they were produced on a rubber
calender by friction calendering. The adhesive was a soft
cohesively weak mass which adhered well to itself, since the
fabric was saturated with the adhesive on both sides, but
adhesion was poor to other surfaces.

        The first truly pressure sensitive industrial tapes

were masking tapes used for automobile body painting and cellophane film tapes for office and household use. It is said that a manufacturer of hospital tapes noticed that the tape sales of one of his distributors was increasing at an unusually fast rate. An investigation disclosed that the distributor was selling the tape to the automotive industry to hold paper sheets employed as masks for automobile body painting. This suggested the need for a special tape.

About the same time Richard G.Drew of the Minnesota Mining and Manufacturing Company developed a masking tape for the automotive aftermarket. The backing used was crimped paper.[6] Drew had joined 3M in 1921 and one of his first projects was the development of this tape. It lead to the development of saturated paper tapes, because unsaturated kraft paper did not have a sufficiently high internal bond to be unwound from a roll without splitting the paper backing.[7] Drew started 3M's first tape laboratory in 1926 and he later developed regenerated cellulose film tape for package closure.[4] Regenerated cellulose film tape was the first clear tape on the market and it found many uses besides package closures.[8,9] While most of the hospital tapes were produced by calendering the adhesive, 3M's strength was in solventborne coatings because of their experience in manufacturing abrasive paper. The solvent coating technique was more suitable for manufacturing industrial tapes. Drew retired in 1964. He was a prolific inventor and has his name on 30 major pressure sensitive tape patents. Transparent film tapes were developed in Germany in 1935 by Beiersdorf.[5]

Further development of various tapes grew at a fast rate and various conceivable backings were used. While hospital tapes were dominantly on fabric backings, saturated paper backed tapes became the most important product for industrial and other non first aid applications. Later paper started to be displaced as a tape backing by various films. Film backing is less expensive than saturated paper. If a thick and bulky tape is required, paper is still the backing of choice, but for applications where the thickness is not important, films are more economical to use. Thus various films became increasingly used as tape backings. Unplasticixed vinyl film tapes became important in Europe and polyester film tapes in the United States. Later biaxially oriented polypropylene film tapes started to replace both films, especially polyester, as the dominant backing for commodity product uses, such as packaging tapes.

LABELS

The use of pressure sensitive adhesives for labels developed separately from that of tapes. Labels are essentially printed products where printing and diecutting are the most important manufacturing steps. The making of label stock, however, is similar to tape making. Labels are also less expensive than tapes. They had to compete with inexpensive gummed paper labels. This price differential also helped to separate these related industries. For a long time the tape and label industries grew side by side. Only recently has the integration of the manufacturing of these two product classes started to take place.

The pressure sensitive label business was started by

R.Stanton Avery in Los Angeles, California. His first contact
with these products was at Ad-Here Paper Co. which invited him
to become production manager. The company planned to apply
strips of adhesive to the edge of paper either before or after
printing. The sheets carrying advertising messages would be
stuck to shop windows and removed again. Stan Avery's first
project was to build the adhesive applicator.[10] The adhesive
was applied, air dried, and faced with strips of removable
glassine. The paper then was cut to the required size for
printing. Ad-Here Paper Co. also had another pressure sensitive
product called Gum-Tack. These were little pieces of double-
faced adhesive paper tape used to secure four corners of a
window sign. This work took place during the depression and Ad-
Here had problems meeting the payroll. Stanton Avery left them
and started on his own in the late 1930s to develop some simple
coating equipment and to print and sell stickers by direct mail.

Kleen Stik was also engaged in a similar effort, producing
printable sheet stock, but suspended their business during World
War II, while Avery continued to develop the label business in
association with Kum-Kleen Products Co. The US Government was
interested in these labels, because of their resistance to water
as compared to gummed labels. In 1942 Avery received an order
from the Navy Department for five million labels. This order
prompted Avery to develop a rotary label die-cutter. Increased
business required additional coating facilities and Fasson
Division was established in 1953 in Painesville, Ohio. Avery
also signed up a manufacturing licensee- the William Sessions
Co. in England. Avery's international business grew from this
undertaking.

The development of the label industry, as we know it today,
became possible with the invention of silicone release coatings,
which became commercially available in 1954.[10] Silicone
release coatings were the result of silicones first introduced
during World War II to protect electrical systems in aircraft.
While there are many polymers and their compounds which are
suitable as release coatings for tapes and labels, silicone
release coatings are unique in several performance respects. It
is difficult to imagine the growth of the label industry without
silicone release coatings.

OTHER PRODUCTS

In addition to the two main product classes of tapes and
labels, pressure sensitive adhesives are also used for many
applications which do not fall into these two categories.
Pressure sensitive adhesives were found to be useful for many
different applications: wallcoverings, automotive trim, note
pads, film lamination, and many other products penetrating
practically all industries. The growth of these miscellaneous
products is faster than that of labels and tapes, which already
have reached maturity.

The importance of pressure sensitive adhesives continues to
grow in the medical product area, which was the first
application of pressure sensitives. Many changes and
developments took place. The introduction of acrylic adhesives
coincided with the work to make tapes which are less irritating
to the skin. In the 1950's and 60's many different porous

breathable tapes on light weight backings were prepared with a
relatively thin adhesive coating. The early work on such
nonocclusive dressings came from Smith & Nephew Ltd. in the form
of porous vinyl film dressing and discontinuously applied
adhesive. The development of nonocclusive tapes gained momentum
with the introduction of nonwoven fabric tape by 3M Co. These
tapes were much less irritating to the skin than earlier
products and they dramatically changed the products used for
surgical and first aid applications. Pressure sensitive
adhesives also expanded into holding means for a variety of
dressings and medical devices. Pressure sensitive adhesives
have also expanded into some unique medical applications:
surgical drapes where the incision is made through the adhered
film; in the 1980's into elastomer film/adhesive dressings where
the adhesive coating is applied directly over the wound;
electrically conductive adhesives to hold electrocardiogram
electrodes or grounding pads used in conjunction with
electrosurgery; and for transdermal drug delivery systems where
the adhesive might also act as a medium for drug diffusion in
addition to holding the drug reservoir.

MATERIALS

     The growth of the pressure sensitive adhesive product
industry greatly depended on the development of raw materials.

     Natural rubber was the only elastomer used for pressure
sensitive adhesives for a long time until rubber shortages
during World War II caused the adaptation of other elastomers.
Polyisobutylene, compounded with vulcanized vegetable oil, ester
resin, and polybutene, was adopted for some hospital tapes.[11]
Natural rubber was tackified with various available resins:
coumarone-indene resins a product of coal and wood rosin. The
latter and its derivatives became the most important tackifiers
for pressure sensitive adhesives. Wood rosin was not stable
and oxidized easily causing loss of tack ("dry out") in a
relatively short time. In 1934, the Hercules Powder Company
started manufacturing hydrogenated rosins. The development of
hydrogenated and esterified wood rosin was an important step
toward improved adhesives of acceptable stability. Terpene
resins became preferred materials for natural rubber based
adhesives in the late 1950's and 1960's. Later development of a
large variety of synthetic petroleum based resins took over many
applications. The increased use of aqueous emulsions increased
the demand for resin dispersions which could be readily added to
aqueous polymer latexes.

     The first attempts to use elastomers other than natural
rubber were forced upon the fledgling industry by the rubber
shortage during World War II. An extensive research program was
financed by the Federal Government to develop a replacement for
natural rubber. Styrene-butadiene copolymer came out as the
preferred substitute. Adhesives were prepared employing SBR and
this elastomer is still used mainly in solventborne
formulations. Styrene/butadiene latex was first sold for
pressure sensitive adhesive applications only in the 1970's.[12]
SBR is more difficult to tackify than natural rubber and it
crosslinks on aging, while aging of natural rubber is
accompanied by chain scission. Blends of the two elastomers
compensate somewhat for the deficiencies of components.

The development of polyisoprene, chemically a similar elastomer to natural rubber, except of lower molecular weight and narrower molecular weight distribution, was thought to replace natural rubber for pressure sensitive adhesive work. Natural rubber is not a uniform product and it was thought that the uniformity of polyisoprene would serve well in the compounding of pressure sensitive adhesives. Chemical similarity notwithstanding, the mechanical differences were such as to make polyisoprene of low level importance for pressure sensitive adhesives.

A great enrichment of the pressure sensitive adhesive industry took place in the 1950's by the introduction of acrylic pressure sensitive adhesives. Acrylic polymers have a long history. Acrylic acid was first synthesized in 1843. By 1901, research was carried out on acrylic acid esters. The first commercial production of methacrylic polymer took place in 1927 by Roehm und Haas AG in Germany. Acrylic emulsions have been made by BASF since 1929 and the first patent application for pressure sensitive acrylic emulsions was submitted by BASF in 1943.[13] Despite all this early activity, pressure sensitive adhesive applications first came in the 1950's and attained their current importance in the 1960's. Polyvinyl ethers were also introduced at about the same time and were used for hospital tapes.

The development of acrylic pressure sensitive adhesives was important in many different ways. Acrylic polymers can be inherently pressure sensitive without any compounding. This put to rest various theories of pressure sensitive behavior thought to be linked to the presence of two phases in rubber/resin systems. It became clear that pressure sensitive properties are closely related to the viscoelastic properties of the adhesive. Carl Dahlquist was one of the pioneers in this important recognition.[14]

Acrylic adhesives are saturated and stable compounds suitable for many applications where stability on outdoor exposure, stability to higher temperature, and clarity are important. These polymers added to the variety of available adhesives and increased the reliability of pressure sensitive products. They replaced other adhesives in most hospital and first aid uses, because of their low level of skin irritation due to allergic reactions. The adhesive consists of one component polymer and does not require various additives, which sometimes may act as irritants.

The advent of acrylic adhesives also contributed to the character change of this industry. Adhesives used to be compounded and consumed captively and there were very few pressure sensitive adhesives on the merchant market for general usage. Acrylic adhesives started to be produced by a number of acrylic polymer manufacturers and they became available to anyone, allowing many coaters to enter the pressure sensitive adhesive field without accumulating the experience of adhesive compounding and acquiring the required equipment.

Acrylic adhesives are available in many different forms: as solution polymers, aqueous emulsions, hot melts, and 100% solids radiation curable coatings. While early acrylic adhesives were mainly solution polymerized, aqueous emulsions started to gain importance in the 1970's, because of air pollution regulations

and currently is the fastest growing adhesive type. Hot melt
acrylic adhesives are available, but not very functional. Ultra
violet and electron beam radiation curable acrylic adhesives are
used only for exceptional applications.

Another important development responsible for the growth of
pressure sensitive hot melts was the introduction of
thermoplastic A-B-A type blockcopolymers. Before the
development of these polymers, pressure sensitive hot melt
adhesives were based on ethylene/vinyl acetate copolymers, which
were deficient in many ways, especially in shear resistance.
The introduction of polystyrene/ polyisoprene/ polystyrene block
copolymers (and blockcopolymers with butadiene midblock) opened
possibilities to compound tacky, high shear resistance at room
temperature adhesives, which could be applied at hot melt
application conditions, because of a sufficiently low viscosity
at elevated temperature. This type of thermoplastic rubber was
first marketed in 1965 by Shell Chemical Company and was
immediately accepted for pressure sensitive adhesive
applications. While these polymers are also suitable for
solventborne and calenderable adhesives, their main impact was
on hot melt adhesive application. Later developments followed:
in 1972 Shell Chemical Company started to manufacture S-EB-S
(ethylene-butylene midblock) polymers which are completely
saturated and have better aging properties.[15] They did not
gain the importance of S-I-S polymers, because of the higher
price.

The advent of these polymers also had an important side
effect on the pressure sensitive adhesive industry, which was
traditionally quite secretive about formulations and the
compounding of adhesives. The manufacturers of thermoplastic
rubbers, however, took it upon themselves to instruct any
interested party on the use of their rubber for pressure
sensitive adhesives and contributed significantly to the
dissemination of compounding information and to the general
understanding of pressure sensitive adhesives.

Practically all available elastomers were compounded into
pressure sensitive adhesives, but only a few found permanent
usage, because of some special properties. Silicone adhesives
are used for their resistance to high temperatures, their
capability to retain tack at very low temperatures, and their
permeability to pharmaceutical compounds. Butyl rubber is used
in corrosion protective tapes for underground oil and gas pipes,
because of its resistance to oxidation and general chemical
stability. Butadiene-acrylonitrile rubbers are used for their
solvent resistance. So far polyurethane based adhesives have
found some limited applications for low adhesion surface
protective tapes.

EQUIPMENT

Pressure sensitive adhesives are applied in relatively
thick coatings: 20-75 g/m^2 (20 to 75 μm in thickness) and
heavier for some tapes. The construction of pressure sensitive
products, however, also requires various light weight coatings;
primers (in the range of 1 g/m^2), release coatings (0.5-1 g/m^2).
The pressure sensitive adhesive industry has broad requirements
for coating and auxiliary equipment. Most of the equipment used

for pressure sensitive adhesives, however, was adapted from other product uses, especially from paper converting and from textile processing. The pressure sensitive adhesive coating business was concentrated in a small number of companies, most of which were capable of doing their own equipment development. Only recently the pressure sensitive adhesive business became large enough to warrant special design effort by the equipment manufacturers.

While early adhesive coatings were applied from solvent solutions, calender coating soon became an important technique for making tapes. The manufacturers of hospital tapes were familiar with rubber calendering and adopted these machines for pressure sensitive adhesive application. The main change came about with the entry of 3M Co. into the manufacturing of industrial tapes. 3M Co. has chosen to work with solvent solutions, because of their experience in the coating of abrasive products. Thus calenders for solid coatings, and knife, reverse roll, and other roll coaters for solventborne coatings became the most widely used machines. Floating knife, gravure and rod coaters were and are used for light weight prime and release coatings.

New requirements for pressure sensitive adhesive coatings came about with the growth of hot melt adhesives. Slot orifice coaters competed for some time with roll coaters for pressure sensitive applications with the former a clear winner. The coater developed by George.C.Park in the 1970s was an important early machine. Increased use of adhesive latexes also resulted in the requirement for different coating equipment. Leveling of the coating can be a problem when coating aqueous adhesives. Knife coaters have a tendency to introduce scratches which are difficult to level and these coaters were found unsuitable for aqueous coatings. Reverse roll coaters could be used, but the adhesive must be carefully compounded to improve its leveling. Reverse and other roll coating introduces longitudinal striation marks, which should level before the coating solidifies in the drying oven. Wire-wound rod coaters found increased usage with the advent of aqueous adhesives, although they are not suitable for higher weight coatings. In the 1980's slot orifice coaters were adapted for aqueous coatings application.

The development of 100% solids silicone release coatings in the 1980's placed a difficult demand on coating equipment manufacturers. Silicone coatings are very light: 0.5-1 g/m^2. If the coating is applied from a dilute solution, typically 20 g/m^2 of wet coating could be applied to obtain a 1 g/m^2 dry deposit. Gravure and wire wound rod coaters are quite suitable for this use. For the deposition of light weight 100% solids coatings, which are thermally or radiation cured, new coating methods were required. Several methods are currently competing for this use: multi roll transfer coater (mainly by Bachofen and Meyer AG), reverse offset gravure (promoted mainly by Black Clawson Co.), and the adaptation of rubber covered roll coater by Bonnier Technical Group SA. Transfer roll coaters are preferred, but the long term verdict is not yet known.

Drying equipment used for pressure sensitive adhesive coatings is an adaptation of equipment used for other applications. The increased use of emulsion coatings has increased the interest in infra red radiation drying.

While slitting is used for many paper, film and foil
applications, pressure sensitive adhesive tape slitting required
the perfection of differential rewinding allowing then
independent slip of each core on the driving shaft. The
continuous strive towards less expensive products, tapes, such
as oriented polypropylene film packaging tapes, required fully
automated slitting equipment. Such equipment has been made
available for the industry primarily by Italian slitting
equipment manufacturers.

TESTING

     Testing of pressure sensitive adhesive properties and
understanding adhesive behavior are closely related.
The standard testing of pressure sensitive adhesives involves
determination of tack, peel adhesion, and resistance to shear.

     Rolling ball tack test measurement is probably the oldest
tack test,being used in the 1940's.[16] Later, probe tack
testing was developed in order to simulate a subjective thumb
test. Wetzel pioneered the probe tack procedure.[17] Hammond
has developed a self contained probe tack instrument in the
early 1960s, which became known as the Polyken Tack Tester and
is widely used in the industry.[18]

     Peel test is the most often used test to evaluate pressure
sensitive adhesives and it yields more information about
adhesive behavior than other tests. The test has been discussed
by many authors and has produced a considerable volume of
literature.[19] Early mathematical analysis of peel tests was
started with aircraft designers. Several authors, especially
D.H.Kaelble, contributed extensively to the analysis of pressure
sensitive adhesive peel.

     Shear (and creep) resistance is the simplest property to
evaluate. The test results are also easily relatable to the
mechanical properties of pressure sensitive adhesives.[20]

     The explanation of the pressure sensitive adhesive
phenomenon changed along with a better understanding of adhesion
in general. During the 1950's and 1960's the current
understanding of adhesion was developed. It was realized that in
addition to the interfacial condition, which is relatable to the
capability of the polymer to wet the adherend, mechanical
properties of the adhesive are even more important, since they
determine not only the shear, but also peel behavior. This
realization prompted increased testing of mechanical
viscoelastic properties of pressure sensitive adhesives. While
dynamic mechanical properties have been investigated since the
mid 1940s in the laboratories of some pressure sensitive
adhesive manufacturers [21], industry wide recognition of the
importance of such test results was delayed for a long time
period. The subject started to be discussed in the the 1970's
(M.Sherriff, D.W.Aubrey, G.Kraus) and became widely discussed
and reported only in the 1980's. Raw material manufacturers,
especially Hercules Inc.(J.B.Class and S.G.Chu) were mainly
responsible for creating a wide interest through their published
material.[22]

110

CONCLUSIONS

The history of the development of pressure sensitive products might have been through its most active and most interesting stages, but the future promises not only a steady growth along the same lines, but also diversification into more product areas and better engineered manufacturing as the products become high volume commodity items. The main changes expected are from the development of new materials adding to the further diversity of materials used for pressure sensitive adhesives.

ACKNOWLEDGEMENTS

The author appreciates the help and comments by Helmut Mueller, BASF AG, Ludwigshafen, Dr.Keiji Fukuzawa, Tokyo, and Carl A. Dahlquist, St.Paul.

REFERENCES

1.   Professional Uses of Tape, 3rd edition (Johnson & Johnson, New Brunswick, N.J. 1972).
2.   Du Xuan, Journal of the Adhesion Society of Japan 4 (2), 73-78 (1985).
3.   W.H.Shecut and H.H.Day, U.S.Patent 3,965 (1845).
4.   Paper Film and Foil Converter 51 (10),90-96 (1977).
5.   100 Jahre Beiersdorf 1882-1982, BDF Hamburg (1982).
6.   R.G.Drew, U.S.Patent 1,760,820 (1930), Reissue 19,128 (1934)(assigned to Minnesota Mining and Manufacturing Co.).
7.   C.Bartell in: Handbook of Pressure Sensitive Adhesive Technology, 2nd edition, D.Satas, ed. (Van Nostrand Reinhold, New York 1989) pp.675-690.
8.   O.J.Hendricks and C.A.Dahlquist in: Adhesion and Adhesives, Vol.2, R.Houwink and G.Salomon, eds. (Elsevier, Amsterdam 1967) pp.387-408.
9.   R.G.Drew, U.S.Patent 2,177,627 (1939).
10.   J.D.Jones and Y.A.Peters in: Handbook of Pressure Sensitive Adhesive Technology, 2nd edition, D.Satas, ed. (Van Nostrand Reinhold, New York 1989) pp.601-626.
11.   M.H.Kemp, Adhesives Age 4(12),22-25 (1961).
12.   A.D.Hickman in: Handbook of Pressure Sensitive Adhesive Technology, 2nd edition, D.Satas, ed.(Van Nostrand Reinhold, New York 1989) pp.295-316.
13.   H.Fikentscher and C.Schuster, German Patent 862,957 (1943).
14.   C.A.Dahlquist in: Adhesive Fundamentals and Practice (McLaren, London 1966).
15.   E.E.Ewins,Jr.,D.J.St.Clair, J.R.Erickson and W.H.Korcz in: Handbook of Pressure Sensitive Adhesive Technology, 2nd edition, D.Satas, ed. (Van Nostrand Reinhold, New York 1989) pp.317-373.
16.   F.H.Hammond, Jr. in: Handbook of Pressure Sensitive Adhesive Technology, 2nd edition, D.Satas, ed. (Van Nostrand Reinhold, New York 1989) pp.38-60.
17.   F.Wetzel, Characterization of Pressure-Sensitive Adhesives, ASTM Bulletin No.221, 64-68 (1957).
18.   F.H.Hammond, Jr. Polyken Probe Tack Tester, ASTM Spec. Publ. 360(1963).

19. D.Satas in: Handbook of Pressure Sensitive Adhesive
     Technology, 2nd edition, D.Satas, ed. (Van Nostrand
     Reinhold, New York 1989) pp.61-96.
20. C.A.Dahlquist in: Handbook of Pressure Sensitive Adhesive
     Technology, 2nd edition, D.Satas, ed. (Van Nostrand
     Reinhold, New York 1989) pp.97-114.
21. Private conversation with Carl A.Dahlquist.
22. S.G.Chu in: Handbook of Pressure Sensitive Adhesive
     Technology, 2nd edition, D.Satas, ed. (Van Nostrand
     Reinhold, New York 1989) pp.158-203.

HISTORY OF DIAGNOSTIC COATINGS

W.J. SCHRENK* AND A.M. USMANI
Boehringer Mannheim Biochemistry R&D Center
Indianapolis, Indiana 46250

ABSTRACT

Coatings have been used for protection and decoration by cave man to present day man. The use of coatings in dry reagent chemistries for rapid quantitative analysis of body analytes e.g., glucose and cholesterol is rather recent, having occurred mainly within the past 30 years.

In a dry reagent chemistry, a suitable coating containing enzymes and an indicator is coated as a thin film. A clinical sample e.g., blood, 10-50 μl in volume, is placed on the miniature, disposable dry reagent. A sequence of rapid reactions results in formation of color that can be quantified. In this paper, we shall describe the chemistry, technology and history of coating based dry chemistries.

INTRODUCTION

Science has not yet discovered an insulin formula that will respond to the body's senses. Since injected insulin does not automatically adjust itself to the current blood sugar level, millions of people with diabetes need to check if change is necessary. Insulin dose required to mimic body's response must be adjusted day to day depending upon diet and physical activity. Thus, self-monitoring of blood glucose levels is essential to people with diabetes to lead a more normal life. These people include women with diabetes who wish to have children. By regular and accurate monitoring of her blood glucose level, an expectant mother can have a normal pregnancy and give birth to a healthy child. Athletes with diabetes also face significant problems and must monitor their blood glucose levels that could vary with the level of physical exertion. Other body analytes also require monitoring in patients with various medical problems.

Prior to the development of dry reagent chemistries, blood glucose and other analytes of clinical importance were quantitatively measured by analysts in a laboratory using wet methods.

In a typical glucose measuring dry reagent, glucose oxidase (GO) and peroxidase (POD) enzymes along with a suitable indicator e.g., tetramethylbenzidine (TMB) are dissolved and/or dispersed in a latex or water soluble polymer. The enzyme-containing coating is applied to a lightly pigmented plastic film and dried to a dry thin film. The coated plastic, cut to about 0.5 cm x 5 cm size is the dry reagent. The user applies a drop of blood and allows it to react with the strip for about 60 seconds or less. The blood is wiped off and the developed color is then read by a meter or visually compared with the pre-designed printed color blocks to precisely determine the glucose level in the blood.

Coatings have played an integral part in the development of dry chemistry reagent diagnostics. In this chapter we will describe background, history, chemistry and technology of dry reagents.

* Present Affiliation: Gen Probe, San Diego, California.

BACKGROUND

Dry chemistry reagent diagnostic tests are used for the assay of metabolites by concentration or by activity in a biological matrix [1]. Many clinical chemistry diagnostic tests are based on the principle of reacting a body fluid e.g., blood or urine containing a specific analyte with specific enzymes. The reactive components are usually present in excess, except for the analyte being determined. This is done to ensure that the reactions will go to completion quickly. Other enzymes or reagents are also used to drive the reactions in the desired directions [2]. Glucose and cholesterol are the most commonly measured analytes. Glucose concentrations deviating from the normal are considered clinically significant, indicating chronic, hyper- or hypoglycemia.

Glucose is determined by conversion to gluconic acid and hydrogen peroxide using glucose oxidase. Hydrogen peroxide is then coupled to an indicator, e.g., 3,3´, 5,5´-tetramethylbenzidine (TMB) by the enzyme peroxidase. The rapidly ($<$ 60 s) developed color is then measured at or around 660 nm. The chemical reactions are shown below:

$$\text{Glucose} + O_2 + H_2O \xrightarrow{\text{Glucose Oxidase}} \text{Gluconic Acid} + H_2O_2$$

$$H_2O_2 + \text{Colorless Indicator} \xrightarrow{\text{Peroxidase}} 2H_2O + \text{Colored Dye}$$

Cholesterol esters present in blood serum can be quantitatively saponified by cholesterol esterase into free cholesterol and fatty acids. Free cholesterol is oxidized by cholesterol oxidase, in the presence of oxygen, to cholest-4on-3one and H_2O_2. There are several indicator peroxidase systems that may be used to detect H_2O_2, e.g., TMB or oxidation of 4-aminoantipyrine (AAP) followed by coupling with naphthol to form a compound that could be read at 490 nm. The sequence of reactions leading to color development is shown below.

$$\text{Cholesterol-Fatty Acid-Lipoprotein Complex} + H_2O \xrightarrow{\text{Cholesterol Esterase}}$$

$$\text{Cholesterol} + \text{Fatty Acid} + \text{Protein}$$

$$\text{Cholesterol} + O_2 \xrightarrow{\text{Cholesterol Oxidase}} \text{Cholest-4on-3one} + H_2O_2$$

$$\text{AAP} + \text{Naphthol} + 2H_2O_2 \xrightarrow{\text{Peroxidase}} \text{Dye} + 4H_2O$$

Color is measured by using reflectance spectroscopy or by visually comparing the reacted strip to a pre-printed color chart [3,4]. Dry reagent test kits are usually available in the form of thin strips. Strips are usually disposable either coated or impregnated and mounted onto a plastic support or handle. The most basic diagnostic strip thus consists of a paper or plastic base, polymeric binder, and reacting chemistry components consisting of enzymes, surfactants, buffers and indicators. Diagnostic coatings or impregnation must incorporate all reagents necessary for the reaction. The coatings can be single or multi-layered in design. A list of common metabolites assayed by dry chemistry diagnostic test kits, although not comprehensive, is given in Table 1 [1,4].

TABLE 1. Common metabolites analyzed by diagnostic test kits.

Bilirubin	Glucose	Theophylline
Blood Urea Nitrogen	GOT	Total Protein
Calcium	Hemoglobin	Triglycerides
Cholesterol	Ketones	Urea
Creatinine	Phosphorus	Uric Acid

It is estimated that the number of dry chemistry measurements which provide clinical information, including all the routine dry chemistry blood glucose and urine tests performed by diabetics and all dry chemistry measurements made in hospitals, now exceeds the number of measurements made by all methods combined.

HISTORICAL

The basic examples of self testing are weight and body temperature measurements. Self testing or home monitoring is rather an old concept. Before the discovery of insulin, Elliott Joslin, some 75 years ago, advocated that diabetics should avoid carbohydrates and monitor it by frequent urine testing with Benedict's qualitative test available at that time [5]. In 1922, insulin became available and the self testing of urine became more popular. Around the mid 40's, Compton and Treneer developed a dry tablet containing sodium hydroxide, citric acid, sodium carbonate, and cupric sulfate [6]. This tablet, when allowed to react with a small sample of urine in a test tube, boiled resulting in reduction of blue cupric sulfate to yellow or orange color if glucose was present. Around 1956, glucose urine strips, impregnated and based on enzymatic reactions using glucose oxidase, peroxidase, and indicator were introduced by Miles and Eli Lilly [7]. Blood glucose test strips, based on the above dry chemistry format and coated with a semi-permeable membrane to which blood could be applied and wiped off, were introduced in 1964 by Miles. Hand-held meters utilizing reflectance for reading glucose were introduced in the late 60's although they were rather heavy and bulky.

Paper matrix dry reagents have a coarse texture, high pore volume and uneven large pore size. This results in color development in the reacted strips that is non-uniform. In the early 1970's, Boehringer Mannheim in Germany, developed a coating film type dry reagent. The film type dry reagent obtained by applying an enzymatic coating onto a plastic support gave a smooth fine texture and therefore, uniform color development [8]. Even today such coated strips are the workhorse of the diagnostics industry.

Discrete multi-layered coatings, a technology developed by the photographic industry, were adapted in the late 70's by Eastman Kodak to coat dry reagent chemistry formats for clinical testing [9,10]. Each zone of a multi-layered coating provides a unique environment to allow sequential chemical and physical reactions to occur. The basic multi-layered coating consists of a spreading layer, separation membrane, reagent and reflective zone, coated onto the base support. The spreading layer wicks the sample and applies it uniformly to the next layer. The separation membrane holds back certain sample constituents e.g., red blood cells and allows only the desired metabolites to pass through to the reagent zone that contains all the necessary reagent components. Other layers maybe included for enzyme immobilization, selective absorption, filtering or reflecting media.

CONSTRUCTIONS OF DRY CHEMISTRY STRIPS

The basic components of typical dry reagents that utilize reflectance measurement are a base support material, a reflective layer, and a reagent

116

layer either single or multi-layer. The functions of the various building
blocks are now described. The base layer serves as a building base for dry
reagent and usually is a thin, rigid thermoplastic film. The function of
the reflectance layer is to reflect light not absorbed by the chemistry to
the detector. Typical reflective materials are white pigment filled
coating, foam, membrane, paper and metal foil. The reagent layer contains
the integrated reagents for a specific chemistry. Typical materials are
paper matrix, fiber matrix, coating film as well as hybrids.

An example of a single-layer coating reagent that effectively excludes
red blood cells (RBC) is shown in Figure 1. In this case, a coating usually
an emulsion type that contains all the reagents for a specific chemistry, is
coated onto a lightly TiO_2 filled thermoplastic film and dried. For glucose
testing, the coating will contain glucose oxidase, peroxidase, and an
indicator e.g., TMB. It may also contain buffer for pH adjustment, minor
amounts of ether-alcohol type organic coalescing agent, and traces of a
hindered phenol type antioxidant to function as a ranging compound. A drop
of blood is applied to this coating and allowed to react for a short pre-
determined time (usually 60 s), the excess blood is wiped off with cotton or
other absorbent material, and then read either visually or by a meter.

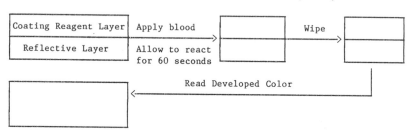

FIG. 1. Coating film type dry reagent.

In the reagent impregnated type (Figure 2), the paper matrix is
impregnated with the reagent. A film membrane, usually an ethyl cellulose
coating, is applied over the paper matrix to help exclude RBC.

Film Membrane
Reagent Impregnated Paper
Reflective Support Layer

FIG. 2. Paper impregnated and overcoated film membrane type dry reagent.

The schematic of a typical multi-layer coating dry reagent for example
BUN is shown in Figure 3. It consists of a spreading and reflective layer.
Sample containing BUN is spread uniformly by this layer. The first reagent
layer is a porous coating film containing the enzyme urease and buffer (pH
8.0). Urease reduces BUN to NH_3. A semi-permeable membrane coating allows
NH_3 to permeate while excluding OH^- from the second reagent layer. The
second reagent layer is comprised of porous coating film containing a pH
indicator wherein the indicator color develops when NH_3 reaches the semi-
permeable coating film. Typically, such dry reagents are slides (2.8 x 2.4
cm) with an application area of about 0.8 cm^2 and the spreading layer
thickness is about 100 μm.

BUN

Spreading and reflective layer (TiO$_2$ pigmented porous cellulose acetate coating)	Sample containing BUN is spread uniformly
Reagent layer 1: Porous gelatin coating film with urease and pH buffer	BUN + H$_2$O $\xrightarrow{\text{Urease}}$ 2NH$_3$ + CO$_2$
Semi-permeable membrane coating (cellulose acetate butyrate)	Permeable to NH$_3$; excludes OH$^-$
Reagent layer 2: Porous cellulose acetate coating film with a pH indicator	HI$_{colorless}$+NH$_3$ $\longrightarrow$ I$^-$(colored)$^{+NH4^+}$
Transparent plastic support	

FIG. 3. Schematic of a multi-layer coating dry reagent.

DIAGNOSTIC COATING POLYMERS

The polymer binder is necessary to incorporate the system's chemistry components in the form of either a coating or impregnation. The reagent matrix must be carefully controlled to mitigate or eliminate non-uniformity in reagent concentrations due to improper mixing, settling, or non-uniform coating thickness. Aqueous based emulsion polymers and water soluble polymers are extensively used. Table II provides a list of commonly used matrix binders [10,11].

TABLE II. Common matrix binders [10,11].

Emulsion Polymers	Water Soluble Polymers
Acrylics	Polyvinyl alcohol
Polyvinyl acetate: homo and co-polymers	Polyvinyl pyrrolidone
	Highly hydroxylated acrylic
Styrene acrylic	Polyvinylethylene glycol acrylate
Polyvinyl propionate: homo and co-polymers	Polyacrylamide
	Hydroxyethyl cellulose
Ethylene vinyl acetate	Other hydrophilic cellulosics
Lightly crosslinkable acrylic	Various co-polymers
Polyurethane	

Polymers must be carefully screened and selected to avoid interference with the chemistry. The properties of polymers e.g., composition, solubility, viscosity, solid content, surfactants, residual initiators, film forming temperature, glass transition temperature, and particle size should be carefully considered [12]. Generally, the polymer should be a good film-former with good adhesion to the support substrate and should have minimal or no tack for handling purposes. The coated matrix or impregnation must have the desired pore size and porosity to allow penetration of the analyte being measured, as well as the desired gloss, swelling characteristics, and surface energetic. Swelling of the polymer binder due to absorption of the liquid sample may or may not be an advantage depending on the overall system. Emulsion polymers have distinct advantages over

118

water soluble polymers due to their high molecular weight, superior
mechanical properties, and potential for adsorbing enzymes and indicators
due to micellar forces [11].

Polymeric binders used in multi-layered coatings include various
emulsion polymers, gelatin, polyacrylamide, polysaccharides e.g., agarose,
water soluble polymers e.g., polyvinyl pyrrolidone, polyvinyl alcohol, copolymers of vinyl pyrrolidone and acrylamide, and hydrophilic cellulose
derivatives e.g., hydroxyethyl cellulose and methyl cellulose [10].

ENZYME CONSIDERATIONS: IMMOBILIZATION AND STABILITY

Dry chemistry diagnostics often take advantage of the immobilization
capabilities of enzymes. Immobilization is a useful tool in designing
multi-layered coatings to reduce interactions between layers. Enzymes can
be bound by covalent coupling, adsorption onto a polymer surface, entrapment
within a polymer matrix or by crosslinking [12-16]. Some immobilized
enzymes show higher thermal stability and lower sensitivity to pH and
oxidation which allows more flexibility in the design and processing of
reagent strips.

A variety of polymeric adsorbents e.g., diethylaminoethyl cellulose,
polysaccharides, carrageenan, polyacrylamide, polyacrylates, polystyrene,
polyvinyl pyrrolidone, polyvinyl alcohol, polyvinylethylene glycol acrylate,
collagen and gelatin are used for bonding with enzymes. Adsorption of
enzymes onto polystyrene is probably the most widely used method of
immobilizing antibodies in solid phase immunoassay.

Enzyme stability is essential in dry chemistry diagnostics. Decrease
in enzyme activity can reduce the shelf-life and performance of the
diagnostic product. Basic approaches used to stabilize enzymes include
addition of a water soluble compound and immobilization. The chemical
binding of the enzymes onto polymers reduces the freedom to undergo
conformational changes and thus reduces denaturing of enzymes [17]. In
another stabilization technique, certain hydrophobic polymers are dispersed
onto a carrier permeable to liquid and oxygen. Although the mechanism is
not well understood, dispersed polymers improve the stability of reagent
peroxidase. Preferred classes of polymers are co-polymers consisting of
hydrophobic recurring units e.g., styrene, alkyl acrylates, and anionic
recurring units e.g., acrylic acid and sulfonate. Recently, Azhar et al
have described stabilization of a glucose oxidase/peroxidase emulsion
coating and proposed several models for predicting shelf stability [18].

CHARACTERIZATION OF DIAGNOSTIC COATINGS

In general, characterization of polymers and coatings is as important
as the designing of these materials. Not much work in characterization of
diagnostic coatings has been done in the past. However, recently Burke et
al have provided an overall characterization outline for dry chemistries
shown in Table III [19]. Noteworthy are their work on thermal analysis of
enzymes, SEM examination of surfaces and interfaces, and x-ray photoelectron
spectroscopy analysis of diagnostic coatings. The color aspects of dry
reagent usually used in clinical analysis have been described by Genshaw
[20].

NEW TRENDS

Electrochemical sensors, biosensors, DNA probes, and implantable sensors are
a few of the new trends in dry reagent diagnostics [21-23]. Electrochemical
sensors such as enzyme electrodes are making their way into the glucose and
cholesterol diagnostic field. The electrochemical sensors work by coating a

miniaturized peroxide electrode with a polymer coating that contains, for example, immobilized glucose oxidase. The amperometric detection of hydrogen peroxide is then proportional to the glucose concentration [22].

TABLE III. Methods for characterization of dry reagent chemistries [19].

Polymer Binder	Enzyme	Coating	Dry Reagent Coated Surface
Viscosity	Activity	Viscosity	Dry film thickness
Solid content	Mol wt (SEC)	Solid content	Surface pH
pH	Thermal stability (DSC)	PVC	UV stability
Particle size and distriution (emulsion)	Michaelis constant, K_m	P/B ratio	Gloss; surface energetics (critical contact angle)
Spectroscopy FTIR NMR	Maximum velocity, V_{max}	Density	Porosity
Mol wt (SEC)	Substrate specificity	Fineness of dispersion	Interfacial aspects (SEM)
T_g, T_m (DSC)	pH stability	Leveling/ sagging	Coatings defects (FTIR)
Tensile strength			Chemical species/ reactions (XPS, FTIR, ATR)
Modulus			Performance characteristic
Elongation			Color development

FTIR   Fourier transform infrared

NMR   Nuclear magnetic resonance
SEC   Size exclusion chromatography

DSC   Differential scanning calorimetry
SEM   Scanning electron microscopy
XPS   X-ray photoelectron spectroscopy
ATR   Attenuated total reflectance

PVC   Pigment volume concentration
P/B   Pigment to binder
T_g Glass transition temperature
T_m Melting temperature

New trends in the dry reagent diagnostic field are bound to be dependent on the application of new technologies from the polymer and coating field. The adaptability of many polymeric systems has not yet been explored for use in the dry reagent area. The ability to tailor make polymers and coatings for specific applications in the diagnostic field is a concept currently being investigated.

120

CONCLUSIONS

We have provided a historical development of dry reagent diagnostic coatings along with the chemistry and technology of single and multi-layered coatings.

Polymers and coatings have played an important role in the development of dry chemistry reagent diagnostics. The market for home use diagnostics is expected to increase in the 1990´s. This trend toward self-testing demand simple yet accurate measurements of analytes e.g., glucose and cholesterol. We hope and expect that contributions from polymer science, specifically coatings science, will continue to aid in the development of diagnostic tests.

REFERENCES

1. H.E. Spiegel, Kirk-Othmer Enc. Chem. Technol, 3rd edition.
2. N.W. Tietz (Ed.), Fundamentals of Clinical Chemistry, W.B. Saunders Co., 1976.
3. A.H. Free and H.M. Free, Lab. Med., 15, 1595 (1984).
4. B. Walter, Anal. Chem., 55, 449A (1983).
5. E.P. Joslin, H.P. Root, P. White and A. Marble, The Treatment of Diabetes, 7th edition, Lea and Febiger, Philadelphia, 1940.
6. W.A. Compton and J.M. Treneer, US Patent 2, 387, 244, 1945.
7. A.H. Free, E.C. Adams and M.L. Kercher, Clin. Chem., 3, 163 (1957).
8. H.G. Rey, P. Rieckmann, H. Wielinger and W. Rittersdorf, US Patent 3, 630, 957, assigned to Boehringer Mannheim, 1971; also see P. Vogel, H-P Braun, D. Berger and W. Werner, US Patent 4, 312, 834, assigned to Boehringer Mannheim, 1982.
9. T.L. Shirey, Clin. Biochem., 16, 147 (1983).
10. E.P. Przybylowicz and A.G. Millikan, US Patent 3, 992, 158, assigned to Eastman Kodak, 1976; also B.J. Bruschi, US Patent 4, 066, 403, assigned to Eastman Kodak, 1978.
11. M.T. Skarstedt and A.M. Usmani, Polymer News, 14, 38 (1989).
12. W. Scheler, Makromol. Chem. Symp., 12, 1 (1987).
13. J.R. Schaeffer, B.A. Burdick and C.T. Abrams, ChemTech, 546 (September 1988).
14. C.E. Carraher, Jr. and L.H. Sperling, Polymer News, 13, 101 (1988).
15. W.H. Scouten, Methods in Enzymology, 135, 30 (1987).
16. R.C. Boguslaski, R.S. Smith and N.S. Mhatre, in Current Topics in Microbiology and Immunology, 1971.
17. A. Freeman, Trends in Biotechnol., 2, 147 (1984).
18. A.F. Azhar, A.D. Burke, J.E. DuBois and A.M. Usmani, in Progress in Biomedical Polymers, C.G. Gebelein (Ed.), Plenum, New York, 1989.
19. A.D. Burke, A.F. Azhar, J.E. DuBois and A.M. Usmani, Pacific Polymer Proceeding, 1989.
20. M.A. Genshaw, Color Res. Appl., 10, 235 (1985).
21. L.B. Wingard, Jr. and E.E. Spaeth, Biotechnol., 7, 116 (1987).
22. S.J. Updike et al., Diabetes Care, 11, 801 (1988).
23. D. DeRossi, Res. Dev., 67 (May 1989).

History of Vinyl Coatings

Raymond B. Seymour
University of Southern Mississippi
Hattiesburg, MS 39406

Vinyl chloride, which was synthesized by Regnault in 1835 by the dehydroclorination of the "oil of the dutch chemists", was one of the first available monomers. The four Dutchmen who were immortalized after obtaining a compound of an "olefiant gas" from the reaction of ethylene and chlorine in 1795 were Dieman,Trotswyck,Bondt and Laurverenburgh. The term vinyl chloride monomers (VCM) was coined by Kolbe and listed in his Lehrbuch in 1854. However, other 19th century chemists such as A. Geuther used the term elayl chloride instead of VCM and Kekule' and Zücke called their monomer monochloroethylene.

It has been assumed by many chemist historians that Regnault produced polyvinyl chloride (PVC) but no reference to vinyl polymerization was recorded in the literature until 1860 when A. W. Hofman accidently produced polyvinyl bromide. In 1872, E. Baumann produced PVC and described the product as a converted product or an isomer.

In 1912, I. Ostromislensky polymerized vinyl bromide, in the presence of sunlight, and isolated polymeric homologues with different solubilities, which he called AB and Γ forms of Kaupren bromide. Klatte and Rollet used organic peroxides as initiators for the polymerization of VCM in 1914.

Because of its insolubility and intractibility, PVC was essentially useless as a plastic or coating until E. Reid and W. Lawson produced copolymers of vinyl chloride and vinyl acetate (Vinylite). His contemporaries Voss and Dickhauser also produced soluble flexible copolymers of VCM and used the term internal plasticization to describe the difference between their products and PVC. Other VCM copolymers which were used as coatings were produced by the copolymerization of VCM with esters of acrylic acid, maleic and fumaric acid as well as with vinylidene chloride.

In spite of the successful use of these VCM copolymers as coatings for textiles, wood and concrete, they did not adhere to unprimed steel. This deficiency was overcome by the incorporation of maleic acid as a third monomer. When blended with phenolic resins, this terpolymer (VMCH) was widely used as a can coating.

Coatings of VCM and vinyl butyl ether (vinoflex) filled with iron oxide are used for magnetic recordings and as coatings for aluminum. However, the most widely used VCM copolymer is Vinylite VYHH which is produced from 87 percent VCM and 13 percent vinyl acetate. As much as 30 percent VYHH is soluble in a 50-50 acetone/toluene solvent. Reed plasticized these vinyl chloride/vinyl acetate copolymers with dioctyl phthalate and with low molecular weight polyesters.

While (PVC) is not very soluble in organic solvents, PVC coatings can be deposited from dispersions of the resin in a liquid plasticizer. The mixture is thermally fused at 165 C after application. Since the PVC in these plastisols is produced by emulsion techniques, it is coated with a layer of surfactant which prevents penetration of the liquid plasticizer into the solid PVC at ordinary temperatures.

Published 1990 by Elsevier Science Publishing Co., Inc.
Organic Coatings: Their Origin and Development
R.B. Seymour and H.F. Mark, Editors

Many chemists were aware of the difficulties of dissolving or plasticizing PVC produced by emulsion polymerization techniques but G. Powell recognized the commercial potentials for these liquid products, which he called plastisols in 1944. When volatile organic liquids were also present, the plasticizer--PVC system was called an organosol.

Most producers of PVC offered PVC emulsions to coatings technologist in the 1950's. PVC coatings deposited from emulsions required a plasticizer and a coalescent agent. The later was compatible with both water and the resin and assured the deposition of a continuous coating. Powered PVC, which can be electrostatically sprayed, was also used as a coating in Germany in 1955.

In addition to being used as a comonomer with VCM, vinyl acetate is also used as a polymer in emulsion systems. Vinyl acetate was synthesized by Klatte in 1912 and polymerized in 1914 by Klatte and Rollett. However, because of its low softening point, there was little interest in the use of polyvinyl acetate (Mowilith, PVAC), as a coating. However, emulsions of PVAC were produced by W. Starck and H. Freudenberger and used as adhesives and coatings. PVAC water-borne coatings have good resistance to hydrolysis. PVAC is also used as a textile coating.

W. Hermann and W. Haehnel produced polyvinyl alcohol (PVAL) in 1927 by the hydrolysis of PVAC. PVAL (Elvanol) has been used as a water soluble coating, which may be stabilized by the application of formalin to the surface of the deposited coatings.

The principal use of PVAL is for the production of polyvinyl acetals (Mowitals), such as polyvinyl butyral (PVB). H. Hopff patented PVB, which he produced by the acid condensation of PVAL and butyraldehyde in the late 1920's. This innovative invention was modified by chemist at Societe Nobel Francaise, Union Carbide, duPont, Eastman and Monsanto.

One of the first applications of these acetals was the use of polyvinyl formal (PVF) as a wire coating by G. E. However, the widest use of PVB (Butvar) is as an inner liner in shatter-resistant glass. Cellulose nitrate which was first used for this application in the 1920's, was replaced by cellulose acetate but PVB became the preferred inner liner in 1940.

The properties of PVB are dependent on the number of residual acetate and hydroxyl groups present in the final product. G. Morrison patented a preferred composition in the 1930's and these polymers were produced at Shawingan Falls, Ontario. These polymers were also used as textile coatings during World War II but were more expensive than other resin coatings. PVB was also used in a "wash primer" formulation, which could be applied to untreated steel before the application of other coatings. This primer (WP-1) is a two component system consisting of a solution of PVB and zinc tetraoxychromate (basic component 0 and an aqueous ethanolic phosphoric acid solution (acid component).

Over a 175 thousand tons of PVC dispersion coatings and almost 25 thousand tons of PVC latex coatings and adhesives were consumed in the US in 1988. The principal use of the dispersed coatings is for flooring (77 thousand tons).

References

Baumann, E., "PVC," Liebig's Ann 163 308 (1972).

Berry, E. E., "Wash Primer," Rev Curr Lit Paint, Colour Varnish 35 777 (1960).

Blaikie, K. G., Crozier, R. N., "Resistance of PVAC to Hydrolysis," Ind Eng Chem 28 1155 (1936).

Brown, C. F., "PVAC Textile Coating," US Pat, 2 381 720 (1948).

Fikentscher, H., Gaeth, R., "VCM, Vinyl Butyl Ether Copolymers," US Pat, 2 100 900 (1945).

Golovoy, A., "Electrostatic Sprayed PVC Polymer Coating," J. Paint Technol 45 (518) 45 (1973).

Gottesman, R. T., Goodman, "PVC," Chapt. 18 in "Applied Polymer Science," R. W. Tess, G. W. Poehlein, PBS, American Chemical Society, Washington, D.C., 1985.

Hermann, W. O., Haehnel, W., "PVAL," US Pat. 1 672 156 (1924).

Hofman, A. W., "Vinyl Polymerization," Liebig's Ann 115 271 (1878).

Hopff, H., "PVB," US Pat. 1 955 068 (1930).

Kaufman, M., "The Chemistry and Industrial Production of Polyvinyl Chloride," Gordon and Breach, New York, 1969.

Klatte, F., "Vinyl Acetate Monomers," US Pat. 1 084 581 (1912).

Klatte, F., Rollett, A., "Free Radical Polymerization of VCM, US Pat. 1 241 738 (1914).

Klatte, F., Rollett, A., "Polyvinyl Acetate," US Pat 1 241 738 (1944).

Kohlbe, H., "Listing of Vinyl Chloride," Ausfurliches Lehrbuchder Organishchen Chemie, Braunschweig 1 347 (1854).

Lawson, W. E., "Vinyl Chloride Copolymers," US Pat. 1 867 014 (1928).

Levinson, S. B., "Powder Coatings," J. Paint Technol 44 570 38 (1972).

Morrison, G. O., Skirrow F. W., Blaikie, K. G., "PVB," US Pat. 2 036 092 (1945).

Nordlander, B. W., Burnett, R. E., "PVF, Wire Coating," US Pat. 2 216 020 (1945).

Levinson, S. B., "Powder Coatings," J. Paint Technol 44 570 38 (1972).

Morrison, G. O., Skirrow F. W., Blaikie, K. G., "PVB," US Pat. 2 036 092 (1945).

Nordlander, B. W., Burnett, R. E., "PVF, Wire Coating," US Pat. 2 216 020 (1945).

Ostromislensky, I., "Polyvinyl Bromide," Chem Zentr 1 1980 (1912) US Pat. 1 712 034 1 791 009 (1911).

Park, R. A., "Vinyl Resins Used in Coatings," Chapt. 50 in "Applied Polymer Science," R. W. Tess, G. W. Poehlein eds. American Chemical Society, Washington, D.C., 1985.

Pfeffer, E. C., deBeers, F. M., "Vinylite Can Coating," US Pat 2 433 062 (1947).

Powell, G. M., Quarles, R. W., "Plastisols," Off Dig Fed Parnt and Varnish Production Clubs (263) 696 (1946).

Quarles, R. W., "Terpolymers of VCM, Vinylacetate and Malaic Anhydride," US Pat. 2 458 639 (1947).

Reed, M. C., "Plasticized VCM Copolymers," Ind Eng Chem 35 896 (1943).

Regnault, V., "Vinyl Chloride Monomer," Ann 15 63 (1835).

Reid, E. W. "Vinylite (Vinyl Chloride--Vinyl Acetate Copolymer)," US Pat 1 935 577 (1928).

Reppe, W., Starck, W., Voss A., "Vinyl Chloride Copolymers," US Pat. 2 118 864 (1928).

Schildknecht, C. E., "Vinyl and Related Polymers," John Wiley, New York, 1952.

Seymour, R. B., "Vinylchloride--Vinylidene Chloride Copolymers," US Pat. 2 348 154 (1944).

Starck, W., Freudenberger, H., "PVAC Emulsions," US Pat. 2 227 163 (1946).

Voss, A., Dickhauser, E., "Vinyl Chloride Copolymers," US Pat. 2 012 177 (1928).

Whittington, L. R., "A Guide to the Literature and Patents Concerning PVC Technology," Society of Plastics Engineers, Stanford, Connecticut, 1963.

HISTORY OF POLYVINYLIDENE CHLORIDE COATINGS

NORMAN G. GAYLORD
The Charles A. Dana Research Institute for Scientists Emeriti
at Drew University, Madison, NJ 07940

ABSTRACT

The rise of supermarkets, the packaging revolution and the development of polyvinylidene chloride, actually vinylidene chloride copolymer, coatings by The Dow Chemical Co. and E.I. duPont de Nemours & Co., were interdependent events in the post-World War II era. These coatings, still widely used today because of their functional characteristics, acted as barriers to moisture vapor and gases and provided flavor retention and grease and oil resistance, as well as gloss and heat sealability. Copolymers with high vinylidene chloride content, generically known as Saran, included copolymers with vinyl chloride, alkyl acrylates and methacrylates and acrylonitrile as well as terpolymers with an acid-containing monomer, were applied as solution or latex coatings onto a variety of substrates. Polymeric films including cellophane, medium and high density polyethylene, oriented poly(ethylene terephthalate) and polypropylene and polystyrene, as well as paper, have been coated with vinylidene chloride-based heat seal and barrier coatings.

HISTORICAL BACKGROUND

Vinylidene chloride (VDC) has been known for more than 150 years. However, it was not until Staudinger and Feisst published their results on the reactions and structure of the homopolymer in 1930, that industrial researchers began to investigate the chemistry and applications of polyvinylidene chloride (PVDC) and VDC copolymers. During the period 1933-1939, R.M. Wiley and colleagues at The Dow Chemical Co. studied the copolymerization of VDC and the plasticization and stabilization of PVDC and VDC copolymers. Their efforts resulted in the commercialization of Saran thermoplastic resins by Dow in 1939.

Since 1940 the development of resins for coatings, extrusion, molding, film, fiber, tape, etc. has been of interest to many companies in the United States and abroad and several hundred patents have issued. The early work of Dow and subsequent developments have been summarized in a number of reviews [1-9].

Saran is a generic term for polymers with a high VDC content in the United States. However, the terms Saran, polyvinylidene chloride and PVDC have been used interchangeably in the literature for VDC copolymers of both known and unknown composition.

CHARACTERISTICS OF PVDC AND VDC COPOLYMERS

Staudinger et al. reported [10,11] that PVDC homopolymer is a crystalline material with a linear head-to-tail structure (I). Reinhardt [1] proposed that the head-to-tail structure had a serpentine configuration (II).

The crystalline homopolymer has m.p. 198-205°C and T_g -17°C. However,

Published 1990 by Elsevier Science Publishing Co., Inc.
Organic Coatings: Their Origin and Development
R.B. Seymour and H.F. Mark, Editors

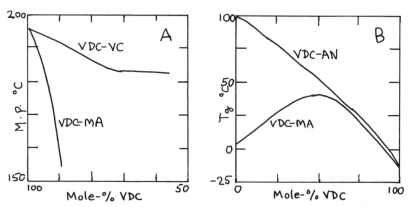

I   II

PVDC has found little use due to its high melting point, rapid crystallization, thermal instability and the tendency to evolve HCl during plastics
working and incompatibility with most plasticizers.

Copolymerization results in a disruption in the sequence of VDC units
and a decrease in the tendency for rapid crystallization. The reactivity
ratios in some VDC copolymerizations are shown in Table I.

TABLE I. Reactivity ratios in VDC (M_1) copolymerization

M_2	r_1	r_2
methyl acrylate (MA)	1.0	1.0
methyl methacrylate (MMA)	0.24	2.53
vinyl chloride (VC)	3.25	0.30
acrylonitrile (AN)	0.37	0.91

To promote the formation of copolymers with a uniform composition, the more
reactive monomer must be added to the polymerizing mixture throughout the
reaction period.

The melting point of the VDC copolymer decreases in the presence of
small amounts of comonomer (Fig. 1A). Although the T_g of the VDC-AN copolymer decreases as the AN content decreases, the T_g of VDC-acrylic ester copolymers increases with decreasing but low comonomer content until the composition is approximately equimolar but then decreases as the acrylic ester
content decreases (Fig. 1B).

FIG. 1. Melting point (A) and glass transition temperature (B) as a
function of VDC copolymer composition.

Copolymers having a high VDC content and retaining the desirable characteristics of the homopolymer are crystalline. The copolymers are amorphous
when the comonomer content is increased, e.g.

> 15 mole-% methyl acrylate
> 25 mole-% acrylonitrile
> 45 mole-% vinyl chloride

However, since the properties which are a function of crystallinity, e.g.
solvent resistance and barrier properties, are desired, commercial copolymers
are made with lower amounts of comonomer.

VDC COPOLYMER COATINGS

The use of supermarkets for shopping, the packaging revolution and the
development of VDC copolymer coatings by The Dow Chemical Co. and E.I. duPont
de Nemours & Co. were interdependent events in post-World War II era.

The objectives in coating substrates, particularly packaging films,
with VDC copolymer coatings, are to provide

> heat sealability
> gas and moisture vapor barrier
> resistance to solvents, fats and oils
> high gloss
> flavor retention

Substrates which have been coated with VDC copolymers for packaging
applications, include paper and board as well as polymeric films, as summarized in Table II.

TABLE II. Substrates coated with VDC copolymers for packaging applications

Regenerated cellulose	Polyamide
Paper or board	Polyamide-coated paper or board
Polyethylene	PE-coated paper or board
Polypropylene	Poly(vinyl chloride)
Polystyrene	Polyester

The excellent heat sealability of films coated with VDC copolymers
results from the low Tg of the coatings and the cohesive strength of the
copolymer.

The water vapor permeabilities of various polymers, including PVDC,
are compared in Table III.

TABLE III. Water vapor permeability of various polymers

Polymer	Density, g/ml		Permeability[a]	
	Amorphous	Crystalline	Amorphous	Crystalline
Polyethylene	0.85	0.96	200-220	10-40
Polypropylene	0.85	0.94	420	
Polyisobutylene	0.91	0.94	90	
Poly(vinyl chloride)	1.41	1.52	300	90-115
Polyvinylidene chloride[b]	1.77	1.96	30	4-6

[a] $g/hr \cdot 100m^2$ at 7.1 kPa (53mm Hg) pressure differential and 39.5°C for
1 mil film
[b] By extrapolation from VDC copolymers

The low water vapor permeabilities for polyisobutylene and PVDC are due to their symmetrical structures and the resultant low free volumes. The gas permeabilities of both polymers are also extremely low. While PVDC is highly crystalline, polyisobutylene is totally amorphous. Butyl rubber, a 98/2 isobutylene/isoprene copolymer, is used where barrier properties and flexibility are needed and grease resistance is not mandatory. VDC copolymers are used where both barrier properties and grease resistance are necessary, e.g. packaging applications.

The gas permeabilities of various VDC copolymers are shown in Table IV. The gas permeability increases as the comonomer content increases.

TABLE IV. Gas permeability of VDC copolymers at $25^{\circ}C^a$

		Permeability x 10^{13}		
Polymer		Oxygen	Carbon Dioxide	Nitrogen
PVDC		1.5	9.0	0.75
VDC/VC	90/10	3.2	22.0	0.75
	85/15	9.0	45.0	
	70/30	8.1		
	50/50	27.0		
VDC/AN	80/20	3.2	8.0	0.4
	60/40	16.0	35.0	1.9

^{a} $(cm^3 \cdot cm)/(cm^2 \cdot sec \cdot kPa)$ kPa x 0.75 = cm Hg

VDC copolymers for coatings are prepared by solution, emulsion or suspension polymerization using free radical catalysts. Solution polymers are generally of lower molecular weight and the polymer is isolated and redissolved in a suitable solvent for lacquer coating applications. Emulsion polymers are of higher molecular weight and the latex or emulsion is either used per se for latex coating or is coagulated to precipitate the polymer which is used in lacquer coating. Suspension polymers are used for extrusion coatings.

The characteristics of VDC copolymer latex and lacquer coating compositions are shown in Table V.

TABLE V. Comparison of VDC copolymer latex and lacquer coating compositions

	Latex	Lacquer
Solids, %	40-60	10-20
Medium	water	THF; MEK-toluene
Viscosity	low	high
Copolymer molecular weight	high	moderate
Application temperature, $^{\circ}$C	20-30	60-80
Shear stability	poor	good
Freeze-thaw stability	poor	good
Wetting agents	present	none
Drying	relatively easy	solvent traces difficult to remove

The advantages of latex coating, i.e. compositions with high solids and low viscosity, low application temperature and ease of drying, are combined with the reduced fire hazards and absence of solvent recovery due to the use of water as medium. However, the aqueous compositions have poor mechanical and freeze-thaw stability. The surfactants, present due to their use in latex preparation and as post-polymerization stabilizers, may exude to the surface of the dried coating, causing slip and blocking problems.

The shelf life of a VDC copolymer latex is related to the copolymer composition. The higher the VDC content, the more rapidly the latex particle crystallizes. Although the latex has the same viscosity and appearance, the crystallized particles no longer have the ability to coalesce to form a pinhole-free coating. Since the barrier properties are related to the co-polymer composition, the latex with the better barrier properties has the shorter shelf life. Consequently, a commercial latex which must have a long shelf life, yields a coating with poorer barrier properties than a latex which can be coated when and/or where prepared. The effect of copolymer com-position on the crystallization time of latex particles is shown in Table VI.

TABLE VI. VDC copolymer latex stability

Polymer		Time for crystallization of latex particles
PVDC		1 hr
VDC/MA	98/2	1 day
	97/3	1 week
	96/4	2 weeks
	95/5	3 weeks
	94/6	5 weeks
	90/10	6 months

The VDC copolymer coating undergoes crystallization on aging. Thus, there is a limited shelf life for a high barrier coated substrate, since the heat sealability is decreased as the coating crystallizes. The tendency for the coating to crystallize also results in exudation of the surfactant. Although this may present a surface problem, it also provides a wettable surface. When cellophane was coated with a VDC copolymer lacquer and the coated film was used to package bakery products which required refrigeration, water which volatilized from the package contents formed droplets on the inner surface of the film, i.e. fogging occurred. When the cellophane was coated with a VDC copolymer latex, the wettable surface caused the condensed water to spread over the inner surface of the package and improved the appearance thereof.

When paper or board is coated with a VDC copolymer latex, the coating contains pinholes due to the wicking action of the substrate. To overcome this problem, multiple coats must be applied to the substrate. Alternatively, a thin layer of polyethylene is applied by extrusion lamination, followed by coating with a VDC copolymer latex.

Due to the great interest in polyethylene extrusion and the commercial installation of extrusion lamination equipment, polyethylene coating was adopted in the U.S. as the preferred coating for paper and board. In con-trast, multiple layers of VDC copolymer were used for paper coating in Europe. A comparison of the properties of polyethylene-coated and VDC copolymer-coated paper are shown in Table VII.

The coating of polyester film, e.g. poly(ethylene terephthalate), with a VDC copolymer provides a heat sealable film for bag making and for vacuum or inert gas packaging. The film may be coated on one side or both sides, e.g. 0.1 mil coating per side on a 0.5 mil base film, for improved barrier properties and/or to provide adhesion for an extruded or laminated coating on the coated side of the base film.

TABLE VII. Paper coating with polyethylene and VDC copolymer

	Polyethylene Coating	VDC Copolymer Coating
Coatings	single	double
Coating thickness, mils	0.5-2.0	0.2-0.5
Heat seal strength	strong	strong
Flexibility	excellent	poor crease resistance
Moisture vapor transmission resistance	good	excellent
Gas transmission resistance	poor	excellent
Grease resistance	poor	excellent
Gloss	low	high
Yellowing	none	slight

The adhesion of VDC copolymer coatings to various substrates is enhanced by corona discharge treatment of the base film or by coating the
film with a suitable prime coat. Notwithstanding surface treatment of the
base film, adhesion of VDC copolymers is promoted by copolymerization of a
carboxyl-containing monomer into the copolymer. Monomers such as acrylic
acid, methacrylic acid and itaconic acid have been incorporated into both
lacquer and latex coating compositions.

The coating of an oriented film such as biaxially oriented polyester
film, in some cases presents a special problem. While adhesion of a VDC
copolymer containing carboxyl groups, to the as-cast unoriented film is
adequate to provide acceptable heat seal strengths, the adhesion is decreased when the base film is oriented. In order to increase the long term
adhesion, e.g. in X-ray film, the unoriented film is coated with the VDC
copolymer composition and then the coated film is subjected to stretching
and orientation.

In addition to the use of VDC copolymer coatings on films for packaging applications, the copolymers are also used in coatings, binders, prime
coats and adhesives in non-packaging applications, as shown in Table VIII.

TABLE VIII. Non-packaging applications for VDC copolymer coatings

Binder for paints
Binder for non-woven webs
Binder for particles on magnetic tape
Laminating adhesive
Primer for bonding silver halide emulsion to
    photographic film
Flame retardant carpet backing
Barrier coating for steel piles
Lining for tank interior: ship, railroad car,
    fuel storage

SUMMARY

Vinylidene chloride copolymer coatings, applied from solvent or latex
compositions or by extrusion lamination, impart heat sealability, grease
resistance and barrier properties to polymeric substrates, making them
suitable for packaging applications. The copolymers generally contain
small quantities of acrylonitrile, methyl acrylate or vinyl chloride. The
vinylidene chloride copolymers are also useful as binders, adhesives and
prime coats in non-packaging applications.

REFERENCES

1. R.C. Reinhardt, Ind. Eng. Chem. 35, 422 (1943).
2. J.J.P. Staudinger, Brit. Plastics 19, 381, 453 (1947).
3. C.E. Schildknecht, Vinyl and Related Polymers (Wiley, New York 1952), Chap. VIII, pp. 446-469.
4. J.F. Gabbett and W.M. Smith in: Copolymerization, G.E. Ham, ed., (Interscience, New York 1964), Chap. X, pp. 587-637.
5. G. Talamini and E. Peggion in: Vinyl Polymerization, Vol. 1, Part 1, G.E. Ham, ed. (Dekker, New York 1967), Chap. 5, pp. 331-367.
6. W.R.R. Park, Plastics Film Technology (Van Nostrand and Reinhold, New York 1969), Chap. 3, pp. 49-74.
7. R.A. Wessling and F.G. Edwards in: Encyclopedia of Polymer Science and Technology, H.F. Mark, N.G. Gaylord and N. Bikales, eds., (Interscience, New York 1971), Vol. 14, pp. 540-579.
8. R.A. Wessling, Polyvinylidene Chloride (Gordon and Breach, New York 1977).
9. D.S. Gibbs and R.A. Wessling in: Encyclopedia of Chemical Technology, 3rd Edition, Kirk-Othmer, eds., (Wiley, New York 1983), Vol. 23, pp. 764-798.
10. H. Staudinger, M. Brunner, and W. Feisst, Helv. Chim. Acta 13, 805 (1930).
11. H. Staudinger and W. Feisst, Helv. Chim. Acta 13, 832 (1930).

JOSEPH J. MATTIELLO

CHARLES H. FISHER
Chemistry Department
Roanoke College, Salem, Virginia 24153

Unfortunately for society and science and technology, Joseph J. Mattiello lived only a short period, namely, 48 years. But during that brief life of 48 years, this internationally-known leader in the paint and varnish industry served his country unusually well and achieved much more than most individuals who live much longer.

Joe Mattiello served in his country's military forces on two continents and in both World War I and World War II. At the age of 16, he served on the Mexican border as a private in the Sixty-ninth Regiment of the New York National Guard. Not many months after returning from the border, he became a sergeant in Company H of the 165th U.S. Infantry. While in France from October, 1917, to March, 1919, he saw action in several major battles. Joe was wounded in the right arm and in both legs on October 17, 1918, and lost his right leg. He was awarded the Regimental Citation and Order of the Purple Heart. In World War II he served three years as a civilian chemical expert assigned to the Quartermaster Corps, for which he received the meritorious civilian service award. During the latter three-year period, he edited a book for the Quartermaster General entitled "Protective and Decorative Coatings" (Government Printing Office, 1945).

Following his service in World War II, Joe earned his B.S. degree at Brooklyn Polytechnic Institute in 1925 and, while holding down a job, continued his work at Polytechnic and in 1931, received his M.S. degree. He attended the Department of Engineering at Columbia University and was awarded the Ph.D. degree in 1936.

Joe co-authored several papers describing his graduate research performed prior to receiving the Ph.D. degree in 1936:

"Potentiometric Determination of Acid Number of Linseed Oil and its Free Fatty Acids," Industrial and Engineering Chemistry, Analytical Edition, Vol. 4 (1932), page 52-.

"Linseed Oil: Changes in Physical and Chemical Properties During Heat Bodying," Industrial and Engineering Chemistry, Vol. 24 (1932), pages 158-162.

"Fish Oil: Changes in Physical and Chemical Properties During Heat Bodying," Industrial and Engineering Chemistry, Vol. 28 (1936), pages 1022-1024.

"Flow Dispersion and Livering Characteristics of Pigment Pastes Made With Bodied Linseed Oil," National Paint, Varnish, and Lacquer Association, Circular No. 502 (1936), pages 31-138.

He began his industrial career with the Hilo Varnish Corp., Brooklyn, in 1925 as a chemist. Active in the affairs of the New York Paint and Varnish Production Club for many years, he served as Secretary-Treasurer in 1937, Vice-President in 1938, President in 1939, and as Chairman of the Technical Committee in 1936, when the Club's paper, "Properties of Linseed Oil, Heat Bodied in Air and Vacuum and Its Behavior With Pigments, With and Without Wetting Agents," was presented at the Federation convention. He

Published 1990 by Elsevier Science Publishing Co., Inc.
Organic Coatings: Their Origin and Development
R.B. Seymour and H.F. Mark, Editors

133

was author and co-author of numerous articles dealing with the paint and varnish industry. Dr. Mattiello was the editor of a five-volume work, "Protective and Decorative Coatings," and at his death was preparing a sixth volume. He was President of the Federation of Paint and Varnish Production Clubs in 1943-44 and also consultant on protective coatings for the Quartermaster.

Additional publications by Mattiello, as author or co-author, include:

"Oiticica and Tung Oils: Changes in Physical and Chemical Properties on Heat Bodying," Industrial and Engineering Chemistry, Vol. 30 (1938), pages 211-215.

"Continuous Laboratory Method for Bodying Oiticica Oil," Industrial and Engineering Chemistry, Analytical Edition, Vol. 12 (1940), pages 77-80.

"Protective Coatings Industry. Paints, Varnishes, Lacquers, Enamels, and Plastics," Chemical and Engineering News, Vol. 23 (1944), pages 136-143.

"Protective Organic Coatings as Engineering Materials" (Edgar Marburg Lecture), Proceedings American Society for Testing Materials, Vol. 46 (1946), pages 493-532.

"Preparation and Study of Synthetic Drying Oils," Journal of the Society of Industrial Research (India), Vol. 8B (1949), pages 29-35.

Joe Mattiello was a member of many organizations, including the Association of Research Directors; Gallows Bird Society; National Paint, Varnish, and Lacquer Association; American Oil Chemists Society; Sales Executive Club; American Chemical Society; American Society for Testing Materials; American Society of Military Engineers; American Institute of Chemical Engineers; Chemists' Club; Fellow of the American Institute of Chemists; the Father Duffy Chapter of Rainbow "42nd" Division Veterans; and the Purple Heart Association. He was also a member of the Phi Lambda Upsilon, national honorary chemical society; Sigma Xi, national honorary scientific society; Alpha Phi Delta and Delta Kappa Pi fraternities; as well as honorary member of the Oil and Colour Chemists' of Great Britain and a Fellow of the New York Academy of Sciences.

In 1946, Joe was honored by being selected to give the Marburg Lecture, feature scientific presentation at the Annual Meeting of the American Society for Testing Materials. His topic was "Protective Organic Coatings as Engineering Materials."

The French Government made him a member of the Legion of Honor in 1947, while he was attending the First International Congress of Paint Technologists in Paris, as a representative of the Federation of Paint and Varnish Production Clubs.

The Federation of Societies for Coatings Technology established the Joseph J. Mattiello Lecture to commemorate one of the outstanding men in the Federation and in the coatings industry.

Joe was born in New York City of Italian parentage on February 28, 1900. His father, Celestino, was a sculptor and died when Joe was four years old. His mother, Elizabeth Bottigleri, was one of a distinguished family and died when he was eleven, an event that placed Joe in the care of an aunt who gave him a good primary education. Joe married Josephine Critelli on September 18, 1922; their children were Margaret Anne (Mrs. Harry Kim, Jr.), Elizabeth (Mrs. Joseph Yozzo), Rosamond, and Barbara Mattiello. Dr. Joseph J. Mattiello died of a heart attack at his home

in Brooklyn, New York, on May 16, 1948. He was 48 and Vice-President and
Technical Director of the Hilo Varnish Corp. at the time of his death.

Joseph J. Mattiello (măt´-ē-ĕl-lo), the son of Celestine and Elizabeth (Bottigliere) Mattiello, was born on Feb. 28, 1900, in New York City. His primary and secondary education was received in New York City public and private schools. His high school education was interrupted in 1916 when he enlisted in the New York National Guard. After serving on the Mexican border, he went to France in October, 1917, with the American Expeditionary Forces in the 165th U.S. Infantry which saw heavy action in a number of historic battles. He was a member of the 69th Regiment in the famous Meuse-Argonne offensive in which he was wounded in right arm and both legs and suffered the loss of his right leg. For his W.W. I service, sergeant Mattiello received the Regimental Citation and the Purple Heart.

Returning home, he finished high school and in 1921 matriculated at the Polytechnic Inst. of Brooklyn, receiving his B.S. in chemistry in 1925. In 1922, he married Josephine Critelli. He joined Hilo Varnish Co. in Brooklyn in 1925. He continued his education at Polytechnic Inst., receiving his M.S. in 1931, then at Columbia Univ. to obtain his Ph.D. in 1936 in organic chemistry.

Beginning in 1932, he authored 11 and coauthored seven publications on analyses and properties of paints, varnishes, printing inks, bodied oils (linseed, fish, oiticica, tung, and synthetic oils); raw materials used in coatings; and engineering aspects of coatings. He contributed items to the Handbook of Engineering Fundamentals.

His most prolific publication period occurred in the 1940s. In addition to 12 journal publications in the 1940s, his editorship of the five-volume work "Protective and Decorative Coatings, Paints, Varnishes, Lacquers, and Inks", Wiley & Sons, N.Y., 1941-5, was a classic reference book for those involved in or with the coatings industry and won him worldwide renown. Volume 6, which he had been working on, never appeared because of his untimely death on May 16, 1948.

Despite the aggravating pain from his prosthetic right leg and his difficulty in walking even with the aid of a heavy cane, he was highly sociable and gregarious as evidenced by his involvement in and activities with numerous technical and professional organizations. He was notably active in the Federation of Paint and Varnish Production Clubs, serving as president; the New York Paint and Varnish Production, serving as president; the National Paint, Varnish and Lacquer Assocn.; American Oil Chemists; Oil and Color Chemists; Oil and Colour Chemists Assocn. of Great Britain (honorary member); and the New York Printing Ink Production. He served on Committee D-1 and its subcommittees and on Committee D-17 of the ASTM, as chairman-elect of the newly formed Metropolitan Long Island sub-section of the New York ACS Section, and vice-president of the Am. Inst. of Chemists. He was a member of the Assocn. of Research Directors, Sales Executive Club, Chemists Club, Am. Inst. of Chem. Engrs., Sigma Xi, Phi Lambda Upsilon, Delta Kappa Pi, and others.

During W.W. 2, beginning in July, 1942, he was a consultant on protective coatings with the Military Planning Division, Office of the Quartmaster General. His major contribution was the development of paints, varnishes, and lacquers for use on Quartermaster Corps equipment and for the development of organic coatings and methods of application in the pre-coating program and for coatings of metallic containers,

Published 1990 by Elsevier Science Publishing Co., Inc.
Organic Coatings: Their Origin and Development
R.B. Seymour and H.F. Mark, Editors

such as rations cans. For these contributions, he was presented
with the Meritorious Civilian Service Award by the War Dept.
on Feb. 21, 1946.

He was the ASTM Edgar Marburg Lecturer in 1946. His talk,
"Protective Organic Coatings as Engineering Materials" was
published in the ASTM Proceedings (99 pages).

In 1947 he was nominated a fellow of the New York Academy
of Science. In October, 1947, he attended the First Technical
International Congress of the Paint and Associated Industry in
France. As the official representative of the Federation of
Paint and Varnish Clubs and the Natl. Paint, Varnish, and Lac-
quer Assocn., he presented a talk on "Functions of Organic
Coatings in Present-Day Engineering Problems." Following the
meeting, he was decorated by the French Government with the
Legion of Honor in recognition of his was activities and
civilian services.

He succumed to a heart attack at 4:30 a.m., Sunday May
16, 1948, at his home in Brooklyn, survived by his wife and
four daughters. At the time of his death, he was director,
V.P., Technical Director, and member of the Executive Committee of Hilo Varnish Corp. He is memorialized by the Mattiello
Lecture presented at the Federation of Societies for Coatings
Technology annual meeting.

Publications by Joseph J. Mattiello:
1. Ind. Eng. Chem, Anal. Ed. **4**, 52-6(1932)-"Potentiometric
   Detn. of Acid No. of Linseed Oil and its Free Fatty Acids"
   (with B. P. Caldwell).
2. Ind. Eng. Chem. **24**, 158-63(1932)-"Linseed Oil Changes in
   Phys. and Chem. Properties During Heat-Bodying"
   (with B. P. Caldwell).
3. Natl. Paint, Varnish & Lacquer Assocn. Circ. #502, 31-138
   (1936)-"A Study of the Flow, Dispersion and Livering
   Characteristics of Pigment Pastes made with Bodied Linseed
   Oils" (with L. T. Work).
4. Ind. Eng. Chem. **28**, 1022-4(1936)-"Fish Oil: Changes in Phys.
   and Chem. Properties During Heat-Bodying" (with L. T. Work,
   C. Swan, and A. Wasmuth).
5. Ind. Eng. Chem. **30**, 211-15(1938)-"Oiticica and Tung Oils:
   Changes in Phys. and Chem. Properties on Heat-Bodying"
   (with S. O. Srenson, C. J. Schuman, and J. H. Schuman).
6. Am. Ink Maker **17**, No. 11, 23, 47(1939)-"Litho Varnishes."
7. Ind. Eng. Chem., Anal. Ed. **12**, 77-80(1940)-"Continuous Lab.
   Method for Bodying Oiticica Oil" (with V. Marchese and Work).
8. Am. Paint J. **24** (1940), a series of six papers in the Mar.
   and Apr. issues based on a lecture "Raw Materials Used in
   Manufacturing Vehicles."
9. Am. Ink Maker **18**, No. 9 (1940)-"Lithographic Varnishes."
10. "Protective and Decorative Coatings: Paints, Varnishes, Lac-
    quers, and Inks," Wiley, N.Y., 5-volumes, (1941-5).
11. Am. Paint J. **31**, No. 10 (1946)-"Protective Org. Coatings as
    Engineering Matls." (based on an address).
12. ASTM Proc. **46**, 493-592(1946)-"Protective Org. Coatings as
    Engineering Matls." (Edgar Marburg Lecture).
13. Paint Ind. Mag. **62**, 298-306(1947)-"Functions of Org. Coat-
    ings in Engineering Problems" (based on an address).
14. J. Sci. Ind. Res.(India) 8B, 29-35(1949)-Prepn. and Study
    of Synthetic Drying Oils" (with H. V. Parekh).

Prepared by: Herman Skolnik
239 Waverly Rd.
Wilmington, DE 19803

ARTHUR KING DOOLITTLE

CHARLES H. FISHER
Chemistry Department
Roanoke College, Salem, Virginia 24153

Nearly everyone has heard of the CBS television game called
"Jeopardy." In this exciting game, the interrogator or host gives the
answer to a question and then challenges the contestants to give the
question corresponding to the answer. Let's play the game; first the
answer:

HOST: A graduate of Columbia University in 1923 with a degree in
chemical engineering, he served in the U.S. Army, 1917-1919, and ended this
service with the rank of Colonel. He was an outstanding chemical engineer
and researcher for several corporations before he became a research leader
in 1932 at the Carbide & Carbon Chemical Company, South Charleston, West
Virginia. He occupied several important positions at Carbide during his
twenty-nine years of outstanding research at this organization. Leaving
Carbide in 1961, he became President of the Arcadia Institute for
Scientific Research and pursued successful careers as a professor and
consultant.

Versatile and creative in research, he distinguished himself by his
successful achievements in both engineering and fundamental chemical
investigations. His principal research activities were concerned with
liquids and their physical properties, coatings, polymers, solvents,
plasticizers, mechanistic theory of solutions, free-space viscosity
equation, and internal force equation. He directed research that created
the vinyl coatings technology that helped launch plasticizers as a major
petrochemical industry.

He was selected as the authority to write the chapter entitled
"Industrial Solvents" in the important Kirk-Othmer Encyclopedia of Chemical
Technology. He was author in 1954 of a widely-used book having the title
"The Technology of Solvents and Plasticizers." Because of these and many
other technical publications, he was widely acclaimed as a distinguished
pioneer in the science of liquids, solvents, plasticizers, coatings, and
polymers.

Now the answer has been given. What was the question?

Many chemists and chemical engineers, particularly those of his era,
would respond with the proper question: "Who was Arthur King Doolittle?"

There is much more to be said about Arthur King Doolittle. His
biographies can be found in many publications, including The Chemical Who's
Who, Who's Who in America, American Men And Women of Science, and World
Who's Who in Science From Antiquity to the Present. The last-mentioned
publication lists Doolittle among the scientists considered to be the most
eminent in the entire world in all history.

Doolittle was a registered professional engineer in West Virginia,
Delaware, New York, and New Jersey. He was a partner with Dorr
Consultants, New York, 1959-1961. His academic assignments included
Professor of Chemistry, Drexel Institute of Technology, Philadelphia,
1961-64; and Member, Advisory Board, Chemical Engineering Department,
Princeton University, 1955-58.

Published 1990 by Elsevier Science Publishing Co., Inc.
Organic Coatings: Their Origin and Development
R.B. Seymour and H.F. Mark, Editors

Arthur Doolittle found time to serve in technical organization as
officer, committeeman, or member: American Institute of Chemical Engineers
(Chairman of the Charleston Section, 1943-44; National Director, 1951-54;
and Vice-President, 1955); American Association for the Advancement of
Science (Chairman of the Management Committee, Gordan Conferences, 1955-56;
Advisory Board member, 1950-58); American Chemical Society (Chairman,
Paint, Plastics, and Printing Ink Division, 1952-53; Division Counselor,
1952-56; Advisory Board, Industrial and Engineering Chemistry, 1954-56; and
Council Committee on National Meetings and Divisional Activities); American
Society for Testing Materials; and the American Association for the
Advancement of Science.

Doolittle was a member also of the Order of Daedalians, Sigma Xi, Phi
Beta Kappa, Tau Beta Pi, Phi Lambda Upsilon, Chi Beta Phi, Alpha Chi Rho,
Columbia University Club (N.Y.C.), Quiet Birdmen Club, Cosmos Club
(Washington), and Army-Navy Club (Washington).

The general nature of Doolittle's research on solvents is indicated by
the title of one of his many publications: "Mechanism of Solvent Action:
Influence of Molecular Size and Shape on Temperature Dependence of Solvent
Ability," Industrial and Engineering Chemistry, vol. 38 (1946), pages
535-540.

Doolittle's paper (with R. H. Peterson, Journal of the American
Chemical Society, vol. 73 (1951), pages 2145-2151) entitled "Preparation
and Physical Properties of a Series of n-Alkanes" is the best source of the
properties of the higher n-alkanes at temperatures up to 300°C.

Some of the additional publications by Doolittle had the titles:
"Critical Appraisal of the Volume-Entropy-Energy Equation for Liquids"
(Journal of Applied Polymer Science, vol. 25 (1980), pages 2305-2315).

"A VSE Equation of State for Liquids: n-Alkanes, Benzene, Mercury, and
Water" (Journal of the Franklin Institute, vol. 301 (1976), pages
241-251).

"Polymer Solution Thermodynamics. State-of-the-Art Survey" (Journal of
Paint Technology, vol. 41 (1969), pages 483-488).

"Development of an Equation for Internal Force in a Liquid" (Preprints
Division of Petroleum Chemistry, American Chemical Society, vol. 10 (1965),
pages 45-52).

"Volume, Entropy, and Energy of Liquids" (Preprints Division Organic
Coatings & Plastics Chemistry, American Chemical Society, vol. 26 (1965),
pages 248-265).

"Molecular Volume and Density of High Molecular Weight Hydrocarbons"
(Preprints Division of Petroleum Chemistry, American Chemical Society, vol.
12 (1967), pages 21-34).

Doolittle was decorated with the Croce di Cavaliere al Merito Della
Republica Italiana in 1967. The Union Carbide Corporation established the
Arthur K. Doolittle Award to honor its one-time employee; this Award is
given to authors of outstanding papers presented at meetings of the
Division of Polymeric Materials: Science and Engineering of the American
Chemical Society.

Born in Oberlin, Ohio, on November 15, 1896, Arthur King Doolittle was
the son of Frederick Giraud and Maud (Tucker) Doolittle. He married Dortha
Bailey on August 8, 1923; their children were Robert Frederick II and

Elizabeth May (Mrs. Donald Charles Peckham). Doolittle received three
degrees from Columbia University; A.B., 1919; B.S., 1920; and Chemical
Engineer, 1923.

Before embarking upon a career in research and development, Doolittle
served in the U.S. Army, 1917-1919. He was a test pilot for Sopwith Camel
airplanes and flew damaged planes from France to England for repairs during
World War I. This experience might have been responsible for the fact
flying was one of his hobbies.

Doolittle worked for several industrial corporations before becoming a
member of Carbide in 1932: Research engineer, Dorr Co., Westport, CT,
1923-1925; plant engineer, Sherwin-Williams Co., Chicago, 1925-1929, and
chief of the lacquer division, Newark, 1929-1931; development engineer,
spray drying, Bowen Research Corporation, New York, 1931; and director,
lacquer research, Bradley Vroom Co., Chicago, 1931-1932.

Arthur K. Doolittle, who with Mrs. Doolittle had lived in their home
in Wallingford, Pennsylvania, died in his sleep at the age of 85 years on
January 21, 1982, in the Broomall Presbyterian Home in Broomall,
Pennsylvania.

James Scott Long--From The Chemistry of Coatings to The History of Science

Brenda H. Mattson and Gerald A. Mattson
University of Southern Mississippi
Hattiesburg, MS 39406

When Ray Seymour asked me to write a paper on the life of James Scott "Shorty" Long, I expected simply to report a list of his scientific papers, of his jobs, and of his honors along with a requisite amount of personal data, "Shorty" and sweet. It did not take me long to see that it was going to be difficult to avoid discussing Long's personal philosophy since it significantly impacted his scientific work. Long was a prolific writer. His credits include more than 100 technical publications and 50 patents, but his writings were not limited to the scientific field. Among other things he wrote about were religion, newspapers, women, kids, purposeful living, diet, pollution, smoking, and the history of science.

James Scott Long was born on August 11, 1892 in York, Pennsylvania. Introduced early in life to science, his father was a dentist and an uncle was an M.D., Long frequently mentioned an incident in his youth which had a profound effect on him. In October, 1905, when Long was 13, his father took him to the York County Agriculture Fair. It must be noted that his father, a Scotsman, did not spend money frivolously, and his spending $4.50 for three tickets to a special attraction taking place at the racetrack was definitely unusual. That day (only 22 months after their first flight) the Wright Brothers were flying their one cylinder airplane around the track. Long wrote of that day and his father's reaction to it, "...many thousands of people applauded, screamed, whistled.... I looked at my father, tears were in his eyes and running down his cheeks. He curled his mustache.... This was a sign that he was thinking...and then he uttered the sentence that DETERMINED MY CAREER IN LIFE.... " He said, "someday THERE WILL BE BETTER AIRPLANES THAN THIS...." He knew instinctively that this, or anything else, could be improved. "This sentence made a research man of me."[1]

His father wanted Long to go to Annapolis. And although he received an appointment to the Naval Academy, it was for the following year. So in the interim, he went to Lehigh University in Bethlehem, Pennsylvania. Until Long enrolled, he had never heard of Lehigh. "Its football teams did not rate much.... I was a little boy 115 pounds, [and 5'3"tall, hence the nickname 'Shorty'], no muscle ability, no conversational ability, could not dance or play cards, no auto, of course, and no money. So I studied constantly."[2] To the Naval Academy's loss, Long remained at Lehigh (as did an impressive list of corporate CEO's of that period).

While Long was an undergraduate student at Lehigh, he worked in the summer as an apprentice painter for his grandfather, a general contractor. His tasks were what he termed "doggy work.... mixing--the white lead paste with additional linseed oil, driers, thinners, [and] sometimes colored pigments to tint them."[3] He worked long hours, including Saturdays. Late one Saturday afternoon, as Long was returning from a paintbrush cleaning shed, a neighbor who was a physician, asked Long to take his wife to a dance. The physician had to deliver a baby. At the dance the attractive young wife of the physician introduced him to a friend, who in turn introduced Long to her husband. After finding out that Long was a chemical engineering student a Lehigh, the friend's husband asked Long to do some work for him, for he had "bought a half interest in a patent leather plant... and the patent leather cracked sometimes the first or second time" a pair of shoes were worn.[4]

The businessman wanted Long "to get a microscope and make THIN sections of the leather with a razor blade... to find out whether the 3 COATS were too thick or too thin

Published 1990 by Elsevier Science Publishing Co., Inc.
Organic Coatings: Their Origin and Development
R.B. Seymour and H.F. Mark, Editors

at places and why the finish on the leather cracked."[5]   When Long visited the manufacturing plant, he found out that the superintendent was not helpful.   The superintendent who bought the materials for making the patent leather, and got a rake-off according to Long, would tell no one what the materials were or how they were used. Cutting samples of the finished product to study them did not accomplish much toward solving the problem, so Long was hired to go to the factory one day each month to try to continue the investigation.   With the businessman's permission and without the superintendent's knowledge, Long went to the plant one night and surreptitiously took samples of the materials. It was then he learned that the main raw materials used were "a special linseed oil...driers such as litharge, umber, and prussian blue...and naphtha."[6] Observing the process, Long saw that the superintendent even carefully controlled the testing of the product, as if he were trying to hide something.   On another of his evening plant trips Long took samples of films from each of the three coats which were applied to the leather, as well as samples of the leather itself, and tested them at Lehigh on a Gardner tension machine to determine the breakage elongation. Although leather from different parts of the hide varied in breakage elongation, elongation was generally 35-40%, whereas none of the linseed oil film coats would stretch more than 27%. Obviously, if in the processing of or in the wearing of the patent leather it was stretched more than 27%, it would crack. Long began to study linseed oil and "how the elongation would vary with 1. the extent of cooking, 2. the driers used, 3. temperatures used in cooking, 4. refining or treating of the linseed oil, 5. iodine number of the oil.... Long also tried "THICKENING the linseed oil by blowing AIR through it at temperatures below $50°$ until it had viscosities..." similar to the higher temperature processed oil.[7] The lower temperature effected polymerization with a lower percentage of oxygen, and although it was superior, it still was not suitable. He later utilized polyurethanes to make a patent leather which did not crack and the patent leather Long developed even outlasted the soles of the shoe.

A number of important lessons arise from this story. First, Long began working at an early age.   We in the teaching profession need to more actively encourage student's working at in-field jobs, as an apprentice, as a gopher, whatever!   Second, do not be reluctant to go to a dance with a physician's wife. Seriously, be prepared to take advantage of a situation. Third, be enterprising and within the rule of law and good ethics, do what is necessary to solve a problem. Fourth, work hard particularly in the evening. As you will learn Long worked hard throughout his life and long after most people are retired.

Long published his first paper at age 19 while an undergraduate at Lehigh. After graduation with Ch. E. degree in 1914 and an MS in 1915, Lehigh hired Long as an instructor, then as an assistant, an associate and a full professor, within a period of 19 years.   In 1922 Long took a one year leave of absence from Lehigh (at 1/2 pay) to meet a residency requirement and earn the Ph.D at Johns Hopkins, where "Long worked for the noted Hopkins' surface scientist Walter A. Patrick, and published the major findings of his thesis as 'Absorption of Butane on Silica Gel'...."[8]

His doctoral research supported his interest in the coatings field. While at Lehigh, Long organized a Paint Research Institute, a cooperative research program between the University and the largest producers of linseed oil and other drying oils such as Archer Daniels Midland. These companies as well as Raybestos-Manhattan, New Jersey Zinc, William O. Goodrich and Armstrong Cork contributed to a research fund of which Long served as Research Director. The cooperative research carried out under his direction led to a number of patents, to sixteen of what we now call refereed papers and to approximately twenty trade publication papers.

In 1934 Long accepted a position as Chemical Director of Devoe and Raynolds (later Celanese), a paint manufacturer since 1754 and according to Long the oldest corporation in America. Long had come to the realization that he had been more of a formulator than a researcher, that his papers on linseed oil were wrongly conceived. Linseed oil is an ester and if you put it on lettuce instead of olive oil and ingest it, it will digest readily in an hour. For applications subject to moisture and exposure Long realized that something more hydrolytically stable would have to be found. Researchers must design and synthesize coatings, balancing all the factors that make a coating desirable. Long had always tried to replace art with science in the coatings field. Recalling his father's statement that everything can be improved, Long now realized that researchers must develop new systems and not just gather data on existing systems. Devoe and Raynolds gave Long a five years carte blanc contract to produce a new type resin, an ether resin, an epoxy.

During Long's tenure at Devoe the laboratory staff increased from 25 to 150. Long hired a brilliant young organic chemist S.O. Greenlee, a student of Roger Adams at Illinois. Shortly thereafter a Shell Chemical man gave Long a sample of epichlorohydrin. In only two weeks Greenlee had made the first sample of an epoxy resin. The commercial epoxy resins developed thereafter were not only financially successful but prevented untold losses from corrosion and other types of environmental degradation.

Credit for the development of the epoxy resin goes to three men. Long for the concept and inspiration, Aaron Rosen, another of Long's young staff at Devoe, for the research outline and Greenlee for the synthesis.

Over the years Long could not help but notice the amount of "monkee," i.e. gelled linseed oil, that had to be discarded in the manufacture of paints. At Devoe he had the laboratory work on two projects to eliminate these losses. A kettle was designed to maintain a constant ratio of oil volume to the area of oil exposed to air. The Devoe laboratory also worked on a solvent process to maintain prescribed pigment volume concentrations. Eliminating the direct fire process which decreased decomposition and utilizing a small quantity of solvent such as xylene which azeotropically distilled off to effect stirring at the molecular level were important developments for the coatings manufacture.

His background in surface science were important in the development of a two coat house paint. Quality painting prior to this development required three coats. The Devoe laboratory developed a primer with controlled penetration to keep the primer in the upper three layers of wood cells. The finish coat which utilized T_iO_2 with nine times the hiding power of white lead was designed to get the optimum balance of twenty properties. Long felt that this purposeful design concept to balance properties in coatings was one of his major contributions to the paint industry. Long also referenced a collaborative patent with GE in 1931 on utilizing an electron stream to cure paint on automobiles as one of his major contributions.[9]

Long's publications in research journals essentially ceased during his 20 years at Devoe (1935-1955). That is not to say he quit writing and talking. He published in the trade journals and in numerous Devoe marketing papers. In the latter part of his Devoe years, he even assumed marketing responsibilities. The quotation in the American Paint Journal that "undoubtably there exists no abler or more knowledgeable a spokesman for the production side of the business"[10] corroborates many citations regarding Long's marketing skills.

In truth he was always marketing since college. He had to sell students on science, sell companies on funding academic research, and sell Devoe on increasing their research activities. Long tried to sell everything he believed in, including religion.

Long was a devoutly religious man. He taught Sunday school for most of his life and supposedly while at Devoe in Louisville, Kentucky, Long taught the largest Sunday school class in America. Long found that the tremendous study of religious matters including the studies of evolution...by great scientists such as Darwin, Asimov, Carson,

Haldane, etc...have been very PROFITABLE IN STIMULATING RESEARCH...."[11]

In the years after retirement from Devoe in 1955 Long had time to write on numerous subjects. On the subject of the news media he wrote:

...read steadily in your field instead of listening to TV or radio over 5 minutes a day. I long ago cancelled the local newspaper and stopped listening to TV commentators because their reports are often just rumors, or if facts are seriously slanted in favor of a political party or socialistic or communistic trend.[12] Do not let athletic facts and figures ever get into your minds for you men are chemists, not athletes.[13]

On the subject of women he wrote:

...women have a variety of abilities such as working, sewing, housekeeping, and cleaning, decorating, washing, raising of children, teaching the children, nurturing children with love and moral concepts, making a house into a home, correspondence and communities, care of health and the family, civic and church activities, and clubs, entertaining, care of pets, and many others.[14]

On the subject of kids he wrote:

...kids should do less talking among...themselves because they do not know much, especially in [the] teens, and therefore they transmit a lot of misinformation.[15]

Long's writings in the later years including the history of science contain examples of bias and he may have been better served by not communicating his personal opinions so often. But, speaking and writing were the way he communicated his science. And speak and write he did. He gave thousands of talks, technical talks, service club talks, schools, etc. Perhaps if Long had not written more than 100 technical papers, numerous trade and industrial papers, a multi-edition classic textbook for "Qualitative Analysis" with D. S. Chamberlain and H. J. Anderson, a problem solving text, "Chemical Calculations", again with H. J. Anderson and co-edited seven books in five volume series titled "Treatise in Coatings", with Professor Raymond Myers of Kent State University, he would not have written what some might now view as inappropriate.

One needs only read the archieval records of the "Treatise on Paints" to see how much effort Long gave to simply editing a book. The correspondence between contributors and between co-editors Myers and Long is overwhelming. After reviewing this extensive amount of correspondence one can only have a great deal more respect for Long and other editors of technical books.

Long did not stop researching when he retired from Devoe and Raynolds. He approached the President of Mississippi Southern University (now called the University of Southern Mississippi) for a faculty position. President McCain responded that he was prevented from paying Long a salary as he was past the mandatory retirement age (Long was 71). Long replied that the Pan American Tung Research League would pay his salary. Needless to say McCain's quick response was "your hired."[16]

Throughout his life, Long was a positive, optimistic person. He foresaw the possible synthesis of 50 million compounds from eleostearic acid, the major component of tung oil. During the period in the mid to late 60's, Long continued to be a motivator and spokesman for the League but relied on a young organic chemist named Shelby Thames for the day-

147

to-day technical guidance of program. Hurricane Camille decimated the tung oil trees in Mississippi in 1969, and Long retired to Clearwater, Florida.

He left a legacy in Mississippi. His work with Shelby Thames helped convince this energetic scientist to press forward for a Department of Polymer Science at USM. Long would be proud of the growth of this department and its soon-to-be dedicated 13 million dollar Polymer Science Research Center.Long left a legacy at Lehigh University as well. The Paint Research Institute referenced earlier evolved into the current Center for Surface and Coatings Research at Lehigh.

Long was formally honored by his profession with the George B. Heckel Award in 1953, the Mattiello Award 1954, the Gold Medal of the American Institute of Chemists in 1954, and honorary doctorates from Lehigh University and from North Dakota State University. The Chicago Section of the American Chemical Society also named him as one of the top ten scientists in the field in 1947.

Reading the writings of James Scott Long caused me to reflect on his purposeful design concept for research and his persistent interest in spreading the word that everything can be improved.I hope this paper on James Scott Long will likewise stimulate you.

REFERENCE

1. James Scott Long, file 706, box 7, James Scot Long Collection, McCain Library, University of Southern Mississippi, Hattiesburg, Mississippi. (hereafter cited as Long Collection).
2. Long Collection, file 006 (not boxed).
3. Long Collection, file 211, box 21.
4. Ibid.
5. Ibid.
6. Ibid.
7. Ibid.
8. Ned D. Heindel, "James Scott Long", (typescript biographical sketch). Copy provided by Dr. Heindel. (hereafter cited as Heindel, sketch).
9. Patent No. 1,818,073 (1931).
10. Long Collection, box 12.
11. Long Collection, file 104, box 1.
12. Long Collection, file 211, box 21.
13. Ibid.
14. J. S. Long, "Parable on the Use of Capabilities", [n.d.],[n.p.], Long Collection.
15. Interview with Dr. William D. McCain, March 8, 1989.
16. Heindel, sketch.
17. The authors gratefully acknowledge the assistance of Shelby F. Thames in preparing this manuscript.

Roy H. Kienle (1896-1957), Polymer Pioneer

George B. Kauffman

Department of Chemistry
California State University, Fresno
Fresno, CA 93740

Kienle was an internationally known authority on resins, pigments, dyes, textile finishes, polymers, and coatings and their physicochemical relationships. He was responsible for some of the first ideas on the colloidal nature of films and for fundamental research on polymers and resins, which resulted, among other things, in the commercial production of the important class of alkyd resins. He also made significant contributions to the elucidation of the complex physicochemical relationships in pigment-vehicle systems and pigment coatings, and he was known for his use of physicochemical methods such as microscopy and spectrophotometry in his research on synthetic resins.[1]

LIFE AND CAREER

Roy Herman Kienle was born in Easthampton, Massachusetts on April 27, 1896, the son of Edmund Frederick Kienle, a businessman, and Emily Mary Kienle (née Hupfer) and the grandson of Gottlieb John Frederick Kienle and Barbara Kienle (née Hofmann). He had a sister, Mildred Kienle. His grandfather emigrated to the United States from Württemberg, Germany, in 1854 and later settled in South Hadley Falls, Massachusetts.

After graduating from elementary school, young Roy attended the Williston Academy (now the Williston Northampton School) in Easthampton, "the one medium-sized New England prep school with an unequivocal and successful program for training the mind and character for learning and leadership, all in a friendly atmosphere" [2]. Its founder, Samuel Williston, the son of an Easthampton minister-farmer, never finished high school; he walked the hundred miles to Phillips Academy in Andover, where he ruined his eyes by studying at night by an oil lamp so that his formal education was cut short. Determined that poor boys in western and central Massachusetts have a school of quality equal to Phillips, in 1841 he and his wife Emily, who together had amassed a fortune in a button manufacturing business, founded the academy known for its rigorous courses. At Williston, from which Kienle graduated as valedictorian in 1912, an outstanding faculty presented chemistry in a realistic but entertaining manner, which strongly influenced him in his choice of a career [3]. From 1951 until his death, Kienle was a trustee of Williston.

A state scholarship enabled Kienle to attend the Worcester Polytechnic Institute, founded in 1865 at Worcester, Massachusetts, where he attained equal academic success and from which he received the Salisbury Award in chemistry and graduated in 1916 with a B.S. degree in chemistry. A lack of funds made graduate work impossible so that same year he accepted a position as chemist at the Pittsfield, Massachusetts plant of the General Electric Company. Hoping that he could work out a plan to combine industrial work with graduate study, a goal that he eventually accomplished, in 1917 he transferred to GE's research laboratories in Schenectady, New York, where he made the acquaintances of later Nobel laureate Irving Langmuir. His plans for graduate study were frustrated by the entrance of the United States into World War I. On May 16, 1918 he entered the U.S. Army Chemical Warfare Service at the American University Experimental Station in Washington, DC, being commissioned as a second lieutenant in an incendiary unit. He was promoted to first lieutenant and chief of the unit on July 13, 1918, and he was discharged on March 15, 1919.

Published 1990 by Elsevier Science Publishing Co., Inc.
Organic Coatings: Their Origin and Development
R.B. Seymour and H.F. Mark, Editors

149

After his discharge, Kienle spent a year with E.I. duPont de Nemours & Co. before returning to GE's research laboratories at Schenectady, where he worked as a research chemist on synthetic resins [4], polymers, coatings, electrical insulation, wire enamels, and plastics. Here he studied insulating material made from linseed oil, proposed some of the first ideas on the colloidal nature of films, and developed the basic theory of the structure of linear and cross-linked polymers. Here he was also influenced by Willis Rodney Whitney (1868-1958), founder and longtime Director of the GE laboratories, and he developed his research credo of "Why? How? What?" According to Kienle, the research chemist should first find out why things are useful and then ask himself how this knowledge could be applied to other uses [3].

On June 17, 1920 in Pittsfield Kienle married Ruth Lynn Hine (October 27, 1897-December 4, 1983), the daughter of Lewis Nelson Hine, a Pittsfield business executive. The couple had two sons--Lawrence Frederick Kienle (born June 16, 1923), who received his M.D. degree from the Albany Medical School and practiced radiology until his retirement, and Robert Nelson Kienle (born October 20, 1927), who followed in his father's footsteps by attending Williston (where he graduated in 1945) and becoming a research and development chemist. Robert received his A.B. (1949) from Princeton University and his M.S. (1951) and Ph.D. (1954) degrees from Yale University, and at the time of his retirement was Director of Original Equipment Material Technology for the Uniroyal-Goodrich Tire Company.

At General Electric Kienle worked on the commercial development of an important class of resins that he called "alkyds" [5,6]. His most basic and fundamental work here involved condensation polymerization and his famous three postulates underlying polymerization reactions, most of which was published in Industrial and Engineering Chemistry and the Journal of the American Chemical Society. He was also awarded numerous patents for his discoveries and inventions, such as those for enameling wire [7] and for producing synthetic resins [8], resinous composition products [9], resinous condensation products [10], and resinous compositions [11]. At GE he was finally able to implement his plan for graduate study on a part-time basis by attending classes at night and carrying out the required experimental laboratory research on weekends. In this way, while at GE, he received his M.S. degree in chemistry from Union College in Schenectady in 1927 and, while working at Calco, his Ph.D. in chemistry from Rutgers University in New Brunswick, New Jersey in 1938. His doctoral dissertation was a continuation of his work on polymers although after 1933 his research involved primarily pigments and dyes. While at GE he also did graduate work on colloids with Professor Ernst Alfred Hauser at the Massachusetts Institute of Technology at Cambridge.

Most of his significant work and international recognition came long before he received his doctorate.

On August 28, 1933 Kienle became a research chemist at the Calco Chemical Company, later a subsidiary of the American Cyanamid Company, to organize a section on physical chemistry and "to work as a chemist in a chemical industry." An important factor in his decision to move from Schenectady was the opportunity to complete the work on his doctorate at Rutgers (Calco's Bound Brook, New Jersey plant was less than ten miles from New Brunswick). During the 1930s there was little opportunity for advancement for bench chemists. Instead, increased remuneration resulted from advancement in management. Thus increasing managerial responsibilities removed Kienle from direct contact with the laboratory bench.

Consequently, after the mid-1930s only a few publications--mainly general review articles--appeared in the literature with Kienle as the primary author. One of his most strongly held managerial principles was to give full credit to the persons working under his supervision. Therefore his name was rarely included as a coauthor on the numerous publications emanating from his laboratories. His impact on the scientific world, however, can be judged from the pioneering contributions of those working under his leadership and the impact of the Dyes and Pigments Division upon the growth of Calco and later American Cyanamid during the period between the 1930s and the 1950s.

Theodore F. Cooke, retired Research Director of Calco and Cyanamid, who had worked under Kienle at Bound Brook and at Stamford, Connecticut, related that he learned two important concepts from Kienle—"to think broadly and not narrowly when trying to solve a research problem" and "to draw an analogy wherever appropriate between the solution of one problem in an attempt to solve another problem" [12].

Of the numerous research projects that Kienle directed at Bound Brook, Cooke cited three as especially significant. One involved the development of a broad line of pigmented resin printing emulsions to complement the dye pastes that had been used to print color patterns on textiles prior to that time. Another project resulted in the formation of a new bright green crystal form of phthalocyanine blue pigment which complemented the known red shade of blue that was the only form of phthalocyanine blue known previously. A commercial process for preparing the new crystal form was developed, and a patent on the product, the licensing of which brought in more than a million dollars, was obtained. The third project concerned the application of formaldehyde resins to cellulosic fabrics in order to impart wrinkle-recovery and durable-press properties. In a landmark paper, "The Mechanism of Imparting Wrinkle Recovery to Cellulose Fabrics," Cooke, Dusenbury, and Kienle reported physical evidence as well as infrared evidence (for the first time) of covalent bonding of melamine formaldehyde with cellulose, which supported the theory of cross-linking of cellulose chains now universally accepted as the mechanism for durable-press properties [13].

During his 24-year stay with Calco and Cyanamid Kienle received several promotions and transfers. From 1934 to 1936 he was Chief of the Physical Chemistry Division and from 1936 to 1949 Assistant Research Director at Calco in Bound Brook. He also simultaneously served as Assistant Research Director of American Cyanamid from 1941 to 1949, commuting two days each week to Cyanamid's Stamford Laboratories. He remained at Bound Brook, where he was Director of Application Research (1949-1952) and Director of Applied Research (1952-1954) of the Calco Chemical Division of American Cyanamid. On August 1, 1954 he transferred to the Stamford Laboratories of the Research Division of American Cyanamid, where he became Director of the newly created Research Service Department, a position that he held until his death of heart failure in Stamford at the age of 61 on September 2, 1957.

After he had received his doctorate in 1938 Kienle considered changing his career by entering academe so that he could return to the laboratory and tackle basic research problems instead of managerial problems. He taught courses on resins, plastics, and polymers as adjunct professor at Brooklyn Polytechnic Institute, Rutgers University, and New York University. However, the advent of World War II and its pressing scientific needs prevented him from leaving industry for university teaching. He, nevertheless, always cherished an aspiration to retire "to teaching and having fun doing research" [3].

During the war Kienle served with distinction at the Office of Scientific Research and Development (OSRD) as a member of its Advisory Committee on scientific and technical personnel to the War Manpower Commission (1941 to 1945). He played an important role in devising a satisfactory procedure for deferring scientists from military service when industry was in danger of losing highly trained technical personnel by indiscriminate application of selective service quotas. From 1944 to 1945 he was a consultant for the War Production Board.

Honors and Awards. In 1947, on the basis of a reader poll conducted by Chemical Bulletin, a publication of the Chicago Section of the American Chemical Society, Kienle was honored by being selected as one of the "10 oldest chemists or chemical engineers" working in the United States in the field of paint, varnish, and plastics. In 1949 he was selected as the first Joseph J. Mattiello Memorial Lecturer, the most outstanding scientific honor in the protective coatings industry, named after his friend, one of the fathers of the science of chemistry in the paint and varnish field. Kienle presented his lecture, "Physical Chemical Research in the Protective Coatings Industry" [14] at the 27th Annual Meeting of the Federation of Paint and Varnish Production Clubs on November 4, 1949 in Atlantic City, New Jersey, where less than four years later, on October 27, 1953, he presented the keynote speech, "The Protective Coatings Industry: A Practical Science," at the federation's 31st Annual Meeting [15]. As the federation's representative, he also presented a lecture, "Observations on Optical Properties of Paint Films," at two international industrial conferences--the FATIPEC (Fédération des Associations des Technicians des Industries des Peintures, Varnis, Emaux, et d'Encres d'Imprimerie de l'Europe Continentale--Continental European Federation of Associations of Paint, Varnish, Enamel, and Printing Industry Technicians) Congress at Noordwijk in the Netherlands on May 18, 1953 and the Biennial Conference of the Oil and Colour Chemists' Association at Eastbourne, England (June 3-5, 1953). The lecture was one of the first to use multiple slide projectors and dual screens for dramatic effect.

Kienle's selection as Mattiello Lecturer was important enough for the American Chemical Society to feature his picture on the cover of the December 19, 1949 issue of its weekly magazine, Chemical and Engineering News, the scientific equivalent of making the cover of Time magazine [3]. C&EN also devoted considerable space to the contents of the lecture, intended to activate and stimulate research by the protective coatings industry, in which Kienle reviewed the manufacture of decorative and industrial finishes, the raw materials used, and the physical chemical principles involved. He also discussed recent developments in solvents, vehicles, and pigments pioneered by himself and by others. Of works on pigments carried out under his direction he singled out the newly developed, very stable copper phthalocyanine blue and the obtaining of a cross-section of chrome green pigmented alkyd enamel film thin enough for study by electron microscopy. He stated that it was the responsibility of the protective coatings industry to obtain scientific answers to problems and to convey them to the consumer just as raw material suppliers furnish the industry with information on their products. Kienle emphasized that almost every technique, instrument, and theory of physical chemistry was being employed by the protective coatings industry. He concluded [14]:

> The challenge before the industry is to exploit physical chemical investigations on the formulation and manufacture of its products and on the protective coating films which they produce....The problems may seem difficult, but by combining curiosity to know, will to experiment, and ability to observe, they can be solved.

On May 5, 1955 Kienle received the 1955 PaVaC Award for "outstanding contributions to the advancement of the protective coatings industry and the New York Paint and Varnish Production Club" at the club's annual dinner in New York City. The award, first presented in 1951, was awarded to Kienle [16]

> in recognition of his fundamental research and development in the field of resins and polymers which resulted in commercial production of the important class of alkyd resins; also for his postulations and basic theory regarding the structure of linear and cross-linked polymers, and his most recent contributions to the complex physical-chemical relationships which exist in pigment-vehicle systems and in pigmented coatings; in addition, for his ability

as a lecturer in the dissemination of knowledge to the Paint Industry, and finally, for his outstanding contributions to technical committees of the Federation and the New York Paint and Varnish Production Club.

On April 7, 1960 the New York Paint and Varnish Production Club officially changed its name to the New York Society for Paint Technology and at the same meeting presented the first Roy H. Kienle awards to honor outstanding members of its Technical Committee in recognition of their service and contributions to the Society and the paint industry. The awards were named after Kienle "to honor a great scientist, whose accomplishments and achievements gave recognition to the important role of the technical and research men in our paint industry" [17]. The first awards were given to forty-four persons in order to recognize those qualifying for them as much as three decades earlier, but in the future they were given annually to one outstanding chemist in the paint industry. The first award was presented posthumously to Kienle and received by his widow, who was the guest of honor at the meeting.

Memberships. Kienle was a member of numerous domestic and foreign scientific, technical, and social organizations. An active member of the American Chemical Society, he served as Secretary (1932-1937) and Chairman (1937) of its Paint, Varnish, and Plastics Division (now the Division of Polymeric Materials: Science & Engineering) and Secretary of its Eastern New York Section. He was also a Fellow of the American Institute of Chemists and of the New York Academy of Sciences and a member of the National Academy of Sciences and the National Research Council (of which he was Permanent Chairman of its Committee on Industrial Chemistry). He belonged to the American Association of Textile Chemists and Colorists and its British counterpart, the Society of Dyers and Colourists, as well as the British Society of Chemical Industry and the Faraday Society. He was also a member of the Federation of Paint and Varnish Production Clubs (Program and Research Committees), the New York Paint and Varnish Production Club (Technical Committee), the Association of Research Directors, the Chemists' Club, the Sigma Xi Scientific Research Society, the Cosmopolitan Club of the Worcester Polytechnic Institute, and the Masonic Order (New Hope Lodge No. 730). He served as a trustee for the Williston Academy (1951-1957), as a Deacon of the Bound Brook Congregational Church (1946-1948), and as a trustee of the First Congregational Church of Old Greenwich, Connecticut, where his funeral was held.

Personality. Kienle is described by his former colleagues as a "hustler," a very enthusiastic, intense, "take-charge" kind of person, on the move most of the time, who chewed on his fingernails. One characterized him as "the best boss I ever had." His natural curiosity led him to read everything in sight. An inveterate cigar smoker, he was not always careful in putting out his spent cigars. Theodore Cooke reports that "more than once I helped extinguish a fire in the wastebasket in his office started by a discarded cigar" [12]. Kienle frequently came to the office with his ever-present cigar in one hand and a bouquet of flowers in the other, for one of his hobbies was gardening [3]. An ardent sports fan, he coached and refereed basketball games, and his recreational interests also included golf, tennis, and travel. At Calco in Bound Brook almost every day after work he joined fellow researchers Ted Cooke, Dick Vartanian, and Charlie Bacon in a round of golf, in which he ran, rather than walked, around the course. After the game, they would repair to his house and play bridge with his wife.

WORK

As mentioned earlier, Kienle was involved in fundamental and applied developments in a number of different areas. However, because of space limitations we shall confine ourselves to the two contributions for which he is most famous--alkyd resins and the functionality concept of polymerization.

Alkyd Resins. A posthumous resolution "incorporated in the minutes of the [New York Paint and Varnish Production] Club as a permanent record of the esteem and the high regard in which he was held," recalled [17,18]

> [Kienle's] great scientific accomplishments, his world-wide recognition as one of the foremost scientists of this generation in the field of Protective Coatings and Pigments, his creation of the first Alkyd Varnish which earned him the resounding title of "Father of the Alkyds," the discovery of which basically changed our industry from an art to a science.

An alkyd is a polyester produced by the condensation of a polyhydric alcohol (such as ethylene glycol, glycerol, or pentaerythritol) and a polycarboxylic acid or anhydride (such as phthalic acid or phthalic anhydride). The polyester chain is usuallly terminated by controlled amounts of an unsaturated monofunctional organic acid (such as oleic or stearic acid) [19-24]. The great Swedish chemist Jöns Jacob Berzelius prepared the first alkyd resin (glyceryl tartrate from glycerol and tartaric acid) in 1847 [25]. However, it was not until 1901 that the English chemist Watson Smith prepared glyceryl phthalate resins by the reaction between glycerol and phthalic anhydride [26], but neither of these investigators seemed to be aware of the film-forming possibilities of these resins. Glyceryl phthalate resins that showed possibilities for commercial applications were patented in the United States in 1914 [27]. Because the secondary hydroxyl group is less active than the two terminal primary hydroxyl groups in glycerol, the first product formed at conversions of less than 70% is a linear polymer; further heating produces a cross-linked polyester [21, p.246]:

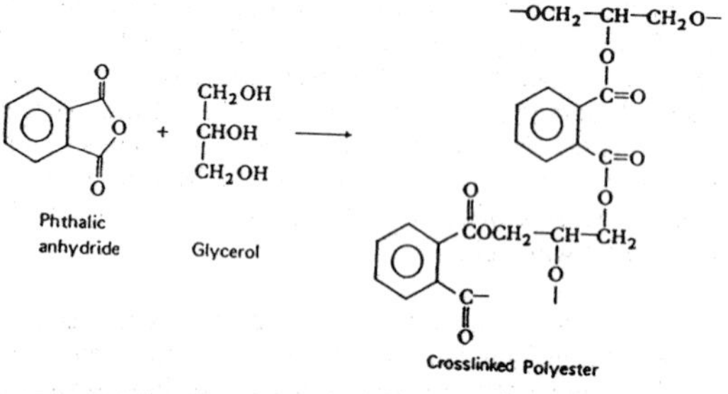

In Germany in 1894 Vorländer prepared glycol maleate [28], and at General Electric in the United States Arsem [29], Dawson [30], and Howell [31] studied the general alkyd reaction, substituting other alcohols for glycerol and other acids for phthalic anhydride.

At the 76th National Meeting of the American Chemical Society (Swampscott, Massachusetts, September 10-14, 1928) Kienle and Ferguson presented a paper, "Alkyd Resins as Film-Forming Materials" [6], which classified them into three general types-- heat-non-convertible, heat-convertible, and oxygen-convertible and described the preparation of solutions of the last two types and the properties of the resulting films. They demonstrated that the oxygen-convertible resins should be seriously considered as film-forming materials because they possess the quick-drying characteristics of nitrocellulose lacquers and the film-building properties of oil-base varnishes.

The first commercial unsaturated alkyds were synthesized from trifunctional alcohols and dicarboxylic acids by Kienle in the mid-1920s. Unsaturated oils (drying oils) were transesterified with phthalic anhydride so that an unsaturated polyester was obtained.

General Electric introduced Glyptal® (glyceryl phthalate) resins in 1926 but did not begin full-scale production of alkyd coatings until 1933 based on Kienle's patent of that year [10]. Other companies began to produce alkyd resins after Kienle's patent was ruled invalid in 1935 because of anticpation by prior art [32], and Kienle obtained numerous other patents [33-36]. In 1949 he reviewed the development of alkyd resins during the past 25 years, showing the growth in their industrial applications and the advances made in the techniques used for their manufacturing and application [37].

The alkyds' toughness and gloss retention on outdoor exposure made them especially adaptable for automobile and household appliance finishes (although they were displaced by acrylic resins after 1957) and made possible the "wrinkle" and "crackle" finishes used on metal products. Although their principal use has been as polymer coatings, marine paints, and baking enamels, they have been available as molding powders since 1948. By the 1950s aqueous alkyd emulsions were being extensively used to finish interior walls [24]. Under various trade names--Glyptal® (General Electric), Beckosol® (Reichhold Chemicals), Duraplex® (Rohm & Haas), and Rezyl® (Koppers Company)--more than 250,000 metric tons of alkyd resins and molding compounds were produced in the United States in 1985.

Kienle coined the word "alkyd" when General Electric's patent department needed a word that would avoid the use of their trade name Glyptal® to describe their new class of polymers. As condensation products of alcohols and acids, they were quickly dubbed "alcids." According to Kienle's son Robert [38], the nomenclature group was unable to decide whether the "c" should be pronounced hard as in alcohol or soft as in acid. The hard "c" was chosen as sounding better. Kienle suggested that the substitution of "k" for "c" would remove the ambiguity in pronunciation. When some members of the group said that the word "alkid" just did not look right, Kienle proposed that the "i" be replaced with a "y," giving the final term "alkyd," which he said had a classical Greek flavor.

The Functionality Concept. In the March, 1969 issue of the Journal of the Oil and Colour Chemist's Association an editorial titled "Who said functionality?" [39] discusses "the striking dichotomy of opinion about the originator of the functionality concept," pointing out that Kienle is favored by the coatings technologists, while academic polymer scientists or rubber, fiber, or plastics specialists favor Wallace Hume Carothers (1896-1937) of nylon and neoprene fame. Both men proposed their ideas in the late 1920s and the 1930s, a time when Hermann Staudinger's covalent macromolecular concept was gaining ascendancy over the opposing micellar concept.

In his previously mentioned paper of 1929 with Ferguson [6], Kienle clearly recognized the reaction of diols with dicarboxylic acids as a means of preparing non-convertible alkyd resins and emphasized the necessity for using trihydric and higher polyhydric alcohols to obtain heat-convertibility. He showed that "lower valent alcohols and acids" retard gel time, which reduces heat-convertibility, decreases acid value, and increases the degree of esterification. This masterful statement of the principles of alkyd formulation, however, stopped just short of presenting a molecular picture, and although it clearly adumbrated the concept of functionality, it did not use the term to indicate the number of active functional groups in a molecule.

In his most famous paper "Observations as to the Formation of Synthetic Resins," presented before the Division of Paint and Varnish Chemistry at the 79th National Meeting of the American Chemical Society, Atlanta, Georgia, April 7-11, 1930, Kienle proposed his three insightful postulates, which provide "a chemical explanation of the formation and properties of synthetic resins and other high polymers" [4]:

1. No high polymer is formed unless the reaction reactivity [functionality] is (2,2) or greater, i.e., poly-reactive molecules must be interacting. If the reactivity is (1,1), (1,2), (1,3), etc., only single compounds result.

2. In high-polymer formation, the reaction should proceed rapidly at first, then more slowly, according to the decrease in the number of effective contacts. At the same time as the molecular complexity increases, the viscosity should increase slowly at first, than more and more rapidly [5].

3. If the distance between the reactive groups in one of the reactive molecules in a polymer is increased, a softer and lower melting product should result....On the other hand, if the number of reactive bonds per-molecular length is increased, the separate parts of the polymeric molecules would be more rigidly and compactly held in position, and hence one would predict an increase in flow point.

Concerning his father's famous three postulates, which "still summarise the picture [of polymerization] more neatly than many a recent lengthy discourse" [39], Robert N. Kienle relates a story told by his father about his travels. Although his lectures at English universities were generally well attended and well received, at one school Kienle was amazed to find the auditorium overflowing with students, literally fighting for seats. At the post-lecture reception he asked a student the reason for the large attendance. The reply was, "Our final exam always includes a question on your three postulates of polymerization. By heaven, we are all going to see the bloke who thought them up" [38].

In his 1930 paper [4], Kienle cited three of Carothers' early papers [40-42] inter alia on "the importance of considering the chemical bonding of the molecules when dealing with the formation, constitution, and properties of high polymers," and he pointed out that of three treatises on resins, only one [43] deals with "the underlying causes of synthetic resin formation." Kienle favored Staudinger's macromolecular concept, but he correctly observed that "formation of chain molecules in itself is not a complete explanation" for the gelation of glyceryl phthalate and other polymers. He also illustrated the cross-linked molecule (using the term "intertwining" rather than cross-linking). Yet, despite his awareness of Carothers' work, he still did not use the term "functionality," preferring to speak of the "reactivity," as "the number of primary valence bonds which are active in the reaction under consideration."

Meanwhile, Carothers had distinguished between condensation and addition polymerization and defined functionality [40], but, while Kienle was concerned primarily with alkyds, Carothers was working mainly with bifunctional reactants. He neglected gelation until 1931, when he first took note of Kienle's work. He declared the difference between the reaction of phthalic acid with glycol and with glycerol to be "obvious" and explicable in terms of one-dimensional as opposed to three-dimensional polymer growth. By 1930 Carothers repeatedly used the expression "cross-linking" [44], in contrast to Kienle, who, as late as 1939, three years after Carothers' famous equation relating the degree of polymerization to the degree of reaction and functionality [45], preferred to explain the gelation of glyceryl phthalate in terms of his own "log-jam" theory [46,47] rather than in terms of the three-dimensional macromolecular theory.

The JOCCA editorial concludes [39]:

Had the distinction between number average and weight average molecular weights been appreciated earlier, it is probable that Kienle and Carothers would have been speaking a common language, each reinforcing the ideas of the other more obviously than they did in the event.

Instead, what emerges from a reading of their publications is a realisation that both Kienle and Carothers were cautious in their interpretation of their own work and in their acceptance of the postulates of others. They worked in different industries, for the most part, with different objectives and with different materials; yet independently they were formulating one of the greatest unifying theories of modern polymer science. Kienle must take credit for his far-sighted three postulates, Carothers for his quantification of functionality theory; but to assign priority for the steps between one is obliged to speculate about their thinking. Who first said "functionality" must remain an open question. Carothers wrote it before Kienle, but the idea is so clearly foreshadowed in Kienle's alkyd paper of 1928 that he may very well have been the first to think it.

ACKNOWLEDGMENTS

I gratefully acknowledge the assistance of Dan R. Bruss, Albany College of Pharmacy; Joseph C. Calitri, Theodore F. Cooke, and William H. Linke, American Cyanamid Company; Dennis H. Grubbs, The Williston Northampton School; Robert N. Kienle, formerly of Uniroyal-Goodrich Tire Company; Louis Navias, formerly of the General Electric Company; William H. Starnes, Jr., Polytechnic University; Thomas C. Werner, Union College; and Mark S. Wrighton, Massachusetts Institute of Technology. I am also indebted to Drs. Robert N. Kienle and Raymond B. Seymour for a critical reading of the manuscript.

REFERENCES

1. The National Cyclopedia of American Biography (James T. White & Co., New York 1968) Vol. 50, pp. 15-16.

2. The Williston Northampton School 1988-89 Catalog (Easthampton, MA 1988) p. 22.

3. Chem. Eng. News $\underline{27}$ (51), 3788-3789 (December 19, 1949).

4. R. H. Kienle, Ind. Eng. Chem. $\underline{22}$, 590-594 (1930).

5. R. H. Kienle and A. G. Hovey, J. Am. Chem. Soc. $\underline{51}$, 509-519 (1929).

6. R. H. Kienle and C.S. Ferguson, Ind. Eng. Chem. $\underline{21}$, 349-352 (1929).

7. R.H. Kienle, U.S. Pat. 1,747,940 (1930).

8. R.H. Kienle, U.S. Pat. 1,878,527 (1932).

9. R.H. Kienle, U.S. Pat. 1,889,923 (1932).

10. R.H. Kienle, U.S. Pat. 1,893,873 (1933).

11. R.H. Kienle, U.S. Pat. 1,898,840 (1933).

12. T.F. Cooke, Recollections of Dr. Roy H. Kienle, typescript, April 28, 1989.

13. T.F. Cooke, J.H. Dusenbury, and R.H. Kienle, Textile Res. J., $\underline{24}$, 1015-1036 (1954).

158

14. R.H. Kienle, Official Dig. Federation Paint & Varnish Production Clubs, 300, 11-52 (1950).

15. R.H. Kienle, Official Dig. Federation Paint & Varnish Production Clubs, 347, 897-899 (1953).

16. J.T. Cassaday, American Cyanamid Co., Stamford, Conn., Interoffice Correspondence, May 3, 1955.

17. Amer. Paint J., 45, 18 (March 28, 1960).

18. E.S. Paterno and M. Bauman, Resolution, The New York Paint and Varnish Production Club, September 12, 1957.

19. C.R. Martens, Alkyd Resins (Reinhold, New York 1956).

20. T.C. Patton, Alkyd Resins Technology (Wiley Interscience, New York 1962).

21. R.B. Seymour and C.E. Carraher, Jr., Polymer Chemistry: An Introduction, 2nd ed. (Marcel Dekker, New York 1988) p. 12.

22. R.B. Seymour, in The Encyclopedia of Chemistry, 2nd ed., G.L. Clark and G.G. Hawley, eds. (Van Nostrand Reinhold, New York 1966) pp. 46-47.

23. C.R. Noller, Chemistry of Organic Compounds, 3rd ed. (W.B. Saunders, Philadelphia, PA 1966) pp. 603-604.

24. K.M. Reese, ed., A Century of Chemistry: The Role of Chemistry and the American Chemical Society (American Chemical Society, Washington, DC 1976) p. 352.

25. J.J. Berzelius, Rapp. Ann. Inst. Geol. Congr., 26 (1847).

26. W. Smith, J. Soc. Chem. Ind., 20, 1075-1076 (1901).

27. M. Callahan, U.S. Pats. 1,101,732 and 1,108,329--1,108,331 (1914).

28. D. Vorländer, Ann., 280, 167-206 (1894).

29. W.C. Arsem, E.S. Dawson, Jr., and K.B. Howell, U.S. Pats. 1,098,777 and 1,119,592 (1914).

30. E.S. Dawson, Jr., U.S. Pat. 1,141,944 (1915).

31. K.B. Howell, U.S. Pat. 1,098,728 (1914).

32. R.B. Seymour, In History of Polymer Science and Technology, R.B. Seymour, ed. (Marcel Dekker, New York 1982) pp. 104-105.

33. R.H. Kienle and H.C. Rohlfs, U.S. Pat. 1,897,260 (February 14, 1933).

34. R.H. Kienle and P.F. Schlingman, Can. Pat. 369,326 (October 19, 1937).

35. R.H. Kienle, U.S. Pat. 2,065,331 (December 22, 1937).

36. R.H. Kienle, Can. Pat. 379,370 (February 7, 1939).

37. R.H. Kienle, Ind. Eng. Chem., $\underline{41}$, 726-729 (1949).

38. Robert N. Kienle, letter to George B. Kauffman, June 6, 1989.

39. J. Oil and Colour Chemists' Assoc., $\underline{52}$ (3), 244-245 (1969).

40. W.H. Carothers, J. Am. Chem. Soc., $\underline{51}$, 2548-2559 (1929).

41. W.H. Carothers and F.J. van Natta, J. Am. Chem. Soc., $\underline{52}$, 314-326 (1930).

42. W.H. Carothers and G.L. Dorough, J. Am. Chem. Soc., $\underline{52}$, 711-721 (1930).

43. J. Scheiber and K. Sändig, Die künstliche Harze (Wissenschaftliche Verlagsgesellschaft, Stuttgart 1929).

44. W.H. Carothers, Chem. Rev., $\underline{8}$, 353-425 (1930).

45. W.H. Carothers, Trans. Faraday Soc., $\underline{32}$, 39-53 (1936).

46. R.H. Kienle, J. Soc. Chem. Ind., $\underline{55}$, 229T-237T (1936).

47. R.H. Kienle, P.A. Van der Meulen, and F.E. Petke, J. Am. Chem. Soc., $\underline{61}$, 2258-2268 (1939).

THE ROY W. TESS AWARD IN COATINGS

SPONSORED BY THE

THE DIVISION OF POLYMERIC MATERIAL: SCIENCE AND ENGINEERING

by

Ronald S. Bauer

Shell Development Company
Westhollow Research Center
Houston, Texas

In 1983 the Division of Organic Coatings and Plastic Chemistry changed its name to Division of Polymeric Materials: Science and Engineering. The purpose for this name change was to better reflect the expanded programming in applied polymer science of the Division, since it was initially established as the Paint and Varnish Division in 1924. Although the intent was to continue programming in the coatings area; Dr. Roy Tess, a former chairman of the Division, became concerned that with no reference to coatings in the Division's new name our earlier roots would be forgotten with the passage of time. With this mind in Dr. Tess and his wife gave the Division a generous gift with which to fund an annual award to be presented to an individual for outstanding contributions to coatings science, technology, and engineering.

In recognition of this gift from Dr. Tess the Executive Committee of the Division established the Roy W. Tess Award in Coatings. The award, which consists of a plaque and a $1000, has been given annually at the fall meeting of the American Chemical Society since 1986. Each fall the Division sponsors a Roy W. Tess Award Symposium at which the award is presented, thus insuring, at least, one half day session on coatings is programmed every year. Recipients of the award have been:

1986   William D. Emmons, Rohm and Haas Company

1987   Marco Wismer (retired), PPG Industries

1988   Zeno W. Wicks, Professor Emeritus, North Dakota State University

1989   Theodore Provder, The Glidden Company

Roy W. Tess, who has long been active in the American Society and was chairman of the Division of Organic Coatings and Plastics Chemistry, received a B.S. degree in Chemistry from the University of Illinois in 1939, and a Ph.D. in organic chemistry from the University of Minnesota in 1944. After obtaining his Ph.D. he joined Shell Development Company in Emeryville, California where he was concerned with research and research management, product development, technical planning, writing, and technical support of marketing. Among his numerous research activities he was involved with the technology, science, and usage of solvents and other petrochemicals, especially in connection with coatings, resins, and polymers. He is the author or co-author of about 30 technical articles or book chapters, 18 patents, and numerous commercial brochures, many of which contain information not otherwise available. He also was editor or co-editor of three books on resins, polymers, coatings, and solvents.

During his career as a research chemist and research supervisor at Shell Development Company he was concerned with the preparation and evaluation of new resins. He was the first Shell chemist to work with the EPON® Resins based on bisphenol A in the middle 40's. His work involved considerable effort on esters of unsaturated and saturated fatty acids of epoxy resins: products including esters modified with styrene, maleic anhydride, cyclopentadiene, silicone, and isocyanates.

Other areas that Tess and his group also investigated were new intermediates (p-tert-butylbenzoic acid; hexanetriol-1,2,6; and 1,3-trimethylene glycol) in alkyd resins and other polyesters as well as the utility of various actual and potential Shell intermediates for aqueous resin dispersions. Some early work on coalescents for latex paint resulted in a presentation before the Federation of Societies for Coatings Technology in 1957 in the first year of the Roon Award. The paper "Use of Hexylene Glycol and Other Solvents in Styrene-Butadiene" by Tess and R. D. Schmitz tied for first place in the open competition.

Published 1990 by Elsevier Science Publishing Co., Inc.
Organic Coatings: Their Origin and Development
R.B. Seymour and H.F. Mark, Editors

161

During 1962-3 he headed a group at the Shell Plastics Laboratory at Delft in the Netherlands involved in applications research to demonstrate how new products could best be used to make superior coatings.

In 1967 Dr. Tess transferred to the Head Offices of Shell Chemical Company in New York City where he was responsible for technical aspects of the solvents business including the structure and oversight of laboratory programs on solvents. For many years Tess served as Shell Chemical Company's representative on the Air Quality Committee of the National Paint and Coatings Association. In 1975 and 1977 under sponsorship of Shell Kagaku K.K. (Shell Chemical Company of Japan) he made lecture tours of Japan where he discussed trends in solvents and coatings in the United States as influenced by air pollution regulations.

His activities in American Chemical Society affairs include many years of service in the Organic Coatings and Plastic Chemistry Division as a member of the Program Advisory Committee and as Chairman of the Division in 1978. In 1974, J. K. Craver of Monsanto and Tess organized a week-long symposium in celebration of the 50th birthday of the Division. The proceedings were published as a book entitled "Applied Polymer Science" in 1975. Considerable attention was devoted to giving a historical perspective to the development of applied polymer science over a period of 50 years. In 1985 the American Chemical Society published an expanded and revised edition of the book, entitled "Applied Polymer Science, second Edition" under the editorship of Tess and G. W. Poehlein, Director of the School of Chemical Engineering at Georgia Institute of Technology.

In addition to his involvement with the American Chemical Society, Dr. Tess was active in the Federation of Societies for Coatings Technology and served on the Board of Directors for several years. He served as President of the Paint Research Institute for three years (1974-1976) and its Board of Trustees from 1972 to 1979. The PRI was the research arm of the Federation and sponsored basic studies at leading universities on topics pertinent to the coatings industry. In 1978 Dr. Tess was given the prestigious Heckel Award for "outstanding contributions to the advancement of the Federations interest and prestige."

After 35 years with Shell, Dr. Tess retired in 1979 with the title of Consultant, the highest technical level at that time in the Company. He and his wife Marjorie moved to Fallbrook, California and established a mini-ranch. He has been active on a part time basis in technical writing, consulting, and participating in the affairs of technical societies, including membership on the Executive Committee of the Division of Polymeric Materials: Science and Engineering.

HISTORY OF AMINO RESINS IN COATINGS

NICHOLAS J. ALBRECHT
ROBERT G. LEES

AMERICAN CYANAMID COMPANY
1937 WEST MAIN STREET
STAMFORD, CT 06904-0060

Amino resins have played and continue to play a
dominant role in the success of industrial
coatings. In this paper the history of amino
resins is chronicled from the introduction of the
first butylated urea formaldehyde resin in the
late 1930s, to the present day use of compliance
acceptable highly alkylated melamine formaldehyde
crosslinking agents. Amino resins based on urea,
melamine, benzoguanamine and glycoluril are
included in the discussion. Application areas,
and the technical reasons for their success are
defined. Some current technical problems, which
amino resins are addressing, are also briefly
described.

INTRODUCTION

Amino resins have been commercially successful for over half a
century, not only in surface coatings but in molding compounds,
adhesives, paper and textile chemicals, and as rubber additives.
They have been referred to as a "mature chemistry" for over
thirty years. But in fact, in the coatings industry they have
been constantly evolving in response to performance demands.

In coatings, "amino resins" is a general term used to identify
thermoset formaldehyde addition products of the following
polyfunctional amines or amides,
> melamine,
> urea,
> benzoguanamine, and
> glycoluril (Figure 1).
Commercially, melamine formaldehyde and urea formaldehyde resins
have generated the greatest interest.

Also included in the category "amino resins" are acrylic resins
containing the formaldehyde addition products of acrylamide or
similar amides. But, for the most part this paper will
concentrate on the former products.

CHEMISTRY

Amino resins' preparation chemistry involves two primary
reactions (Figure 2). The first is an acid or base catalyzed
addition reaction of formaldehyde and the amine (in this
instance, melamine) forming a methylol, or hydroxymethyl
compound. The second reaction is the condensation of either the
methylol on one molecule of melamine with an imino site or
methylol site on another, forming a polymer, or the acid
catalyzed condensation of the methylol with an alcohol to form

Published 1990 by Elsevier Science Publishing Co., Inc.
Organic Coatings: Their Origin and Development
R.B. Seymour and H.F. Mark, Editors

an alkoxy methyl compound. The number of moles of combined
formaldehyde and alcohol can be controlled, as can the molecular
weight distribution to some extent. A wide range of
compositions (Figure 3), and film performance properties are
possible.

Unalkylated amino resins have found use in adhesives, laminating
resins, and molding compounds. However, although soluble in
water, they are insoluble in organic solvents, and have
inherently poor stability. Stability is of such concern, that
even as an adhesive for wood they are sometimes marketed as a
powder, to be dissolved in water prior to use.

Virtually all amino resins used in surface coatings are
alkylated to some extent. The formation of alkoxymethyl sites
provides the stability, compatibility with other binders, and
organo solubility properties required for coatings applications,
and yet still maintains a potential for reactivity. This is
particularly important, since amino resins are not the primary
binder in a coating, but are used to crosslink other binders,
such as alkyds, polyesters, acrylics, urethane polymers, and
epoxy resins. The specific alkylation alcohol used can vary,
although methanol, n-butanol, and iso-butanol are most common.
All other compositional variables being equal, methylated resins
have advantages in cure response, and exterior durability, while
butylated aminos have relatively better wetting characteristics,
and superior intercoat adhesion properties. Typically, surface
coating urea resins have combined formaldehyde contents ranging
from 1.7 to 2.5 moles per mole of urea, and 0.5 to 1.8 moles of
combined alcohol per mole of urea. Surface coating melamine
resins' combined formaldehdye content ranges from 3 to 6 per
mole of melamine, and combined alcohol from 1.5 to 5.5. Degrees
of polymerization are as low as 1.3 and as high as 3-5.

They cure by a condensation reaction, usually acid catalyzed,
involving either self-condensation of methylol, imino and
alkoxymethyl sites, or co-condensation of methylol and alkoxy
with nucleophilic sites (hydroxyl, carboxyl, amide, and thiol)
on other resins (Table I). Hydrolysis and de-methylolation can
also be significant contributing reactions. Reaction volatiles
are predominantly the alkylation alcohol, along with minor
concentrations of water and formaldehyde.

In general, amino resins with degrees of alkylation lower than
80-85% of available amine functional sites on a molar basis
respond well to weak acid catalysis (Table II). In many coating
formulations the carboxyl functionality of the other binders
present is sufficient to catalyze effective extents of cure.
These low degree of alkylation amino resins will also cure well
in buffered systems where basic additives or basic pigment
surface treatments might reduce catalyst effectiveness.
However, they have a higher viscosity than their more alkylated
counterparts, and a higher tendency for self-condensation.
Higher viscosity results in lower solids, higher emission
solvent based formulations, a current day problem.
Self-condensation can be an aid in developing film hardness,
particularly if high concentrations of amino resins are used in
the formulation. Yet, poorer film adhesion to metal substrates
and poorer film flexibility could also result. It is believed
that the rate of self-condensation versus crosslinking or

co-condensation is dependent not only on amino resin composition, but on numerous other formulation variables.

Aminos, in particular those based on melamine, benzoguanamine and glycoluril, with greater than 85% alkylation on a molar basis require strong acids for acceptable extents of cure, even at conventional bake temperatures of 125-150°C. However, they have a relatively lower viscosity and a low self-condensation tendency. These latter two properties make them desirable in low emission high solids solvent based formulations, and in high performance systems where effective crosslink density must be achieved at low amino resin levels.

For a more detailed discussion of the chemistry of amino resins please see references [1-4].

COMMERCIAL SIGNIFICANCE

Since 1940, surface coating amino resins in the United States have had an annual growth rate of 5-7% (Figure 4). But, their volume has consistently been less than 10% of the total amino resin useage. The highest useage has been in adhesives and laminating applications.

Coating application areas for amino resins have changed continually throughout their history. The United States market, 1988, is dominated by melamime resins, which in turn is dominated by those resins containing some methylation. Automotive (OEM) and general metals finishes, are the largest application areas, followed by wood finishes, container coatings, coil coatings, appliance finishes, and coatings for paper and plastic.

EARLY AMINO RESIN HISTORY

For some of us, who have spent most of our profession involved with amino resins in coatings, there is an abnormal reverence for them, a reverence which demands the belief that a triazine ring was first drawn by the Neanderthal in the Rhine valley in Germany. Yet, amino resin chemistry can only be traced as far back as 1884 (Table III), when the first mention of a urea formaldehyde reaction was by Tollens and Holzer, who isolated methylene urea [5]. In 1908 Einhorn further identified the chemistry of amino resins [6]. But, their commercial potential was apparently not recognized until 1920 when a patent issued to Hanns John [7] described one urea formaldehyde resin as a syrup which can be used as an adhesive, for casting dental work, or as a lacquer, a surface coating. The first commercial introduction, a molding compound, is attributed to Rossiter in England in 1926 [8]. The compound was made from a formaldehyde addition to an equimolar mixture of urea and thiourea, and was reinforsed with cellulose fiber. When commercialized, it was given the trademark BEETLE, supposedly, since it "beat all" others. Certainly, the initial commercial successes for amino resins were in molding compound and adhesive applications, and for the most part, the resins were unalkylated.

In the late 1920s, work at I.G. Farbenindustrie investigated the preparation of alkylated amino resins, which ultimately were of interest in "stoving lacquers" or surface coatings [9]. There were numerous other early contributors [10-15].

The first general utilization of amino resins in industrial surface coatings occurred in the late 1930s in both the U.S. and Europe, and involved a butylated urea formaldehyde resin as a modifying or co-reactive resin for solvent borne alkyds. Binder ratios of 85/15 to 70/30 on a solid weight basis of alkyd to amino resin were common. The combination with alkyds was necessary, for if used alone, the urea resin was brittle and lacked adhesion to metal substrates. However, the combination outperformed, in most film properties, the use of either resin alone [16]. Prior to amino resin's introduction, the baking schedule for an alkyd resin enamel was as long as two hours at 250°F. Higher bake temperatures discolored the alkyd film, and shorter schedules resulted in soft, tacky films. The combination of urea resin and alkyd shortened the schedule to 30 minutes at 250°F, and in some instances to 15 minutes at 300°F. Incidently, this short bake schedule probably marked the birth of the modern general industrial finish. The resulting cured films had higher film hardness, and superior chemical and abrasion resistance, but were poorer in exterior durability than the alkyd resin alone (Table IV).

Then, in 1940 entered melamine formaldehyde resins.

MELAMINE FORMALDEHYDE RESINS

Melamine itself was first sythesized on a laboratory scale by Justus Von Liebig in 1834 by heating together ammonium chloride and potassium thiocyanate. However, it was thought to be too difficult to prepare, too costly, and like many discoveries in their initial stages, judged to be of no practical value. It remained a laboratory curiosity for over a hundred years, until 1939 when it was prepared from cyanamide and commercially introduced in the United States.

In contrast to alkyd/butylated urea formaldehyde resins, alkyd/butylated melamine formaldehyde resins did provide exterior durability or ultra-violet radiation resistance, superior to the alkyd resin alone. The melamine resin also improved film color, hardness, chemical resistance, water resistance, heat and mar resistance to a greater extent than the urea resin. Although more expensive than urea resins, overall improved performance properties made melamine resins cost effective. Similar to urea formaldehyde resins, they could not be used alone to protect metal substrates due to film brittleness and lack of adhesion.

Amino resin useage in coatings increased during World War II. They might even have been considered a "war hero", as judged from the following early 1940's statement by one of the leading amino resin suppliers of the time [17],
> " (alkyd/amino resins)... used in
> paints by our arsenal of democracy to
> protect the United Nations war machine,
> (alkyd/amino resins) are synonymous with
> unexcelled surface protection under
> severe conditions of usage."

Over the next 20-25 years, butylated melamine formaldehyde
resins along with their urea resin counterparts developed into a
standard of the industry. Compositionally, these initial resins
were undermethylolated, and underalkylated, and with degrees of
polymerization of 4-6, were considered polymeric. Their
functional sites consisted of a combination of imino, free
methylol and butoxymethyl, with ratios varying from product to
product to emphasize a specific performance property such as
compatability, cure response or film appearance due to
differences in flow and leveling.

The present day dominant melamine resin or crosslinking agent is
based on hexamethoxymethylmelamine (HMMM), which differs
significantly in composition versus the above. HMMM was first
prepared by Gams, Widner,and Fisch [18] in 1941 by etherifying
hexamethylolmelamine with methanol using hydrochloric acid as
the alkylation catalyst.

The first commercial introduction in the United States was in
1960 [19]. It was not an immediate success. Compared to the
less alkylated, butylated, more polymeric melamine resins it was
competing with or attempting to replace, it was more monomeric,
and contained predominantly one type of functional site, methoxy
methyl (Table V). Initially it was more expensive than other
amino resins, and was available only as a waxy solid, since it
crystallized at room temperature. It required the addition of
a strong acid catalyst for acceptable cure response on
conventional 250-300°F schedules. When melted or dissolved in
solvent, it had a much lower viscosity than other amino resins,
and on a weight for weight substitution in the formulation, it
resulted in an unacceptably high solids system at application
viscosity. Too high solids was judged to be a disadvantage in
these "pre-compliance" years. Paint was sold and applied by
volume. Therefore the cost per gallon of paint, was
significantly higher when HMMM was used, and the application
techniques of the day could not consistently take advantage of
the greater coverage or mileage of the alkyd/HMMM formulation.
HMMM was a little ahead of its time.

Attempts were made to encourage its success by recommending
lower levels of HMMM addition to the formulation [20]. Since
it was highly functional, and had a low tendency to
self-condense, the recommendation was valid, and helped to
off-set some of the poor economics. But, HMMM still required
catalyst addition to the formulation, which was too demanding a
change for a conservative coatings industry. Paint
manufacturers were afraid that if the catalyst was mistakenly
omitted from a specific batch of paint, inadequate cure response
would result, and if the required small concentration of
catalyst was over-charged, poor paint storage stability would
occur. It was just too different to be considered for single
package systems.

But in early 1962, HMMM's first commercial surface coating
success occurred in an application well suited to it, coil
coatings, or prefinished metal. It was noted that when HMMM was
used in combination with linear or slightly branched polyester
resins, advantages in film flexibility or formability at high
film hardness and chemical resistance were possible (Table VI).
This unique performance property for an amino resin was
attributed to HMMM's high extent of alkylation, which resulted

in an effective, low self-condensation crosslinking agent. This enabled lower HMMM concentrations in the formulation than when other amino resins were used. Moreover, HMMM's high monomer content contributed to better distribution of the HMMM throughout the film network. Hence higher film formability was possible, a critical property in post-forming coated metal. HMMM did not add to the film formability of the polyester resin, it just did not detract from it as other amino resins had. The performance versus economics relationship of polyester/HMMM systems propelled their growth in coil coatings in the '60s and '70s, a growth which continues.

At about the same time, the early 1960s, liquid versions of HMMM became available [19]. They were not as highly methylated as the solid, crystallized HMMM, and were slightly more polymeric, having degrees of polymerization of 1.5-1.7. These changes destroyed the crystallization potential, assured fluidity at temperatures above 0°F, and obviously improved ease of handling. Furthermore, performance properties for the most part were not compromised.

Also, paint manufacturers were becoming accustomed to catalyst addition to single package systems, and to additive addition in general for that matter. It became less of a concern, and a reasonable "trade-off" for the added performance benefits of HMMM.

The second commercial coatings success for HMMM was in anodic electrodeposition in 1965. In addition to the amino resin, these water borne formulations contained carboxyl functional backbone resins, primarily acrylics, solubilized with base. Binder ratios ranged from 80/20 to 70/30 acrylic/HMMM, on a solid weight basis. Here again, HMMM was well suited, with a minor compositional change. This change was to ethylate, or butylate, as well as methylate the melamine crosslinker, in order to increase its hydrophobicity. This was the first U.S. commercial introduction of a "mixed ether" melamine resin. The increased hydrophobicity enhanced the uncharged amino resin's ability to migrate to the anode within the charged acrylic backbone resin aggregate particle, and to co-deposit with the acrylic resin at the anode. Unmodified HMMM was too water soluble to co-deposit at the anode at the desired concentration. The standard butylated, highly polymeric melamine resins were too hydrophobic to prevent precipitation from the E-coat bath, and too chemically unstable in the bath (pH of 7.0 to 8.0) to be acceptable in the application. High extents of alkylation and high monomer content were desired, both of which were provided by the modified HMMMs.

Success in coil coating and anodic E-coat applications were encouraging but were only an indication of the extent of future accomplishments. In 1966, with the introduction of California's Rule 66, interest in HMMM and methylated melamine resins in general increased. Rule 66 significantly limited the organic solvents emitted into the atmosphere. Suitable high solids backbone resins, alkyds, polyesters, acrylics, which would reduce the solvent content of the coating system, were limited in selection at the time, and therefore water borne sytems were chosen to comply with Rule 66. Methylated amino resins were well suited (Table VII). Compositionally, most of these systems were similar to the anodic E-coat systems discussed previously, except in most instances the acrylic contained hydroxyl as well as carboxyl functionality, and an acid catalyst was added to the formulation. There was also some emulsion technology available, but again it too was crosslinked with methylated melamine resins.

HMMM type crosslinking agents were not the only amino resins used. Partially methylated, slightly more polymeric melamine resins were and still are common. Their use enabled lower bake temperatures (Table VII). Their major deficiency was storage stability of formulated systems, which was poorer than those containing HMMM types. Furthermore, amine stabilizers or solubilizers had to be carefully chosen for optimum stability. Tertiary amines were recommended. Yet, both the HMMM and the partially methylated product line increased in number of product introductions in response to this new anti-pollution demand.

A significant improvement over the partially methylated melamine resins were the high imino methylated melamines introduced in 1976 [19]. Their performance characteristics were similar to the older partially methylateds, except that they were faster curing, lost less weight on curing, had lower free formaldehyde, and emitted less formaldehyde on baking. The technical reasons behind the above were based on the reaction mechanism of partially alkylated melamine resins. It was found that the initial step during cure was demethylolation of the free methylol site, resulting in an imino site and formaldehyde generation. By designing a resin with extents of alkylation similar to the partially alkylateds, but with low free methylol and high imino sites, the initial demethylolation reaction step need not occur. Essentially, this initial step was carried out in the preparation kettle and not in the film during baking. Hence, these resins provided faster cure response and lower formaldehyde generation. Formaldehyde content, odor, and emissions, which were never a concern in the first thirty years of amino resins commercial existance, were becoming a concern in the late '70's. The high imino melamines addressed this concern.

Rule 66 did not end in California. Over the next ten years, compliance demands on organic emissions spread nationally and affected the majority of industrial applications. Water borne coatings were not the only solution, and in time not the best solution. In the early to mid 1970s, high solids systems began to dominate. Water borne systems had application problems, and in many instances required environmental humidity control. They also necessitated slightly higher bake temperatures than the older solvent based systems, primarily because of the need to evaporate or overcome the inhibiting effect of the amine added to the formulation to maintain the pH above 8 and assure adequate single package stability. With the development of lower molecular weight alkyds, polyesters and acrylic resins, so-called "high solids systems", crosslinked with high solids amino resins, became common place. Many of the high solids low viscosity melamine resins already existed. Certainly, in this respect HMMM's time had come. High solids systems were not without their own application problems, but they were closer in handling characteristics and bake requirements to the conventional systems they were replacing than were water borne systems.

In response to high solids' demanding application requirements, and the specific needs of the automotive topcoat industry, the highly alkylated, highly monomeric, mixed ether melamine resins, products similar in composition to those used in anodic electrodeposition, increased in interest. These were HMMM type crosslinkers, except that they contained two alkylation alcohols, the most common combinations being either methanol/n-butanol or methanol/isobutanol (Table VIII). In the acrylic systems used in automotive topcoats, they had superior recoat adhesion versus HMMM, had superior flow and leveling, higher electrostatic spray resistivity, and were acceptable in durability, resistance to ulta-violet radiation, and in their reduction of organic emissions.

UREA FORMALDEHYDE RESINS

You might think that we have abandoned the history of urea formaldehyde resins. Not entirely so. During the time when melamine resins were evolving into high solids crosslinking agents, urea formaldehyde resins were responding likewise.

More monomeric, highly methylated urea resins appeared commercially in 1969 [19]. They were targeted for both solvent and water based general industrial metal and wood finishes to comply with anti-pollution regulations. They were expected to be economical, and to address those interior end-use applications where ultra-violet radiation resistance was not necessary. Disappointingly, commercial success was less than predicted. Their solubility in a broad range of organic solvents was limited, which restricted their use primarily to water borne systems. Over the next several years their film performance/ecomomics versus methylated melamine crosslinking agents became less attractive, as the relative price of melamine crosslinking agents, those based on HMMM in particular, decreased. In 1988, the price of a typical highly methylated, high solids urea formaldehyde resin was 85-90% of its melamine counterpart.

A high solids butylated urea formaldehyde resin [19] was introduced in 1974, unique due to its low molecular weight and high extent of butylation. It was introduced in response to the increasing low emission demands. Its high extent of butylation necessitated the addition of a strong acid catalyst to the formulation for acceptable cure response at 120-140°C bake temperatures. This was judged to be a disadvantage. However, similar to highly alkylated melamine resins, the resin did have a slight advantage in film flexibility versus more polymeric less alkylated urea resins. It has found commercial success, particularly when used in combination with other amino resins.

But in general, urea resin useage in coatings has decreased over the last 10 years in favor of the superior performing high solids melamine resins. Yet, in 1989, urea formaldehyde resins remain the dominant amino technology in wood furniture applications. The primary technical reason is their faster cure response at low bake temperatures versus other amino resins. In highly catalyzed, two package formulations, where the acid catalyst is added to the formulation just prior to use, they cure well at temperatures ranging from 25°C to 70°C. Contrary to use on metal substrates, the high catalyst concentrations

required for cure at these temperatures have only a slight
effect on water resistance properties in wood applications.

BENZOGUANAMINE FORMALDEHYDE RESINS

Of course there are other amino resins. Those based on
benzoguanamine were introduced in the late 1940's [21]. Higher
solids, high alkylation, high monomer content versions appeared
in 1967 [19], specifically in answer to the same demands as
their melamine counterparts.

In respect to performance versus melamine resins, benzoguanamine
resins are slower curing, are not as stable to ultraviolet
radiation, and were initially and still are more expensive.
But, they are generally more organo-soluble and resin compatible
than their melamine counterparts, and they have superior
hydrolysis resistance under basic conditions, commonly referred
to as detergent resistance. For this reason, their major
success has been in appliance finishes.

They also have a lower functionality per molecule than melamine
resins. Therefore, highly alkylated versions of benzoguanamine,
those with a low tendency for self-condensation, have excellent
film flexibility, a property desirable in metal prime coats, and
in applications where post forming of the metal after coating is
desired.

Benzoguanamine resins' high cost, more than twice as high as
similar melamine resins, and their poor U.V. stability, have
limited their usage to those non-exterior applications where
melamine formaldehyde resins have inadequate performance.

GLYCOLURIL FORMALDEHYDE RESINS

Another family of amino resins of technical and commercial
interest are those based on glycoluril. The first U.S.
commercial introduction to the coatings industry occurred in
1975-76, when both unalkylated methylol glycoluril, and highly
alkylated versions became available [19].

The unalkylated version is soluble primarily in water, and
therefore suitable only for water borne systems. But, it is
unique. Unlike other amino resins, it is relatively stable in
water under acidic (pH 4-6) conditions. This allows for very
rapid cure response versus melamine and urea resins, and is the
key to their success [22]. The glycoluril systems do not
require amine addition to adjust the pH to the basic side and
therefore achieve formulation stability. For all amino resins
in water borne systems, the lower the pH, the faster the cure
response.

The highly alkylated glycoluril crosslinkers perform similarly
to their HMMM counterparts. They require a strong acid catalyst
for cure on a general industrial schedule, are high in monomer
content and low in viscosity, have a low self-condensation
tendency and have ultra-violet radiation resistance. Their film
flexibility and formability in polyesters and adhesion to metal
substrates is superior to melamine crosslinkers, which therefore
has led to their success in coil coatings. Yet, they are also
more sensitive to cure response buffering of basic pigment

surface treatments, and basic additives. Therefore, care must
be taken in formulation component selection. Also, they were,
when introduced, and still are more expensive the similar
melamine chemistry, a fact which has hindered commercial
success.

One significant advantage of glycoluril amino resins is that on
curing they emit less formaldehyde than their melamine and urea
counterparts. This property alone predicts increased future
usage for glycoluril resins.

Also, tetramethoxymethyl glocoluril is a solid, melting at
approximately 100°C, and as such has potential as a crosslinking
agent for powder coatings.

MISCELLANEOUS AMINO RESINS

There were a number of miscellaneous amino resins introduced
over the last 10-15 years, which continue to have technical and
commercial interest.

Methylated melamine and benzoguanamine formaldehyde crosslinking
agents containing carboxyl functionality along with alkoxyl
functionality appeared in 1976 [23]. Advantages in film
adhesion to metal substrates, and corrosion resistance were
identified. However, the crosslinking agents were recommended
primarily for dark pigmented coatings, since they could form
yellow and pink complexes with heavy metals such as iron or
zinc.

Several attempts, as early as 1976, were made to introduce
blends of a polyol and amino resin [24]. A number of polyols
were considered, and in general, performance enhancement of
selected properties was noted. However, in most instances the
major value of a blend was one of convience. The paint chemist
had only to add one product to the formulation, rather than
separate additions of amino resin and polyol. The early blends
had little commercial success, primarily do to too narrow a
formulation latitude and poor economics. The most commercially
succcessful blend appeared in the early 1980s. It was based on
an oligomeric styrene allyl alcohol copolymer, and a methylated
melamine formaldehyde resin. A family of products was
commercialized. They did have a cure response and corrosion
resistance advantage versus the melamine resin alone, but
depending upon the level of polyol and the specific formulation,
could also result in lower film flexibility, reduced adhesion to
metal substrates, and decreased U.V. resistance. They again
provided the convience of polyol addition to the formulation.

Also appearing in 1981 were acrylated melamine formaldehyde
resins [25]. They possessed the ability to not only cure by the
traditional acid catalyzed condensation chemistry route, but
also by either thermal initiated radical cure, or oxidative
cure. Advantages in film hardness and chemical resistance,
especially in low temperature cure applications, were claimed.
The importance of these crosslinking agents to the evolution of
amino resins has yet to be determined.

PROBLEMS TO BE ADDRESSED

In order for amino resins to be the dominant coatings
crosslinking technology in the 21st century the following three
areas of concern must be addressed,

- FORMALDEHYDE EMISSION   Improvements not only in
the residual "free" formaldehye content of amino resins are
necessary, but also in the formaldehdye emission on curing. The
elimination of all formaldehyde is the ultimate objective.

- VOLATILE EMISSION (elimination of reaction
volatiles) The development of low viscosity amino crosslinking
agents has already been demonstrated, and viscosities in the
200-300 centipoise range at non-volatiles before reaction of
100% are possible. What needs to be addressed is the further
reduction of reaction volatiles on curing. Reaction volatiles
from amino resins' condensation reaction can add an additional
0.2 to 0.4 pounds per gallon of volatile organic content to an
industrial coating.

- FILM PERFORMANCE (corrosion resistance,
crosslink stability) This is a particular concern in systems
cured at temperatures less than 100°C. As noted previously, the
high acid catalyst concentrations required at this temperature,
have a negative effect on film performance. Improvements in
this area might be realized not only from amino resin design,
but also from more effective catalysts for curing amino resins.

The major amino resin suppliers are pursuing the above concerns,
and some progress is being made. It will be interesting to see
how a "mature" technology will continue to respond.

REFERENCES
[1] Paul S., SURFACE COATINGS SCIENCE AND
TECHNOLOGY, pp. 167-199, Wiley and Sons, New York, 1985
[2] Blank, W.J., J. COATINGS TECHNOLOGY, 1979,
51, 656.
[3] Blank, W.J., J. COATINGS TECHNOLOGY, 1982,
54, 26.
[4] Santer, J.O., PROGRESS ORGANIC COATINGS,
1984, 12, 309.
[5] Tollens B. (quoted by) Ber. dtsch. chem.
Ges., 17, 659 (1884)
[6] Eihorn, A. and Hamberger, A., BER. DTSCH.
CHEM. GES., 41, 24.
[7] Hans John, U.S. Patent 1,355,834
[8] British patents 248,477; 258,950; 266,028
(1925) E.C. Rossiter
[9] I.G. Farbenindustrie, A.G., B.PP 260,253
(1925), 261,029 (1925), 262,818 (1925), 301,696 (1926), 319,251
(1928)
[10] Goldschmidt, H. and Neuss, O., British
Patents 187,605 (1922), 202,651 (1923), 208,761 (1922).
[11] Pollak,F. British Patents 171,096 (1921),
181,014 (1922), 193,420 (1923), 201,906 (1923), 206,512 (1923),
213,567 (1923), 238,904 (1924), 240,840 (1924), 248,729 (1925).

174

[12] Ellis, C., US Patents 1,482,357 (1922), 1,482,358 (1922), 1,536,881 (1922), 1,536,882 (1922).
[13] Ripper, K., US Patent 1,460,606 (1923)
[14] Henkel, British Patent 455,008 (1935)
[15] Hentirch, W. and Kohler, R., German Patent 647,303 (1937)
[16] Norris, W. and Bacon, J., Official Digest, Fed. of Paint and Varnish, Oct. (1948)
[17] American Cyanamid Company, Technical Literature, (1943)
[18] A. Gams, G. Widner, W. Fisch; Helv. Chim. Acta; 24E, 302 (1941)
[19] American Cyanamid Company product. "High Solids Amino Crosslinking Agent", CRT 557.
[20] Cymel® 300 Resin, Technical Brochure CRT70, American Cyanamid Co.
[21] Rohm and Haas Product Introduction
[22] Parekh, G., J. COATINGS TECHNOLOGY, 1979, 51, 658
[23] Cymel® 1141 Resin, Cymel® 1125 Resin Technical Literature, American Cyanamid Company.
[24] Resimene® 797 Resin Technical Literature, Monsanto Co.
[25] AM-300, AM-325 Technical Literature, Monsanto Co.

FIGURE 1

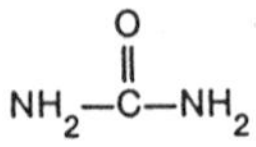

UREA

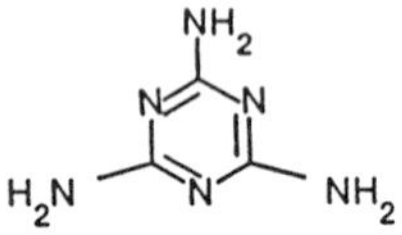

MELAMINE

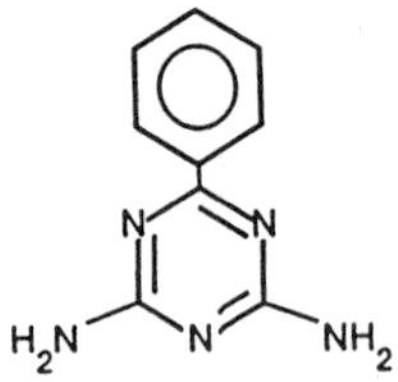

BENZOGUANAMINE

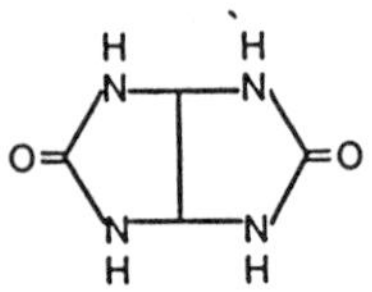

GLYCOLURIL

FIGURE 2

MELAMINE FORMALDEHYDE
RESIN PREPARATION

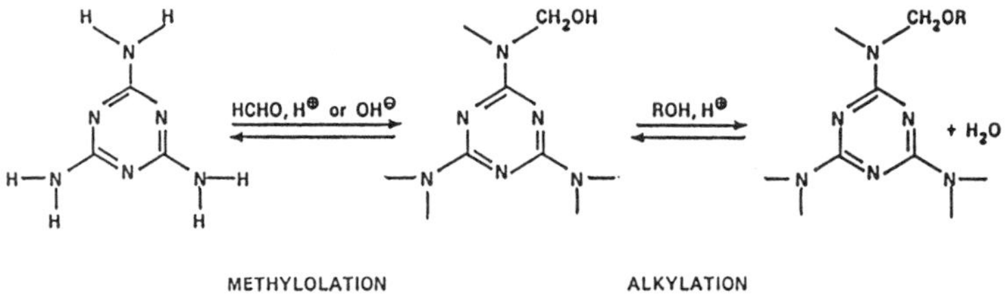

METHYLOLATION   ALKYLATION

FIGURE 3

ALKYLATED MELAMINE FORMALDEHYDE RESIN
FOR SURFACE COATINGS

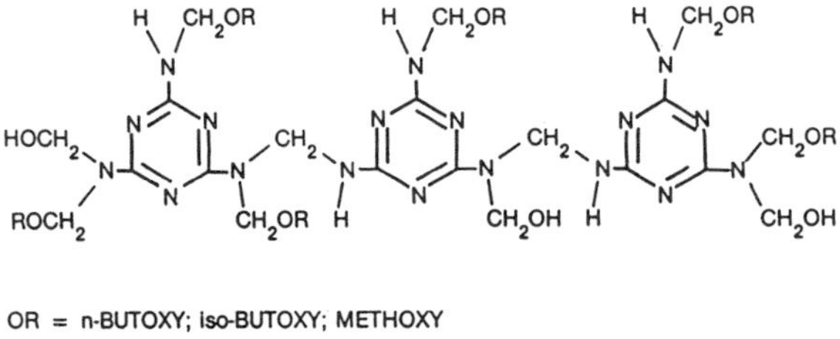

OR = n-BUTOXY; iso-BUTOXY; METHOXY

TABLE I

REACTIONS OF AMINO RESIN

CROSSLINKING (CO-CONDENSATION)

$$-NCH_2OR + R'OH \rightleftharpoons -NCH_2OR' + ROH$$

$$-NCH_2OH + R'OH \rightleftharpoons -NCH_2OR' + H_2O$$

SELF-CONDENSATION

$$-NCH_2OH + -NH \rightleftharpoons -NCH_2N- + H_2O$$

$$2 -NCH_2OH \rightleftharpoons -NCH_2OCH_2N- + H_2O$$

$$-NCH_2OR + -NH \rightleftharpoons -NCH_2N- + ROH$$

$$2 -NCH_2OR \rightleftharpoons -NCH_2N- + ROCH_2OR$$

DEMETHYLOLATION

$$-NCH_2OH \overset{\triangle}{\rightleftharpoons} -NH + HCHO$$

$$-NCH_2OCH_2N- \overset{\triangle}{\rightleftharpoons} -NCH_2N- + HCHO$$

HYDROLYSIS

$$-NCH_2OR + HOH \rightleftharpoons -NCH_2OH + ROH$$

FIGURE 4

AMINO RESIN USAGE
SURFACE COATINGS
(UNITED STATES)

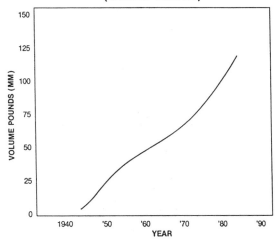

REFERENCE - U. S. TARIFF COMMISSION

U. S. INTERNATIONAL TRADE COMMISSION

TABLE II

MELAMINE, BENZOGUANAMINE RESINS
EFFECT OF ALKYLATION ON
PERFORMANCE PROPERTIES

	EXTENT OF ALKYLATION (MOLE %)	
	< 80—85	< 80—85
CATALYST REQUIRED [1]	pK = < 5	pK = <2
RESIN VISCOSITY	3000—6000 cps @ 85—90% NON-VOLATILE	300 cps @ 100% NON-VOLATILE
SELF-CONDENSATION TENDENCY	HIGH	LOW

[1] FOR ACCEPTABLE CURE RESPONSE AT 125°C

178

TABLE III

AMINO RESIN HISTORY

1884		METHYLENE UREA REACTION IDENTIFIED
1920	U/F	COMMERCIAL POTENTIAL IDENTIFIED
1926	U/F	COMMERCIALLIZATION (MOLDING COMPOUND)
1936-38	U/F	COMMERCIALLIZATION (COATINGS)
1939-40	M/F	COMMERCIALLIZATION (COATINGS)
1941	HMMM	FIRST PREPARED
1960	HMMM	COMMERCIALLIZED
1962	HMMM	SUCCESS IN COIL COATINGS
1965	MODIFIED HMMM	SUCCESS IN ANODIC E-COAT
1966	METHYLATED M/F (RULE 66 WATER BORNE SUCCESS)	
1967	HIGHLY ALKYLATED BG/F COMMERCIALIZATION	
1969	METHYLATED U/F COMMERCIALLIZATION	
1970-74	M/F SUCCESS IN HIGH SOLIDS COATINGS	
1975-76	GU/F COMMERCIALLIZATION (COATINGS)	
1976	HIGH IMINO M/F COMMERCIALIZATION	
1976	CARBOXYL FUNCTIONAL M/F COMMERCIALLIZATION	
1976	M/F RESIN POLYOL BLENDS COMMERCIALLIZATION	
1981	ACRYLATED M/F COMMERCIALLIZATION	

TABLE IV

FILM PROPERTIES OF
ALKYD/AMINO RESIN COATINGS

	ALKYD	ALKYD + UREA RESIN	ALKYD + MELAMINE RESIN
CURE RESPONSE	●	O	O
FILM HARDNESS	◓	O	⊗
CHEMICAL RESISTANCE	◓	O	⊗
WATER RESISTANCE	◓	O	⊗
FILM ADHESION	⊗	O	O
FILM FLEXIBILITY	O	◓	◓
COLOR (INITIAL)	O	⊗	⊗
(OVERBAKE)	◓	O	⊗
EXTERIOR DURABILITY	O	●	⊗

● POOR
◓ FAIR
O GOOD
⊗ EXCELLENT

TABLE V

HEXAMETHOXYMETHYL MELAMINE
COMMERCIAL PRODUCTION *

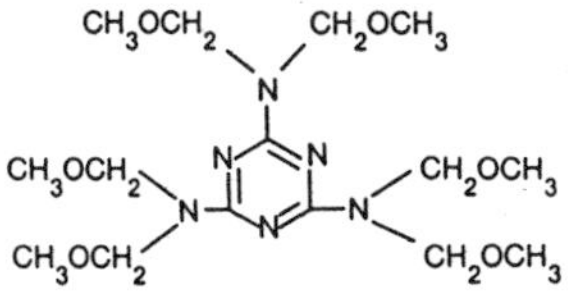

$$CH_3OCH_2 \quad CH_2OCH_3$$

(triazine ring structure with substituents: CH_3OCH_2, CH_3OCH_2, CH_2OCH_3, CH_2OCH_3 on nitrogen atoms)

M/F/METHANOL	1/6.0/5.4
DEGREE OF POLYMERIZATION	1.3—1.4
NON-VOLATILE [1]	>98%
VISCOSITY	WAXY SOLID
	1000—1500 cps (melt; 25°C)

LOW SELF-CONDENSATION TENDENCY

STRONG ACID CATALYST REQUIRED

* AMERICAN CYANAMID COMPANY

[1] DETERMINED AT 45°C FOR 45 MINUTES

TABLE VI

HEXAMETHOXYMETHYL MELAMINE IN
SOLVENT BASED COIL COATINGS

FORMULATION PARAMETERS

POLYESTER RESIN/HMMM	90/10 - 80/20
ACID CATALYST; PTSA	0.1 - 0.4% ON BINDER
SUBSTRATE	PRE-TREATED CRS
BAKE SCHEDULE	500-800°F FOR 45-90 SEC.
FILM THICKNESS	0.8 - 1.2 MILS

PERFORMANCE PROPERTIES

PENCIL HARDNESS	F - 2H
TUKON HARDNESS	6 - 10
REVERSE IMPACT STRENGTH	100 - 160 INCH POUNDS
T-BEND	T0 - T2
MEK RUBS	100 - 200+

180

TABLE VII

METHYLATED MELAMINE FORMALDEHYDE RESINS IN WATER BORNE COATINGS

FORMULATION PARAMETERS

BACKBONE RESIN [1]/AMINO RESIN		80/20—60/40
NEUTRALIZING AMINE		DMAE, AMP
pH		8.0—9.0

	HMMM	MMM
CATALYST	PTSA	—
CURE TEMPERATURE REQUIRED [2]	300—350°F	250—275°F
STABILITY, 23°C	12+ MONTHS	3—4 MONTHS
FILM FLEXIBILITY [3]	60—120	< 20

[1] ACRYLIC, POLYESTER, ALKYD RESIN CARBOXY, HYDROXY FUNCTIONALITY
[2] 20 MINUTE BAKE, 1 MIL DRY FILM
[3] REVERSE IMPACT, INCH-POUNDS; TREATED CRS SUBSTRATE; POLYESTER RESIN BACKBONE

TABLE VIII

HIGHLY ALKYLATED MIXED ETHER MELAMINE RESINS PERFORMANCE IN HIGH SOLIDS SYSTEMS

	MBMM	HMMM
CURE RESPONSE	◐	○
LOW VOC LOW VISCOSITY HIGH WEIGHT RETENTION	◓	◓
FILM FLEXIBILITY	○	○
INTERCOAT ADHESION	○	◐
CORROSION RESISTANCE	◐	◐
FLOW LEVELING	○	◐
ELECTROSTATIC SPRAY RESISTIVITY	○	◐
EXTERIOR DURABILITY	◐	○

○ EXCELLENT
◐ ACCEPTABLE
● POOR

MBMM = METHOXY BUTOXY METHYL MELAMINE
HMMM = HEXAMETHOXY METHYL MELAMINE

History of Phenolic Resin Coatings

Raymond B. Seymour
Department of Polymer Science
University of Southern Mississippi
Hattiesburg, MS 39406-0076

Adolph Baeyer produced an undesired solid when he reacted phenol and acetaldehyde in 1872. Kleeburg repeated this synthesis using formaldehyde instead of acetaldehyde in 1891 but had little interest in this noncrystalline insoluble product. However, Lederer and Manasse isolated the intermediate methylolphenol in 1894 and Arthur Smith patented the solid resinous product in 1899.

In 1905, Leo Baekeland utilized the results of previous investigations and his own ingenuity to produce useful resinous products under controlled reaction conditions. Unlike his predecessors, he recognized that as shown in the following figure phenol was a trifunctional reactant i.e., products could be produced by reaction in the 2, 4 and 6 positions with reactants, such as formaldehyde.

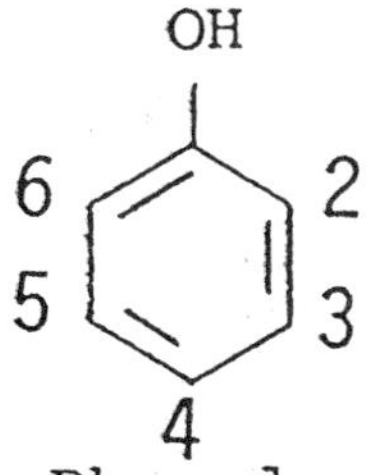

Phenol

Accordingly, he used mild alkaline reaction conditions to condense aqueous formaldehyde with phenol to produce a linear resole or A stage resin as shown below; which could be crosslinked in the presence of weak or strong acids. A segment of the repeating unit in the C stage or in infusible resole resin is also shown below.

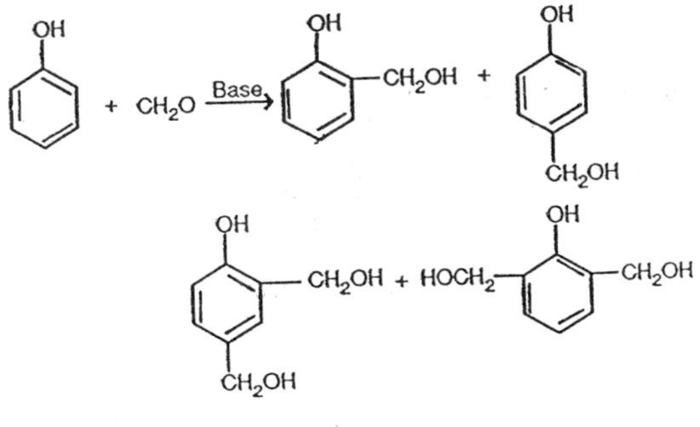

Published 1990 by Elsevier Science Publishing Co., Inc.
Organic Coatings: Their Origin and Development
R.B. Seymour and H.F. Mark, Editors

Resole or A-stage resin

HCHO

Cross-linked infusible resin

Baekeland also produced a linear polymer by condensing phenol with an insufficient amount of formaldehyde under acidic conditions and this novolac resin was then converted to an infusible solid by the addition of more formaldehyde. The formaldehyde was usually supplied by the thermal decomposition of hexamethylenetetramine $((CH_2)_6N_4)$. However, the novolac resin was used primarily for molding rather than as a coating resin.

Baekeland reviewed the history of the development of phenolic resin (PF) in an article in the first volume of Industrial and Engineering Chemistry in which he criticized his contemporaries, such as Kleeburg, whose myopic view was restricted to crystalline and distillable organic compounds. Baekeland's principal interest was in moldable plastics which could compete with shellac but he did investigate resole coatings for pipe and containers. In spite of his belief that phenolic polymers were ring structures, consisting of six phenylmethylol repeating units, he was able to produce a useful commercial plastic which he named Bakelite.

Leo Hendrich Baekeland (1863-1944), who has been called "the father of the plastics industry" was born in Ghent, Belgium on November 14, 1863. He received a Ph.D degree from the University of Ghent in 1884. After serving as a Professor of Chemistry at the University of Bruges, Leo married Celine Swarts, the daughter of his research professor at Ghent and utilized his three year traveling fellowship to honeymoon at Columbia University in New York.

He accepted Professor Chandler's invitation to remain in the USA and was soon recognized as the inventor of Velox photographic paper for which he was awarded one million dollars by George Eastman. This invention was followed by the Townsend Electrolytic cell for Hooker Chemical in 1904. This cell was named after his coinventor , who also served as his patent attorney. These patents and that of Baekeland were numbered among his over 400 patents. He was the first recipient of the Chandler medal and served as president of the American Chemical Society in 1924.

In spite of his success with phenolic adhesives and molding resins, Leo was unsuccessful in his attempts to incorporate phenolic resins in oleoresinous varnishes. This problem was solved by L. Blumer, who produced a soluble resin called Laccain by the condensation of phenol and formaldehyde in the presence of organic acids in 1902 and by

Kurt Albert who produced compatible resins by heating PF with rosin and glycerol in 1926. Rosin--modified PF resins continue to be used in the production of oleoresinous varnishes.

Oil soluble (linear) PF resins were produced in 1928 by replacing phenol by ortho or para--substituted phenols. In 1980, Fry patented this bisphenol A--formaldehyde resin which could be used both as an ethanol solution and as a powder coating. Novotny produced commercial phenolic resins by using furfural in place of formaldehyde. Lebach patented the acid curing of PF resole resins in 1907.

Protective coatings count for less that 1 percent of the 1.2 million tons of PF resins produced annually in the US. However, over 75 thousand tons of PF resins are used for coating abrasives and over 600 thousand tons of these resins are used annually in the US as adhesives for plywood.

PF coatings, such as heat reactive phenolic and substituted phenolic resins, based on p-t-butylphenol and bisphenol A are produced by the alkaline condensation of the reactants with formaldehyde. Heat resistant phenolic coatings are produced by the acid condensation of formaldehyde with phenol or substituted phenols. The major use of the pioneer ethanolic solutions of PF resins was as a baked coating for brass bedsteads. Bisphenol A is a multifunctional reactant which produces coatings which are more flexible than PF coatings. 4, 4-bis (4-hydroxylphenyl) pentanoic acid resins are multifunctional resins which have been used to protect oil well drill pipe, metal drums and backing for copper printing plates. Micaceous iron oxide-pigmented tung oil-modified resole PF resins are used as spar varnishes and for the protection of structural steel.

Substituted heat reactive PF resins have been used in the formulation of neoprene contact adhesives. Commercial oil soluble PF coatings (Albertol, Amberol), which are now available from Reichold, were introduced in 1924. The heat resistance of this type of PF coating was improved by Turkington who used p-phenylphenol as the reactant in 1929. Patents on coatings for other para-substituted phenols were issued to Honel and Turkington in the mid 1930's. It is customary to use these soluble phenolic resins as high temperature blends with drying oils, such as tung oil. However, Richardson developed a cold mixing technique in 1954. Water-borne PF resins were patented by Harding in 1974. PF resins dispersed in volatile aliphatic solvents were introduced by Fry in 1978.

Since PF resins were developed before scientists recognized the importance of polymer science, many of the innovations were based on empirical investigations. In spite of their commercial success many improvements in PF coatings will be made in the future as polymer chemist transfer known information from other polymer systems to PF resins.

References
Baekeland, L. H., "Biographical Memoirs Vol. 124," Natural Academy of Science, Columbia University Press, New York, 1968.

Baekeland, L. H., "History of Phenolic Resins," Ind Eng Chem 1 149 (1909).

Baeyer, A., "Reaction of Phenol and Acetaldehyde," Ber 5 1095 (1871).

Barth, B. P., Chapt. 23 in "Handbook of Adhesives," I Skeist ed., Van Nostrand Reinhold Co., New York, 1962.

Bender, H. L., "Phenolic Resins," Modern Plastics 30 136 (1953).

184

Boyd, W., Merriam, C., Fry, J., "Phenolic Resin Coatings," Chapt. 47 in "Applied Polymer Science," R. W. Tess, G. W. Poehlein eds., American Chemical Society, Washington, D. C., 1985.

Brode, G. L., "Phenolic Resins" in Kirk Othmer Encyclopedia of Chemical Technology," John Wiley, New York, 1985.

Carswell, T. S., "Phenoplasts," Interscience, New York, 1947.

Cohoe, W. P., "Biography of Baekeland," Chem Eng News 23 228 (10 Feb 1945).

Fry, J. S., "Bisphenol A--Formaldehyde Coating," US Pat. 4 180 732 (1981).

Fry, J. S., "PF Dispersion Resins," US Pat. 4 124 5 54 (1978).

Greth, A., "Albertol," Kunststoffe 31 345 (1914).

Harding, J., "Water--borne PF Resins" US Pat. 3 823 103 (1974).

Honel, H. p--t.--Butylphenol-Formaldehyde Coatings," US Pat. 1 996, 960 (1935).

Knopf, A., Scheib, W., "Chemistry and Applications of Phenolic Resins," Springer-Verlag, New York, 1979.
Knopf, P. L., "Phenolic Resins," Springer-Verlag, Berlin, 1985.

Kopf, P. W., "Phenolic Resins," in Encyclopedia of Polymer Science and Engineering Vol. 11, John Wiley, New York, 1988.

Kopperl, S. J., "Biography of Baekeland" in "American Chemists and Chemical Engineers."

Ledder, L., "Methylolphenol" J. Probt Chem. 50 (2) 223 (1894).

W. D. Miles ed., American Chemical Society, Washington, D. C. 1976.

Manasse, O., "Methylolphenol" Ber 27 2409 (1894).

Mann, A. A., Fontobert, E., "Oil Soluble PF Coatings," US Pat. 1 623 901 (1917).

Martin, R. W., "The Chemistry of Phenolic Resins," John Wiley, New York, 1956.

Megson, V. J. L., "Phenolic Resin Chemistry," Academic Press, New York, 1958.

Richardson, S. H., "Cold Blend of Drying Oils and PF," Paint Varnish Prod. 8 (1955).

Sandler, S. R., Caro, W., "Phenol Aldehyde Condensations," Academic Press, New York, 1977.

Sandler, S. R., Caro, W., "Phenol Aldehyde Condensations," Academic Press, New York, 1979.

Seymour, R. B., "History of Thermoses," Chapt. 7 in "History of Polymer Science and Technology," Marcel Dekker, New York, 1982.

Seymour, R. B., Chapt. 7 in "Pioneer in Polymer Science," H. F. Mark, L. Pauling, C. S. Marvel et al eds., Kluwer Academic Publishers, Dordrecht, The Netherlands, 1989.

185

Smith, A., "Phenolic Resins," US Pat. 643,012 (1900).

Turkington, V. H., Butler, H., "p--Phenylphenol--Formaldehyde Coatings," US Pat. 2 017 877 (1935).

Turkington, V. H., Butler, H., "p--Phenylphenol--Formaldehyde Coatings," US Pat. 2 173 346 (1939).

HISTORY OF DIENE/STYRENE BLOCK COPOLYMERS

HENRY L. HSIEH
Phillips Petroleum Company, Research and Development
Bartlesville, OK 74003

ABSTRACT

A block copolymer, or block polymer, is defined as a polymer comprising molecules in which there is a linear arrangement of blocks. A block is defined as a portion of a polymer molecule in which the monomeric units have at least one constitutional or configurational feature absent from the adjacent portions. Block copolymers having branched structures are referred to as star-block or radial block copolymers. The literature, especially patents, contain numerous references to making block copolymers in free radical systems. However, for one or more reasons the methods described are not attractive.

M. Szwarc in 1956 described the use of sodium-naphthalene complex and incremental monomer addition for the synthesis of block copolymers. The term "living polymer" was coined. About the same time, there was a growing awareness, as first indicated by K. Ziegler in the early 1930s that polymerizations initiated with organolithium compounds proceed by step-wise addition and can be conducted with negligible termination or transfer steps. The developments for making thermoplastic elastomers based on ordered copolymers of diene and styrene is the subject of this article.

INTRODUCTION

Block Copolymers

Copolymers are polymer products formed by polymerizing two monomers, M_1 and M_2, into the same polymer chains. The copolymer chains may be a mixture of units of the two monomers with both randomly distributed throughout the polymer chain. Alternatively, individual segments of the chain may contain only one monomer unit followed by a segment containing only the other monomer unit. The first is a random copolymer whereas the second is a type of block copolymer. Molecules of these two types of polymers may be visualized as follows:

$-[A\text{-}A\text{-}B\text{-}A\text{-}A\text{-}B\text{-}A\text{-}B]_x-$ $-[A\text{-}A\text{-}A\text{-}A\text{-}A]_x\text{-}[B\text{-}B\text{-}B]_x-$

A/B Random Copolymer Molecule   A-B Block Copolymer Molecule

The difference between the two can be seen by examination of a section of the polymer chain. In truly random copolymer molecules, any section of the chain containing >10 monomer units would show a constant composition of monomer unit A and B reflecting the ratio of the two monomers in the original charge. By contrast, if the A and B units of the polymer chain are arranged in segments of all A units connected to segments of all B units, then we have a simple A-B block copolymer

Published 1990 by Elsevier Science Publishing Co., Inc.
Organic Coatings: Their Origin and Development
R.B. Seymour and H.F. Mark, Editors

molecule. An examination of any section of the chain (except that containing the A-B joining bond) would show either all A or all B units depending on which part was examined. An example of this would be a copolymer consisting of a block of polystyrene (A-A-A-A-A) connected to a block of polybutadiene (B-B-B). For simplicity, this polymer would be designated as an A-B block copolymer recognizing that A and B now refer to blocks of monomer units rather than to single monomer units. The dash between the A and B says the two segments are connected to each other with little or no mingling of the two. Block copolymers can be a simple two block, A-B structure or a more complex A-B-A tri-block or even more complex A-B-A-B-A multiblock types. Block copolymers also can contain, for example, molecules which contain a block of A, and a block of a mixture of copolymerized A/B monomer. Such a block copolymer molecule can be represented by A-(A/B).

Structure Nomenclature

Structure Feature	Nomenclature
A - B	A-B Diblock copolymer* or Poly (A-b-B) or Poly A-block-Poly B
A - B - A	A-B-A Triblock copolymer* or A-B-A Teleblock or Poly (A-b-B-b-A) or Poly A-block-poly B-block-polyA
A - B - C	A-B-C Triblock copolymer* or Poly (A-b-B-b-C) or Poly A-block-poly B-block-poly C
A - B - A - B - A - B - A	A-B multiblock copolymer* or Segmented poly A-poly B block copolymer*

$$
\begin{array}{c}
A \\
| \\
B \\
| \\
A - B - X - B - A \\
| \\
B \\
| \\
A
\end{array}
$$

A-B Star block copolymer* or

A-B Radial block copolymer*

*Many times the word polymer is used in place of copolymer.

Simple block copolymers, since they contain segments of two or more totally different polymer types, will have some properties common to each type of homopolymer. However, when the two polymer segments are connected, some new characteristics will be exhibited. For example, polystyrene is not appreciably soluble in polybutadiene. A mixture of the two polymers will be translucent at best and perhaps opaque depending on the percent polystyrene in the mixture as well as it's molecular weight. By contrast, a block copolymer of the same overall composition will be transparent. The appeal of block copolymers and their commercial importance is usually due to that rare combination of properties created by the hybridization of the homopolymers. The morphology or phase-separation plays a key role in the properties of the block polymers.

Block copolymers can have very special properties by virtue of the particular arrangement of, for example, glassy blocks of polystyrene and rubbery blocks of polybutadiene. The polystyrene blocks at room temperature are some 75°C below their glass transition temperature while the polybutadiene blocks are some 110°C above theirs. The block copolymer exhibits both glass transitions when taken through the -100 to 150°C range.

The glassy polystyrene block is not soluble to any extent in the rubber portion of the polymer and a simple mixture of the two homopolymers would tend to separate to produce an opaque mixture of resins. However, when the blocks are tied together on a molecular level, as in an A-B block copolymer molecule, they cannot separate on a macro scale and therefore tend to separate on a micro scale.

Simple butadiene-styrene (B-S) or A-B block copolymers appear to be homogeneous and indeed exhibit great clarity when molded into sheet. Nonetheless, the two blocks are still basically incompatible and while massive separation may be prevented by the carbon-carbon links between the blocks, separation will still occur. On a micro-scale, the polystyrene blocks tend to separate from the rubbery polybutadiene blocks and associate with similar blocks from other polymer chains. The polystyrene blocks concentrate to form into domains of glassy polystyrene in a rubbery block matrix. The domain size is in the 100-400Å size which cannot be seen by the naked eye. Therefore the molded polymers appear to be homogeneous and transparent. The products are rubbery and yet exhibit two glass transitions typical of the two homopolymers. Such products will exhibit no green strength (raw strength) and for most applications will have to be vulcanized to develop the strength needed in molded articles.

B-S-B block copolymers also exhibit phase separation and again the polystyrene center blocks associate to form the domain structure. The only difference is that each polystyrene block has two rubbery blocks hanging into the rubbery matrix. Again, the polymer must be crosslinked or vulcanized to develop tensile strength.

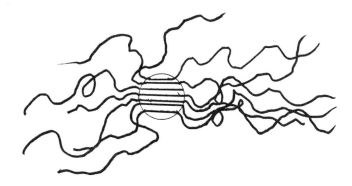

190

Styrenic Thermoplastic Elastomer

If the blocks are arranged with the rubbery (soft) block in the middle of two polystyrene (hard) blocks, S-B-S type, a really unusual, commercially useful property results, namely green tensile strength. Again, as with the other block copolymers, the polystyrene blocks associate in domains, but since there is one at each end of a long rubbery block, a single molecule may be connected to two different domains. Indeed most of the blocks appear to be arranged in this fashion. As long as the polystyrenes are in their glassy, rigid state, the product will appear as a network structure with the rubbery chains anchored between two fixed domains.

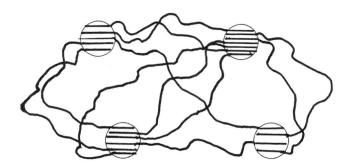

S-D-S Triblock Copolymer (Thermoplastic Elastomer[a])

Two phases
Two Tg's
Characterized by:
    High raw strength
    Complete solubility
    Reversible thermoplasticity

(a) S-D-S = Styrene-diene-styrene, thermoplastic elastomer (TPE) and thermoplastic rubber (TPR) are used interchangeably in publications.

In this form, as long as the polystyrenes are below their glass transition temperature, the polymer has good tensile strength without chemical crosslinking or vulcanization. When the temperature of the polymer is raised above the glass transition temperature of the polystyrene segments (usually 100-110°C), the glassy domains soften and the resin loses its strength because the chains are no longer anchored at each end of the rubbery blocks. When the resin temperature falls below 100°C, the block structure reforms and green strength returns.

Elastomeric block copolymers (thermoplastic elastomers (TPE), or thermoplastic rubber (TPR)) are useful in a variety of fields. The unique behavior of A-B-A and $(AB)_x$ star block copolymers combines the features that were previously considered to be mutually exclusive, namely, thermoplasticity together with elastic behavior.

For this article, the history of the discoveries and developments of styrene-diene-styrene thermoplastic elastomers (styrenic TPE) are reviewed.

Anionic Polymerization

Anionic polymerization, under properly selected conditions, can achieve nearly instantaneous initiation, no chain transfer reaction and no premature termination. In other words, the active chain ends (negatively charged ions, or anions, which can exist in various forms of ion pairs as well as free ions) are formed nearly all at once, remain active during the addition of different monomers and are stable enough to remain unterminated for months even in the absence of monomer. A better block copolymerization process than that afforded by the anionic process could scarcely have been designed. Indeed, the process, often called a "living polymerization", allowed the production of an extremely wide range of well characterized homopolymers and copolymers including block copolymers.

Initiators for most anionic processes are often organometallic compounds. The best known and most extensively investigated are those based on alkyls of the alkali metals of Group I of the Periodic Table as initiators for vinyl or diene addition polymerization. Particularly, the polymerization of dienes and vinyl aromatic compounds with alkyllithium compounds have been studied in great detail. The process is capable of making essentially monodisperse polymers ideally suited for accurate characterization and fundamental structure/property studies. Indeed, the resins lend themselves to the classical characterization techniques needed to provide accurate structure assignments.

Anionic polymerization by organolithium compounds is actually organometal chemistry applied to giant molecules. Both extremely simple and complex polymer structures can be created at will.

HISTORY

Early Events

Alkali Metals as Pzn. Initiator

1910-1914
Methews and Strange [1]
Harris [2]
Schlenk [3]

1930-1940
Midgley [4]
Ziegler [5-9]

1940-1945 (B/S Copzn.)
Marvel, Bailey and Inskeep [10]
Schulze and Crouch [11]

Early Industrial Applications
Germany (Buna 85, Buna 115) [12]
USSR (SK-B) [13]

Butyllithium as Initiators
(Non-Termination Polymerization)

1929-1936
Ziegler [5-9]

192

$$2\ Li^0 + M \longrightarrow LiMLi$$

$$LiMLi + nM \longrightarrow LiM_nLi$$

$$BuLi + M \longrightarrow BuMLi$$

$$BuM_2Li + M \longrightarrow BuM_3Li$$

$$BuM_nLi + M \longrightarrow BuM_{n+1}Li$$

Discovery of "Living Polymers"

This section can just as well be named as rediscovery of non-termination polymerization. Terminationless polymerization was considered by Mark and Dostal in 1935[14] and also by Flory[15]. Ziegler had proposed that polymerization of butadiene or isoprene by lithium metal or alkyllithium is a stepwise addition of monomer with no termination step as shown earlier. Polymer molecules continue to grow, so long as monomer is present. However, it was M. Szwarc in his dramatic article entitled "Living Polymers"[16] and his convincing publication with Levy and Milkovich[17] that conveyed the concept and ramifications of non-termination polymerization.

In the spring of 1955, in a conversation with Prof. Sam Weissman, Szwarc learned the results of Weissman's investigations on electron-transfer reactions. It occurred to Szwarc[19] that electron-transfer to styrene ought to convert its C=C bond into what "naively" could be described as C-C⁻. Then, the radical end should initiate radical polymerization while anionic polymerization would be simultaneously induced by the carbanionic end. Actually, as Szwarc and his associates learned later, the radical ends are short lived. These ends quickly coupled to form a dianion. Nevertheless, the "living" nature of this process i.e. that macromolecules resume their growth whenever a suitable monomer is supplied, is clearly demonstrated. Szwarc visioned that with this "living" process, polymers with a Poisson-type molecular weight distribution, well defined block copolymers, and polymers with functional end groups can be synthesized. Later publications demonstrated these possibilities.[19-22]

The use of highly reactive sodium initiator in THF limited its operability at very low temperature. Side reactions led to premature termination. It was apparent to many of the researchers, that the process of choice for "living" polymerization is the use of lithium-based initiators in a hydrocarbon solvent.

Sam Weissman (pre 1955)

Electron-Transfer

Formation of Sodium Naphthalene Complex

Initiation of Pzn. of Styrene by Electron Transfer

194

Related Events

• 1955 - Firestone group announced the synthesis of cis-polyisoprene by means of lithium metal in a hydrocarbon solvent[23]

• 1957 - Foster and Binder of Firestone published a more detailed article on the preparation of cis-polyisoprene[24]

• 1957 - Hsieh and Tobolsky described the polymerization of isoprene leading to cis-polyisoprene with butyllithium[25]

• Firestone and Shell produced polyisoprene rubber with lithium metal and alkyllithium respectively in the later '50's and early '60's.

Solprene® Rubbers and Phillips Petroleum Company

In the summer of 1957, immediately after the completion of my study at Princeton University under the direction of Prof. A. V. Tobolsky, I joined Phillips Petroleum Company in Bartlesville, Oklahoma. I was assigned to the Solution Polymerization Section of the Synthetic Rubber Branch, R&D. The section was formed only about a year earlier and was headed by J. N. Short. The mission was to explore solution polymerization with organometallic initiators to prepare new and novel stereo-elastomers and elastomers with unique properties. By the time I joined the section, R. P. Zelinski, who joined Phillips a year earlier, and his co-workers had already discovered cis-polybutadiene rubber.[26] Zelinski then turned his attention to the copolymerization of butadiene and styrene with butyllithium. They soon discovered that the mixture of these two monomers will form a tapered block copolymer in a hydrocarbon solvent and a random copolymer in the presence of a polar compound.[27-29] The tapered block copolymer has one block consisting mostly of butadiene (or isoprene) with small amounts of styrene and a polystyrene block (B/S-S type). Both the random and the tapered diblock copolymers were quickly scaled-up in the pilot plant and put in commercial production in 1962.[30-32] This B/S-S type diblock copolymer has outstanding extrusion characteristics, low water absorption, low ash, and good electrical properties. As a result, it is useful in such diverse applications as wire and cable coverings, shoe soles, and floor tiles.[33]

I was assigned to produce elastomers with functional groups at both ends of the chain. The initial objective of this work was to prepare elastomers of which the free chain ends can be incorporated into the network by chemical means during vulcanization.[34-35] This work also through serendipity, led to the development and commercialization of liquid polybutadiene with carboxy or hydroxy end groups (Butarez CTL and Butarez HT) which are used as solid rocket fuel binder.[35] They were also extensively tested as coating material. The reason this work is mentioned here is that it was a direct link to the first preparation of A-B-A triblock copolymer during 1958-1959 at Phillips laboratory.

Our strategy to prepare polymers with functional end groups was to use a dilithium initiator to first form α,ω-dilethiopolybutadiene, followed by functionalization. In 1957, there was no readily available dilithium compound. Thus, much of our work involved exploring the synthesis of the initiator. By early 1958, we had several dilithium initiators available for use. With these initiators, we made "telechelic" polymers and "teleblock" polymers. The word "telechelic" was coined by C. A. Uraneck at Phillips (greek telos, end, chele, claw) to describe polymers possessing two reactive terminal (end) groups. When we made A-B-A type triblock copolymers, we conveniently called them teleblock. We used these two home-made terms extensively in our reports, notebooks, patent applications and later, journal publications. Using the dilithium initiator we made in the laboratory, Zelinski systematically prepared teleblock copolymers of S-B-S, S-I-S, B-S-B, I-S-I, S-B/S-S, -- -- types in 1958.

Encouraged by earlier evaluation in our Compounding Section, G. Kraus, then the manager of Reinforcement Research Section in the Synthetic Rubber Branch, carried out a study of the stress-strain behavior of a series of S-B-S teleblock copolymers having variable S/B ratios. Kraus found that block polymers of styrene and butadiene, consisting of a center block of polybutadiene with polystyrene ends, exhibited high elasticity and tensile strength without the benefits of cross-linking or filler reinforcement, and that the S-B-S teleblock copolymers exhibit "a type of physical cross-linking by association of styrene blocks". These samples were prepared by Zelinski, using a dilithium initiator with incremental addition of monomers (in this case, butadiene was polymerized first and then the addition of styrene).

Later, similar teleblock copolymers were prepared with butyllithium by incremental addition of monomers (in this case, it was styrene first, butadiene following, and then styrene in final step).

By use of organolithium initiators, Zelinski constructed four basic types of ordered copolymers from dienes and styrene. These copolymers differ in monomer distribution within the polymer molecule, and were identified as symbolized below:

DDDD-S-DDDD-S----DS
Random Copolymer

DDDDDDD-SSSSS
Simple Block Copolymer

SSS-DDDDDDD-SSS
Teleblock Copolymer

SSS-(DDDD-SS-)n DDDD-SSS
Multiblock Copolymer"

Zelinski found that teleblock copolymers containing butadiene as the center block were particularly notable for their raw strength, as well as thermoplasticity.

During the period of 1960-1962, we were busy trying to bring carboxy and hydroxy telechelic liquid polybutadienes, random S/B copolymers (solution SBR), medium-cis and medium-vinyl polybutadiene rubbers, and B-S diblock copolymer to commercial production. All of these polymers are initiated with alkyllithium. We recognized that many of these solution rubbers made with an alkyllithium initiator had excess cold flow and poor processability. We worked on the processes of broadening the molecular weight distribution and introducing long chain branching to reduce the cold flow and improve the processability. Both coupling with multifunctional compounds such as $SiCl_4$, $SnCl_4$, polyepoxides, and polyisocyanates, and copolymerization with difunctional monomers such as divinylbenzene were extensively studied. The former approach resulted in "radial" or "star" polymers and the latter approach gave random branching. Both approaches were immediately put in use in the production of many of our Solprene rubbers.

Using the same coupling method, we synthesized "radial teleblock" copolymers in 1962. I described the synthetic steps of making $(S-B)_n-X$ type thermoplastic elastomers in various publications.[37,38]

The radial thermoplastic elastomers were produced in development quantity in our pilot plant in 1964 and put in commercial production in 1973. These TPE's were marketed also under the trade name of Solprene (400 series). By 1978, Phillips and its licensees were producing nearly 50 variations of Solprene rubbers (polybutadiene, solution SBR, S-B diblock and TPE) including over 15 versions of radial teleblock copolymers of the (S-B)$_{3-4}$-X or (S-I)$_{3-4}$-X type. The total annual capacity reached 350,000 tons per year.[32]

Phillips stopped domestic production of all synthetic rubbers between 1982 and 1986, and sold the foreign Solprene rubber interests to our joint venture partners in Belgium, Spain, and Mexico in 1986. The TPE plant located in Borger, Texas was sold to Synthetic Rubber Corp. in Taiwan in April, 1986. The plant was moved to Taiwan and has been in full production since January, 1988.

Phillips, meanwhile, expanded the production of a family of clear, medium-impact styrene-butadiene thermoplastic resins (K-Resins®) in its Houston Chemical Complex.

Kraton® TPE and Shell Chemical Company

It is not possible for me to report on Shell's discovery and development of S-D-S types of TPE with the same degree of intimacy as I did on Phillips' events. Fortunately, Norman R. Legge of Shell (retired) has done this for us. Legge directed, guided, championed and managed this project from the first discovery to the commercial success. The 1987 Charles Goodyear Medal address entitled "Thermoplastic Elastomers" by the medalist Norman R. Legge[39] at the 131st Meeting of the Rubber Division, ACS, Montreal, Quebec, Canada, May 26-29, 1987, described in detail the developments at Shell.

Here is a brief summary from Legge's address:

• The discovery was made while their researchers were engaged in means to impart some green strength to cis-polyisoprene rubber made with alkyllithium, a product of Shell's at that time.

• At the time of their discovery of S-I-S copolymer possessing green tensile strength, Legge was the Director of Elastomer Research Laboratory in Torrance, CA. He transferred from Shell Development in Emeryville to the Director position at Torrance in January, 1961.

• Legge and his senior staff felt the discovery was a major scientific breakthrough and received immediate and strong support from the managers of Synthetic Rubber Division and Shell Development.

• Obtaining broad coverage of the triblock copolymers through patent applications was considered to be the major task. The basic U.S. Patent on the triblock copolymers of styrene and dienes was filed in January, 1962, and issued in 1964.[40] G. Holden and R. Milkovich were the inventors. Milkovich received his M.S. degree in 1957 from the University of Saracuse, and with M. Levy developed the all-glass, high-vacuum technique and established the living electron-transfer polymerization in Prof. Szwarc's laboratory.[17] Subsequently he moved to Prof. M. Morton's laboratory at the University of Akron and received his Ph.D. degree in 1959. He joined Shell following his graduation.

• The thermoplastic elastomers based on block copolymers of styrene and butadiene were announced in 1965.[41,42] A series of papers were presented in 1967-1968 describing the details of the triblock structure, the theory of the thermoplastic elastomer behavior, and some structure limitations.

- The trade name, Kraton, was named by Tom Baron, then Manager, R&D in the Synthetic Rubber Division, later President of Shell Development Company. From a book on greek mythology he was reading, he selected the name of a god of strength, Kratos, and changed the last letter to give a technological sound.

- In 1972, Shell announced the Kraton G, a S-EB-S triblock copolymer. It is a product of hydrogenated S-B-S copolymer.

- Legge considered the successful innovation of TPR came primarily from technological "push", with very little market demand "pull".

- Legge concluded that Shell's success was due to several favorable elements:

  a) Patentable technical breakthrough

  b) Ample pilot plant capacity at the time of discovery

  c) By happy chance, there was thermoplastic processing and testing equipment on hand

  d) A small and very capable research and development staff

  e) Management was oriented toward new products and overall company management was willing to support new ventures in polymers

Past and Future

Commercially, Shell introduced its Kraton in 1965 as the first producer of styrenic TPE. It started at a modest 2,000 metric ton production in 1965 and reached around 30,000 metric tons in 1972 (Chem System estimates). Phillips started the commercial production of its radial copolymers in 1973. From 1973 to 1982, Shell and Phillips provided the bulk of styrenic TPE to the market place. The total production by 1982 reached nearly 80 thousand metric tons. After Phillips stopped its production in 1982, Shell became the dominant producer with a small portion of the market being served by Firestone since 1981. I should mention, Firestone had an active research group (Tate, Halasa, Schulz and others) in the 60's and 70's in anionic polymerization.

Several major rubber and chemical companies in the U.S., Western Europe and Japan also had active R&D programs on styrenic block copolymers during this period.

According to a recent article, Royal Dutch/Shell will spend $170 million through 1991 to boost worldwide capacity 39% to 320,000 m.t./yr.[43] According to another article, new plants in the U.S. are planned by leading importer Enichem, in a joint venture with Arco, and newcomers Dow and Exxon in their Dexco joint venture.[44] Enichem, the second-ranking producer worldwide with around 70,000 m.t./yr capacity, Petrofina, and Repsol Quimica - all once licensees of Phillips - have added capacity or are doing so. Taiwan Synthetic Rubber Corp., as mentioned earlier, has restarted the ex-Phillips plant it shipped from Borger, TX with an annual capacity of 25,000 m.t./yr. Japan has a production capacity of 50,000 m.t./yr at Asahi, Nippon Zeon, Japan Elastomer Company (joint venture of Asahi and Showa Denko), and JSR.

Forecasts by Chem. Systems (Tarrytown, NY) show U.S. demand of 145,000 m.t./yr and exports 30,000 m.t./yr in 1991. Europe's and Japan's demands and exports should reach 163,000 m.t./yr and 34,000 m.t./yr, respectively.

198

CONCLUSIONS

• Ziegler, Marks, Flory, Szwarc and others provided the background, wisdom and vision.

• Still, the discoveries of styrenic TPE were serendipitous, not targeted. It was discovered and made commercially successful by people recognizing a breakthrough.

• Once discovered, in a relatively short time, broad advances were made by scientists both in academics and industry based on the understanding of the two-phase polymeric system. "Domain size", "domain boundary", "boundary layer", "microphase separation", "aggregation", "critical end block molecular weight", etc. were the buzzwords in the 60's and 70's. No other elastomer ever discovered created so much excitement in the scientific community. A few selected references are listed for the readers.[45-72]

• The complexity of the process (requiring incremental addition of monomers, end coupling, etc.) and the unique product (an elastomer acts like thermoplastic) challenged the skills and imagination of engineers and marketing persons alike.

• In the final analysis, one can say the styrenic TPE was more than an extension of an existing synthetic rubber industry; it created an entirely new technology.[73]

REFERENCES

1. F. E. Mathews and E. H. Strange, Brit. Pat. 24,790 (1910).
2. C. D. Harris, U.S. Pat. 1,058,056 (1913).
3. W. Schlenk, J. Appenrodt, A. Michael and A. Thal, Chem. Ber., 47, 473 (1914).
4. T. Midgley, Synthetic and Substituted Rubbers, in Davis and Blake, "The Chemistry and Technology of Rubber" Reinhold, New York, 1937.
5. K. Ziegler and K. Bahr, Ber., 61B, 253 (1928).
6. K. Ziegler, H. Colonius and O. Schater, Ann. Chem. 473, 36 (1929).
7. K. Ziegler and O. Shater, Ann. Chem., 479, 150 (1930).
8. K. Zielger, L. Jakob, H. Wolltham and A. Wenz, Ann. Chem. 511, 64 (1934).
9. K. Ziegler, H. Gilman, and R. Willer, Ann. Chem., 542, 90, (1940).
10. C. S. Marvel, W. J. Bailey, and G. S. Inskeep, J. Polym. Sci., 1, 275 (1946). Rubber Chem. and Tech. 20, 1 (1947).
11. W. A. Schulze and W. W. Crouch, J. Am. Chem. Soc., 70, 3891 (1948).
12. R. L. Bebb and L. B. Wakefield, German Synthetic Rubber Developments, in Whitby, Davis and Dunbrook, "Synthetic Rubber", John Wiley and Sons, Inc. New York and Chapman & Hall, Ltd., Londo, 1954.
13. W. K. Taft and G. F. Tiger, Diene Polymers and Copolymers Other than GR-S and The Specialty Rubbers in Whitby, Davis and Dunbrook, "Synthetic Rubber", John Wiley and Sons, Inc., New York and Chapman & Hall, Ltd., London, 1954.
14. H. Dostal and H. Mark; Z. Phys. Chem. B, 29, 299 (1935).
15. P. J. Flory; "Principles of Polymer Chemistry", Cornell University Press, Ithaca, New York, 1953.
16. M. Szwarc; Nature, 178, 1168 (1956).
17. M. Szwarc; M. Levy; and R. Milkovich, J. Am. Chem. Soc., 78, 2656 (1956).
18. M. Szwarc; Living and Dormant Polymers: A Critical Review; in McGrath "Anionic Polymerization: Kinetics, Mechanisms, and Synthesis", ACS Symposium Series 166, 1981.
19. M. Szwarc; Advances Chem. Phys., 2, 147 (1959).
20. M. Szwarc, Makromol. Chem., 35, 132 (1960).

21. M. Szwarc, Proc. Roy. Soc (London, Ser A, 279, (1964).
22. M. Szwarc, Advances in Polym. Sci. 12, 127 (1966).
23. F. W. Stavely, F. C. Foster, J. L. Binder and L. E. Forman; Ind. Eng. Chem., 418, 778 (1956), paper presented to the Div. Rubber Chem., ACS, Philadelphia, PA., Nov. 1955.
24. F. C. Foster and J. L. Binder, Adv. in Chem., Ser. No. 19, ACS, P-26 (1957).
25. H. L. Hsieh and A. V. Tobolsky, J. of Polym. Sci., Vol. XXV, 109, 245 (1957).
26. British Pat. 848,065 (to Phillips Petroleum Company; April 16, 1956).
27. (To Phillips Petroleum Company) Brit. Pat. 895, 980, Published May 9, 1962, Filed May 9, 1962.
28. R. P. Zelinski (to Phillips Petroleum Company) U.S. Pat. 2,975,160; U.S. Pat. 3,251,905 ; U.S. Pat. 3,287,333; U.S. Pat. 3,301,840.
29. R. N. Cooper (to Phillips Petroleum Company) U.S. Pat. 3,030,346.
30. W. W. Crouch and J. N. Short, presented at 2nd Int. Syn. Rubber Symp. London, October, 1960, Rubber & Plastics Age, 42, 276 (1961).
31. H. L. Hsieh, presented at 3rd Int. Syn. Rubber Symp., London, October, 1964, Rubber & Plastics Age, 46, 394 (1965).
32. H. L. Hsieh, R. C. Farrar and K. Udipi, Anionic Polymerization: Some Commercial Applications, in McGrath "Anionic Polymerization"; ACS Symp. Ser. 166, (1981) Chemtech, 626, October, 1981.
33. H. E. Railsback, C. C. Baird, J. R. Haws and R. C. Wheat, Rubber Age, 94, 583, 1964.
34. C. A. Uraneck, H. L. Hsieh and O. G. Buck; J. Polym. Sci., 46, 535 (1960).
35. C. A. Uraneck, H. L. Hsieh and R. J. Sonnenfeld, J. Appl. Polym. Sci., 13, 149 (1969).
36. D. M. French, Rubber Chem. & Tech., 42, 71 (1969).
37. H. L. Hsieh, Synthesis of Block Copolymers Via Anionic Polymerization in Burke and Weiss "Block and Graft Copolymers", Syracuse University Press, 1973.
38. H. L. Hsieh; Rubber Chem. and Tech., 49, No. 5, 1305 (1976).
39. Norman R. Legge, Rubber Chem. and Tech. Rubber Reviews, 60, No. 3 G3, (1987).
40. G. Holden and R. Milkovich (to Shell Oil Company) U.S. Pat. 3,265,765 (1964).
41. J. T. Bailey, E. T. Bishop, W. R. Hendricks, G. Holden and H. R. Legge, Rubber Age, 98, 69 (1966).
42. M. A. Luftglass, W. R. Hendricks, G. Holden, and J. T. Bailey, Machine Design, 38, 1914 (1966).
43. Chemical Week, May 24 (1989).
44. European Rubber Journal, 19, March (1989).
45. G. Kraus and J. T. Gruver, J. Polym. Sci., 11, 2121 (1967).
46. G. Holden et al, J. Polym. Sci., C, 26, 37 (1969).
47. R. P. Zelinski and C. W. Childers, Rubber Chem. and Tech., 41, 161 (1968).
48. G. Kraus et al, J. Appl. Polym. Sci., 11, 161 (1968).
49. S. L. Cooper and A. V. Tobolsky, Textile Res. J., 36, 800 (1966).
50. R. J. Angelo et al, Polymer, 6, 141 (1965).
51. J. F. Beecher et al, J. Polym. Sci., C, 26, 117 (1969).
52. C. W. Childers and G. Kraus, Rubber Chem. and Tech., 40, 1183 (1967).
53. G. L. Wilkes and R. S. Stein, J. Plym. Sci., A-2, 7, 1525 (1969).
54. G. Uchida, et al, J. Polym. Sci., A-2, 10, 101 (1972).
55. T. Miyamoto et al, J. Polym. Sci., A-2, 8, 2095 (1970).
56. G. Kraus et al, J. Polym. Sci, A-2, 10, 2061 (1972).
57. M. Matsuo et al, Polymer, 9, 415 (1968).
58. A. Keller et al, Kolloid-Z., Z. Polym., 238, 385 (1970).
59. E. T. Bishop and S. Davison, J. Polym. Sci., C, 26, 59 (1969).
60. D. J. Meier, Polym. Prepr., 15(1), 171 (1974).
61. D. G. Fesko and N. W. Tschoegl, J. Polym. Sci., C, 35, 51 (1971).

62. M. Morton, J. E. McGrath, and P. C. Juliano, paper presented at the Rubber Division Meeting, ACS, May, 1967.
63. G. Kraus, C. W. Childers, and J. T. Gruver, J. Appl. Polym. Sci., 11, 1581 (1967).
64. A. Rembaum, F. R. Elias, R. C. Morrow, and A. V. Tobolsky, J. Polym. Sci., 61, 155 (1962).
65. P. Rempp and H. Benoit, paper presented at the Rubber Division, ACS, May (1967).
66. D. J. Meier, J. Polym. Sci., C, 26, 81 (1969).
67. R. J. Angelo, R. M. Ikeda and M. L. Wallach, Polymer, 6, 141 (1965).
68. A. Noshay and J. E. McGrath, (Ed.) "Block Copolymers: Overview and Critical Survey", Academic Press, New York, (1977).
69. D. J. Meier, (Ed.), "Block Copolymers - Science and Technology", MMI Press by Harwood Academic Publishers Chur, London, New York 1983).
70. S. L. Aggrawal (Ed.), "Block Copolymers", Plenum Press, New York (1970).
71. R. J. Ceresa, "Block and Graft Copolymers", Butterworth Publishers, London (1962).
72. J. J. Burke and V. Weiss (Ed.) "Block and Graft Copolymers" Syracuse University Press, Syracuse, New York (1973).
73. N. R. Legge, G. Holden, and H. E. Schroeder, "Thermoplastic Elastomers, A Comprehensive Review", Hanser Verlag, Munich, (1987).

HISTORY OF SILICONES IN COATINGS

Edwin P. Plueddemann, William A. Finzel
Dow Corning Corporation
Midland, Michigan 48686-0995

ABSTRACT

Organosilicon compounds are not found in nature due to the strong affinity of silicon for oxygen. The first organosilicon compounds were prepared in 1863, and the first commercial silicone resins were produced in 1945 in energy-intensive reactions from silica. Silicones ultimately degrade back to silica, but phenyl- and methyl- silicones have remarkable thermal-oxidative stability. Silicones are used as sole resins in coatings, or may be used to modify organic polymers to obtain coatings with intermediate heat- and weather-resistance. Resistance to high temperatures and weathering is generally proportional to the ratio of silicone in the total coating. Specialty silicon compounds are also used in small proportions as additives to coatings to impart special effects, such as adhesion, flow, pigment dispersion, etc.

INTRODUCTION

Although silicones are based on silicon dioxide, the most abundant constituent of the earth's crust, they are a relatively recent development compared to polymers of carbon compounds. In terms of electronegativity, silicon is assigned a value of 1.8 compared with 2.5 for carbon[1]. The result is that all silicon compounds tend to revert to the oxide, and no organosilicon compounds are found in nature.

Friedel and Crafts[2] in 1863 prepared the first compounds with silicon-carbon bonds by the reaction of diethylzinc with silicon tetrachloride. Ladenburg, an organic chemist, joined with Friedel to continue this work and concluded[3] "that the so-called inorganic elements are capable of forming compounds which are analogous to those of carbon." He later showed that hydrolysis of $(C_2H_5)_2Si(OC_2H_5)_2$ gave a stable oil[4] instead of a simple volatile compound analagous to diethyl ketone formed from $(C_2H_5)_2C(OC_2H_5)_2$ but the term "silicone" has been retained for all siloxanes even though there is no silicon analog of a ketone $R_2Si = O$.

A great impulse to silicone chemistry came from Kipping's 54 papers published during the period 1899 to 1944. He made use of the Grignard reagents to prepare organosilicon compounds. He was a theoretical organic chemist and was interested in pure compounds that could be isolated by distillation or crystallization. The oils and glues that he often obtained seemed uninviting to him, but he correctly described them as macromolecules[5].

Corning Glass Works employed Dr. J. F. Hyde to investigate hybrid polymers - a cross of organic polymers and glass - in

Published 1990 by Elsevier Science Publishing Co., Inc.
Organic Coatings: Their Origin and Development
R.B. Seymour and H.F. Mark, Editors

1940. He used the reactions developed by Kipping to prepare
many new organosilicon compounds that could be converted into
polymers that had outstanding heat stability. At the same time
Corning was developing fiberglass and needed a polymeric binder
to impregnate glass tape for high-temperature electrical
insulation. Similar studies were soon underway at the General
Electric Laboratories under E. G. Rochow and W. I. Patnode, and
at Mellon Institute under R. R. McGregor. In 1942, the work at
Corning had progressed to the point where commercial production
was considered. Since they were glass manufacturers, they
approached Dow Chemical Company for assistance in manufacturing.
The results was the formation of Dow Corning Corporation in 1943
as a joint venture by Dow Chemical Company and the Corning Glass
Works. In 1946, the General Electric Company announced their
first production of silicones. The Plaskon Division of Libby-
Owens-Ford Glass Company advertised silicone products for sale
in 1949 - especially in the area of silicon modified alkyd
coatings. About the same time, the Linde Division of Union
Carbide Corporation commenced pilot-plant work on silicon
chemicals.

Organic varnishes are commonly prepared by heating drying oils
in the presence of air to cause polymerization of the oils.
This procedure was tried with silicone oils at the Mellon
Institute. After several unsuccessful attempts to get resins in
this manner, Dr. McGregor finally exclaimed, "How foolish can we
be! We have the world's most stable oil and are trying to
convert it into a varnish. Why not find a use for it as an
oil."[6] The first commercial uses for silicones were in
military applications as damping fluids in aircraft engines to
prevent corona discharge at high elevations. Resinous silicones
prepared from trifunctional organosilicon intermediates were
used with glass tapes as insulating medium in electric motors.
In 1945, both Dow Corning and General Electric announced the
development of silicone rubber that was useful at both high and
low temperatures. When hostilities concluded in the summer of
1945, the military demands ceased and silicones were left
without a market, but the unique properties of silicones soon
were adapted to a peacetime economy. Demand for peacetime
applications soon surpassed the earlier requirements for
military use, and expansion of facilities became necessary for
all producers. As production increased, prices dropped until
silicone fluids and certain silicone resins are now commodity
chemicals with wide ranges of applications. The early days of
silicone manufacture, and their many uses were well documented
by Rochow[7] and McGregor[8].

MANUFACTURE OF SILICONE INTERMEDIATES

Silicones are made from organo-substituted chlorosilanes, which
are prepared from elemental silicon. A metallurgical grade of
silicon formed by the reduction of silica with carbon may be
used:

$$\overset{>1420°C}{SiO_2 + 2C \xrightarrow{\hspace{1cm}} Si + 2CO}$$

Silicon is usually converted to organo-substituted chlorosilanes by reaction with alkyl or aryl halides in a direct process developed by Rochow[9]. This reaction is especially important in manufacturing methylchlorosilanes in the presence of copper catalyst.

$$Si + MeCl \overset{Cu}{\underset{>300C}{\xrightarrow{\hspace{1cm}}}} MeSiCl_3 + Me_2SiCl_2 + Me_3SiCl \text{ (plus other minor products)}$$

Silver is sometimes employed in the direct process for phenylchlorosilanes from silicon and phenylchloride. Phenylmethyldichlorosilane and other specialized monomers are made by Grignard alkylation:

$$MeSiCl_3 + PhMgCl \xrightarrow{\hspace{1cm}} MePhsiCl_2 + MgCl_2$$

A third important preparative method employs silicon hydrides such as trichlorosilane ($HSiCl_3$) or methyldichlorosilane ($MeCl_2SiH$) in addition reactions to terminally unsaturated organic compounds. Reaction may be initial by UV radiation, peroxides, or elevated temperature, but the most active catalysts are soluble platinum compounds such as $H_2PtCl_6·6H_2O$ first described by Speier et al[10]. The mild conditions of the platinum-catalyzed addition allow reactions with unsaturated organofunctional compounds containing heat-sensitive groups.

$$Ch_2=C(CH_3)COOCH_2CH=CH_2 + HSiCl_3 \xrightarrow{\hspace{1cm}} Ch_2=C(CH_3)COOCH_2CH_2CH_2SiCl_3$$

Organofunctional silanes have been used extensively as adhesion promoters for coatings to metals and other inorganic surfaces [11]. More recently it was observed that they are also effective promoters for organic to organic adhesion[12]. Their application in coatings for intercoat adhesion has not been studied extensively.

POLYORGANOSILOXANES

Chlorosilanes are readily hydrolyzed by water to form silanols and hydrogen chloride. Silanols react further with the elimination of water to form siloxanes. Silicone is a generic term including substances constituted of alternating silicon and oxygen atoms in which most of the silicons are also linked to at least one monofunctional organic radical through stable carbon-silicon bonds.

The final product of hydrolysis of monochlorosilanes are disiloxanes such as hexamethyldisiloxane:

$$2Me_3SiCl + H_2O \xrightarrow{\hspace{1cm}} Me_3SiOSiMe_3 + 2HCl$$

Dichlorosilanes give polysiloxanes:

$$n\text{-}R_2SiCl_2 + n\ H_2O \longrightarrow [R_2SiO]_n + 2\ nHCl$$

High molecular weight dimethyl(polysiloxanes) are vulcanized to silicone elastomers:

$$[Me_2SiO]_{>1000}$$

When linear poly(dimethylsiloxanes) are end-blocked by use of hexamethyldisiloxane, silicone fluids result:

$$Me_3SiO[Me_2SiO]_{0\text{-}1000}\ SiMe_3$$

Hydrolysis of trichlorosilanes result in silsesquioxanes, although this term is usually reserved for completely condensed cubical octomers and similar structures:

$$nRSiCl_3 + 3/2nH_2O \longrightarrow [RSiO_{3/2}]_n + 3nHCl$$

Copolymers of R and R_2 monomers are branched because they contain trifunctional structures and form silicone resins:

```
      Me   Ph   Me
     -Si-O-Si-O-SiO
      Me   OH   O  n
```

The relatively high cost of silicone resins is partly explained by the necessary removal of chlorine atoms that constitute a large part of chlorosilane weight. Chlorosilane intermediates may be by-passed by reacting activated silicon directly with dialkyl ethers in the presence of bromide catalysts at moderate temperatures [13]. Thus a crushed silicon/copper alloy with excess dimethyl ether and a trace of methyl bromide in a pressure reactor at 261°C gave a product containing 64% dimethyldimethoxysilane along with other methylmethoxy silanes and siloxanes.

SILICONE RESINS

The process for the preparation and cure of silicone coating resins consists of first the formation and then the condensation of silanol groups.

$$SiCl + H_2O \longrightarrow SiOH + HCl \quad \text{(reaction 1)}$$
$$SiOH + HOSi \longrightarrow SiOSi + H_2O \quad \text{(reaction 2)}$$

Mixtures of chlorosilanes are dissolved in sufficient toluene or xylene to give a 30-50% silicone resin solution after hydrolysis. This solution is mixed with enough water to yield 17% hydrochloric acid. Hydrolysis is rapid at room temperature, but the water may be warmed to 70°C to modify the nature of the hydrolysis products. Simultaneously, the more reactive silanol groups condense to begin the formation of siloxane structures.

The organic layer is separated from the hydrochloric acid layer and washed with water (sometimes a bicarbonate solution) to obtain an acid-free product. The solvent may be completely or partially removed by stripping at 90-150°C under reduced pressure. At this point the resin contains 3 to 6 wt% silanol (calculated as hydroxyl) and has a solution viscosity (60% toluene) of about 20 cP.

The molecular weight can be increased to improve coating properties and physical strength. Again, this involves condensation (reaction 2), but the residual silanols are less reactive. Azeotropic removal of water, sometimes catalyzed by metallic salts, is performed on the 50-90% solids solution. The finished resin contains 0.5-1.2% OH and may have a 50% viscosity of 150 cP.

Pure silicone coatings may be prepared by mixing this silicone resin solution with pigments and a more active catalyst to condense silanols further according to reaction 2. Elimination of the last traces of water is effected without bubbling or film rupture because of the high water vapor permeability of the polysiloxane.

Thus it is seen that utilization of the silicone resin merely continues the basic reaction that occurred in its preparation. This condensation may be interrupted at any stage depending on the particular application desired, and the user may specify and purchase silicone resins that have greater or lesser silanol content, molecular weight, and crosslink density.

A silicone resin may be blended with other silicone or organic resins to obtain desired flexibility, hardness, and degree of crosslinking. Melamine resins commonly used to crosslink organic coatings are also effective crosslinkers for active silicone resins, greatly reducing the baking temperatures required for crosslinking. Most organic and inorganic pigments can be blended with silicone resins by conventional methods to prepare protective coatings.

THERMAL AND CHEMICAL STABILITY

Major applications for silicones in coatings are based on their thermal stability and their resistance to outdoor weathering. These properties are proportional to the percent silicone in silicone organic copolymers and cold blends. The ultimate, of course, are pure silicone resins.

Thermal stability in air is related to the fact that the silicone polymer backbone is completely oxidized, and is therefore not subject to oxidative degradation $[Si-O-Si-O-]_n$. Ultimate oxidative degradation of silicones involves the removal of organic substituents by oxidation of the silicon-carbon bond:

$$Si-C + O_2 ----> SiO + -O-C$$

The useful half life of a silicone coating may be defined as the
time at elevated temperature at which half of the organic groups
are removed from silicon. This varies widely with different
groups on silicon, with phenyl being the most resistant to
oxidation. The half lives of certain groups in air at 250°C
were calculated[14] from weight loss during 100 hour exposure as
shown in Table I.

It is more difficult to rationalize the outstanding weather
resistance of silicones since a siloxane bond is less resistant
than carbon-carbon bonds to attack by aqueous reagents. From
Pauling's scale of electronegativities it is calculated that the
siloxane bond has about 50% ionic character. This bond is
readily hydrolyzed, especially when catalyzed by an acid or a
base. The secret of coating durability in weathering is
probably due to the reversible nature of siloxane hydrolysis.
Siloxanes in the presence of moisture are in true equilibria
that strongly favor condensation of silanols to siloxanes[15]
(Table II). The result is that a photon of absorbed high-energy
radiation in the presence of moisture will hydrolyze siloxane
bonds, but they reform spontaneously. In this way no permanent
damage is done to the film. In practice it has been observed
that cured silicone films resist exposure to sunlight and
moisture for many years. Although silicones are highly
permeable to gases, including water vapor, they protect surfaces
against liquid water, and waterborne ions; so they provide
excellent corrosion resistance to underlying metals.

APPLICATIONS OF POLYORGANOSILOXANE POLYMERS

Silicone or polyorganosiloxane polymers can be soft and rubber-
like or hard and very brittle, depending upon the variables
discussed earlier. Silicone coatings as a class of materials
consume a large percentage of all silicone products produced.

Typical products include solvent and water-based elastomers for
flexible and durable clear and pigmented coatings; silicone
elasto-plastic or conformal coatings for protection of
electronic components from moisture, moderate temperature, and
environmental contaminants; electrical varnishes for high
temperature applications; and release coatings for paper, metal
and other substrates. Release coatings use small amounts of
polydimethylsiloxane for optimum release. Other applications
include casting, laminating, cloth-coating, impregnating, mica
bonding, and encapsulating.

Almost all high temperature applications utilizing silicone
resins are pigmented with colored pigments. Pigment choice is
very important as well as pigment volume concentration[14][16] for
optimum coating performance. Pigmented protective coatings
containing silicone binders serve the needs of high temperature
service; however, a coating must meet a myriad of other
requirements. These requirements are: adaptability to its

substrate, resistance to corrosion and chemicals, flexibility and/or hardness, stability during temperature cycling, color stability, water resistance, and most recently, acceptable volatile organic compound content on application. Since these coatings must withstand a variety of harsh conditions, surface preparation, primer consideration, film thickness, application conditions, and dry film thickness are extremely important. Table III shows the temperature range of silicone and silicone-modified organic resins. Temperatures above 600°F (316°C) utilize pure silicone resin binder. Typical applications for silicone coatings for high temperature are stacks, incinerators, manifolds, boilers, ovens, furnaces, heat exchangers, wood and coal-burning stoves, barbecue grills, and food-release coatings.

APPLICATIONS OF SILICONE-ORGANIC COATINGS

Many higher temperature applications and need for coatings with improved exterior weatherability do not require pure silicone. Silicone organic systems can be prepared from copolymerization of silicone reactive intermediate with carbinol rich organic resins or by cold blending high phenyl containing silicone resins with organic resins. Silicone content of silicone organic binder systems can vary from 15-90% and depend on the specific requirements of the applied coating system. As shown in Table III, organic resins modified with silicone contain 15-50% silicone resin on the total resin system, whereas silicone resins modified with oganic resins contain silicone levels above 50%.

An example[17] of a silicone modified organic is silicone modified alkyd copolymers for air dry maintenance applications. The silanol functional silicone is copolymerized with a long oil alkyd at 30% silicone yielding a copolymer that air dries like a medium oil alkyd yet has excellent outdoor weatherability as compared to an alkyd of equal oil length.

Another example is a silicone modified saturated polyester copolymer where the methoxyfunctional intermediate is copolymerized with the polyester at silicone levels of 30-50%. Pigmented paint systems are applied to primed galvanized steel or similar substrates as coil coatings, heat cured and formed into exterior metal building siding with excellent outdoor weatherability. The coating must have good hardness, flexibility, and toughness.

Copolymerization is used where large volumes of resin are needed in addition to the fact that the cost of the intermediate is much less than typical resins. Copolymerization allows for improved compatibility with a wide range of organic resins; however, the organic resin must have carbinol functionality available for copolymerization or gellation to occur.

In addition to typical solvent-based copolymers, water based[18,19] and high solids[20,21] copolymers have been prepared and are commercially available.

Cold blending of silicone and organic resins can result in an unlimited number of cold blend variations and is a popular method to upgrade both thermal stability and weatherability of organic resins. Good compatibility between the silicone resin and organic resin is necessary for optimum coating film properties.

Examples of cold blended silicone modified organics (see Table III) are process equipment, recreational stoves and lanterns, space heaters, light reflectors, and generators.

Some applications of high temperature silicone coatings require up to 50% organic modification for specific application properties. Examples are acrylic modification for improved tack-free dry time and epoxy resin modification for improved corrosion resistance. Examples of these cold-blended organic modified silicones are heat resistant touch-up paint and vehicle muffler coatings.

Silicone organic copolymer and cold blend protective coatings offer a variety of coating systems for the protective coating industry. New silicone technology under development include radiation cure, low temperature cure systems, solventless and low solvent content based polymers.

SILICONE ADDITIVES

Silicone additives were first used in paints in the late 1940's when it was found that polydimethylsiloxanes had unique properties as antifloating and antisilking additives to reduce color separation. Since silicone additives are effective at low concentrations, most formulators add them to the finished blending or let-down. Polydimethylsiloxane was also used to eliminate problems of cratering, pinholing, edge migration, and orange peel[22]. Low viscosity (20-100 centistoke) polydimethylsiloxanes were used for these applications and usually diluted to low concentrations in aromatic solvents for easier use. These products are known today as "fish-eye killer" and "anticrater agent." As polydimethylsiloxane viscosity is increased, compatibility is decreased to the point that hammertone additives result which is based on the incompatibility of the silicone with the hammertone paint, thus forming the pattern known as hammertone.

Few silicone additives were available to the paint and ink industry before 1960 other than polydimethylsiloxane. As new organosilicon based products were developed for other applications, these chemicals were evaluated as additives for flow-out, pigment dispersion, pigment floating, mar resistance, defoamers, and other surface active properties. Because of air pollution regulations, newer paints and inks require additives for acceptable performance. New silicone additives are being formulated to solve specific paint and ink problems.

At about the same time polydimethylsiloxane was being introduced to the paint and ink industry, the fiberglass-reinforced composite industry was developing new and exciting products, but one problem was poor wet adhesion between the resin and glass fiber[23]. Organofunctional silanes were found to improve wet adhesion of a variety of composites and plastics which are used today. Similar organic/inorganic "coupling agents" were tested and found acceptable as adhesion promotors in paint and inks, both as additives or primers. Thus an aminofunctional silane was found to be an effective adhesion promoter for epoxy or urethane paints to steel and aluminum[24], either as a primer or an (0.2%) additive. Silanes were especially effective in improving recovery of adhesion after a water soak (Tables IV and V).

REFERENCES

1. L. Pauling, The Nature of the Chemical Bond, Cornell University Press, Ithaca, NY (1942).
2. C. Friedel and J. M. Crafts, Compt. Rend. 56 590 (1863).
3. C. Friedel and A. Ladenburg, Liebigs Ann. 143 118 (1867).
4. A. Ladenburg, Liebigs Ann. 164 300 (1862).
5. F. S. Kipping, J. Chem. Soc. 130 104 (1927).
6. R. R. McGregor, Private Communication at a Dow Corning Symposium (1960).
7. E. G. Rochow, An Introduction to the Chemistry of the Silicones, John Wiley and Sons, Inc., New York (1946). Second edition (1951).
8. R. R. McGregor, Silicones and their uses, McGraw Hill Book Co. Inc., New York (1954).
9. E. G. Rochow, U.S. Pat 2,380,995, August 7 (1945).
10. J. L. Speier, J. A. Webster, and G. H. Barnes, J. Am. Chem. Soc., 79 974 (1957).
11. E. P. Plueddemann, Chapter 9 in Treatise on Coatings, Vol I, Part III, R. R. Meyers and J. S. Long, Eds., Marcel Dekker, New York (1972).
12. E. P. Plueddemann, Adhesion Aspects of Polymeric Coatings, pp. 363-378, K. L. Mettal, Ed., Plenum Press, New York (1983).
13. J. R. Malek, J. L. Speier, and A. P. Wright, U.S. Patent 4,088,669, May 9 (1978).
14. L. H. Brown, Chapter 11 of ref. 11.
15. E. P. Plueddemann, J. Adhesion Sci. Technol., Vol 2, No. 3, pp. 179-188 (1988).
16. W. A. Finzel, Properties of High Temperature Silicone Coatings, Journal of Protective Coatings & Linings, Vol 4, No. 8, pp. 38-43 (1987).
17. W. A. Finzel, Silicone Coatings: Popular Choice for the Tough Jobs, American Paint and Coating Journal, January 19 (1981).
18. W. Finzel, K. Fey, New Water-Based Silicone Alkyds for Low-Polluting Coatings, American Paint and Coating Journal, June 28 (1982).

210

19. W. A. Finzel, Low VOC Silicone Alkyd Copolymers for Maintenance Paint Applications, Journal of Water Borne Coatings, Vol. 5, No. 3, August (1982).
20. W. A. Finzel, Higher Solids Silicone Alkyd Copolymers for Lower VOC Maintenance Coatings, High Solids Coatings, Vol 11, Spring (1986).
21. L. M. Parr, High Solids and Water-Reducible Silicone Polyesters, presented at the Eighth Water-Borne and Higher Solids Coating Symposium, February 26 (1981).
22. L. H. Brown, Chapter 11 (Silicone Additives) in Handbook of Coating Additives, L. J. Calbo, ed., Marcel Dekker, New York (1987).
23. W. T. Collins, Chapter 10 (Adhesion Promoters) of ref. 22.
24. P. Walker, J. Coatings Technology, 52 (668) 33 (1980).

Table I. Thermal Life of Various Substituents on Silicon

Group Covalently bonded to Si	Approximate half-life (hrs) at 250°C in air
Phenyl	>100,000
Methyl	> 10,000
Vinyl	101
Ethyl	6
Propyl	2
Cyclohexyl	40
Octadecyl	26

Table II. Equilibrium Hydrolysis of Siloxane Bonds

$$SiOSiR_n(OH)_m + H_2O \longleftrightarrow SiOH + HOSiR_n(OH)_m \quad (n + m = 3)$$

Structure of Siloxane	K eq
$(HO)_3SiOSi(OH)_3$	10^{-5} (estimate)
$(HO)_2RSiOSiR(OH)_2$	10^{-4} (estimate)
$HOR_2SiOSiR_2OH$	2.75×10^{-3}
$R_3SiOSiR_3$	5×10^{-2}
$Si-OR$	10^{-1} (estimate)

Table III.   Temperature Resistance of Coating Systems

Temperature*	Coating System Type
250-400°F (121-204°C)	Silicone modified organic[+] (colored)
400-600°F (204-316°C)	Silicone modified organic (aluminum) Silicone modified organic (colored) Organic modified silicone[++] (colored)
600-800°F (316-427°C)	Organic modified silicone (black and aluminum) Silicone (colored)
800-1000°F (427-538°C)	Silicone (black and aluminum)
1000-1400°F (538-760°C)	Silicone (ceramic and aluminum)

* 100 hours minimum
[+] Silicone modified organic = 15-50% silicone
[++] Organic modified silicone = 50-90% silicone

Table IV. Wet and Recovered Adhesion Values on Mild Steel

Paint	TREATMENT/PRIMER (2% silane in water)	Wet Adhesion MPa/	% Detached	Recovered Adhesion MPa/	% Detached
Polyurethane	Degreased only	5.6	100	6.8	100
	Degreased/primer	7.4	90	12.5	90
	Sandblast only	11.8	95	20.8	60
	Sandblast/primer	22.7	30	29.1	0
Epoxy	Degreased only	7.2	100	10.9	100
	Degreased/primer	28.1	0	29.2	10
	Sandblast only	9.2	100	20.9	100
	Sandblast/primer	26.3	50	27.8	40

Table V. Wet and Recovered Adhesion Values on Aluminum

Paint	TREATMENT/ADDITIVE Treatment/Additive	Wet Adhesion MPa/ % Detached		Recovered Adhesion MPa/ % Detached	
Polyurethane	Degreased only	1.3	100	peeled while drying	
	Degreased/ 0.2% additive	9.5	100	14.5	100
	Sandblast only	10.4	70	15.2	40
	Sandblast/ 0.1% additive	15.6	70	25.2	30
Epoxy	Degreased only	2.2	100	12.9	100
	Degreased/ 0.2% additive	25.3	0	26.9	0
	Sandblast only	7.4	100	22.2	20
	Sandblast/ 0.2% additive	28.2	0	28.7	0

Silane   = 2-(aminoethyl)-aminopropyl trimethoxysilane
Wet   = 1500 hours in water at room temperature
Recovered = after 48 hours in air at room temperature

A HISTORY OF POLYURETHANES

Kenneth N. Edwards

Dunn-Edwards Corporation
Los Angeles, California 90040

Introduction

Polyurethane coatings are, by and large, based upon a condensation chemistry that was an off-shoot of research into areas other than coatings. The basic chemistry existed well before World War II, however the class of resins would probably never have been developed to the extent they are now had Germany not needed rubber latex and other natural oleoresinous oils so desperately during the war years.

In this discussion three basic time periods are explored: first, the prewar years which set the development stage; second, the war and immediate postwar environment; and third, the developments leading up to the early 1970's at which time most of the developmental history had taken place and acceptance started to grow for isocyanate based coatings systems.

The present discussion is a history of polyurethane coatings, not polyurethane products which is an immense field. While reference is made to a number of non–coatings patents and other research, the attempt is to place coatings in time relative to other related works. The effort is to trace where and when various types of coatings evolved without digressing heavily into the chemistry or marketing segments. Elastomeric coatings are treated separately although like their high performance counterparts it is impossible to separate moisture cured and two–component systems.

Discussion

In order to start any discussion of a chemistry as diverse as that of the isocyanates, it is necessary to first define some of the primary reactions that make urethanes so important to the field of coatings and plastics.

To oversimplify, one might state that urethanes are polymers built upon nitrogen bridge reactions excluding urea–formaldehyde types which create their own special presence in the field of coatings.

The reactions which are of specific interest are those which give polyurethanes their name: i.e., those reactions which create chains based upon the repeating linkage "urethane." The two main reactions are, the "urethane reaction" of an isocyanate with a hydroxyl group, and the "urea reaction" of an isocyanate with water. Also included is the amine reaction which is used either by itself or as a first stage in the urea reaction.

$$R-N{=}C{=}O \; + \; R'OH \; \longrightarrow \; R-\overset{\displaystyle H}{\underset{\displaystyle |}{N}}-\overset{\displaystyle O}{\overset{\displaystyle \|}{C}}-O-R'$$

Urethane Reaction

Published 1990 by Elsevier Science Publishing Co., Inc.
Organic Coatings: Their Origin and Development
R.B. Seymour and H.F. Mark, Editors

213

$$R-N{=}C{=}O \ + \ R'NH_2 \ \longrightarrow \ R-\overset{\overset{\displaystyle H}{|}}{N}-\overset{\overset{\displaystyle O}{||}}{C}-\overset{\overset{\displaystyle H}{|}}{N}-R'$$

Urea Reaction

$$R-N{=}C{=}O \ + \ H_2O \ \longrightarrow \ R-NH_2 \ + \ CO_2$$

Amine Reaction Leading to Urea Reaction

If we look upon the isocyanate radical (–N=C=O) and the hydroxyl (–OH) as being attached to difunctional compounds, their reaction products will create linear chains apropos of filaments or elastomers. Such was the early work of Dr. Otto Bayer which will be discussed later. If however the functionality of the NCO and OH containing compounds become three or greater, crosslinked materials of high chemical resistance and inertness are created. High performance coatings as we use them are generally drawn from this end of the spectrum and tend to be very highly crosslinked.

Diamines or more highly functional amines will also crosslink with isocyanates to form chain– or crosslinked structures very similar to the urethane reaction structures.

The urea reaction, as applied to coatings, on the other hand, depends on the presence of water in the atmosphere to crosslink or "moisture cure" coatings. In the formation of foams, the addition of water is deliberate and the liberated CO2 providing all or part of the blowing agent.

Because water is present to some degree in pigments, polyols, and solvents even after drying, all coating applications from isocyanates probably contain traces of all three reactions.

Although some people attribute the start of urethanes to Wohler (1828) who studied the basic chemistry of carbonic acid and urea, isocyanate chemistry probably had its theoretical start with the early work of Wurtz and Hoffman. Prior to the mid–eighteen hundreds these two had already explored the chemistry of monobasic isocyanides. Although they were little better than laboratory curiosities at that time, both aliphatic and aromatic isocyanates were investigated

Early work languished because of a lack of good isocyanate materials. However, when in 1884, Hentschel developed his preparation technique through phosgenation of primary ureas, interest increased thanks to the excellent yields. No real work was done on isocyanate coatings precedents until the early 1930's. W. H. Carothers of duPont was interested in creating synthetic fiber and had been working with linear condensation polymers of various types especially polyester[1]
This work led to nylon, certainly the most successful purely synthetic polymer of its day. Patent 2,071,250 (1937) entitled Linear Condensation Polymers sums up this work and talks about what he called, "linear condensation superpolymers and by this I mean linear condensation polymers capable of being formed into useful fibers," as well as manufacturing techniques for doing so.

A.S.T.M. has divided the five types of urethane coatings created by isocyanate reactions. See Figure 1 below.

FIGURE 1.

ASTM CLASSIFICATION OF POLYURETHANE COATINGS INTO FIVE BASIC TYPES

	One Package (Prolonged Storage Stability)			Two Package (Limited Pot Life)	
ASTM Coding	1	2	3	4	5
Description	Uralkyd	Moisture Cure	Heat (Un-blocking Cure)	Catalyst Cure	Polyol Cure
Curing Mechanism	Oxidation	Moisture	Unblocking with Heat	Catalysis	Hydroxl Groups of Polyol
Pigmentation	Conventional	Difficult*	Conventional	Difficult*	Pigment in Polyol
Overall Rating	Good	Very Good	Excellent	Excellent	Excellent

*Dry, Nonalkaline Pigments Normally Specified

Meanwhile in Germany, Dr. Otto Bayer was starting to experiment with products based upon diisocyanates as a path to fibers competitive to nylon and on March 26, 1937, the German experimental team discovered the diisocyanate polyaddition process which led to German patent 728,981. Linear polyurethanes had advanced German synthetic fiber chemistry to such a point that by 1941 two commercially successful versions called "Perlon U" and "Igamid U" were available. "Perlon U" was used for bristles and other synthetic fiber applications, and "Igamid U" for other plastics applications using linear fibers.

This same year, Dr. Bayer applied for permission from the war time government to construct an isocyanate plant capable of producing 200 tons of aliphatic isocyanate and 100 tons of aromatic isocyanates. In the application he referred to potential applications in the fields of adhesives, foams, lacquers, coatings, synthetic leather, etc.

The use of urethanes in German aircraft coatings was particularly impressive. Allied investigators were perplexed when American performance tests of captured German aircraft did not correspond by a significant percentage to the tests done on these same aircraft by their British counterparts. It was then determined that the only difference between the two sets of tests was that the aircraft had been repainted prior to shipment to the U.S.A. Stripping the new aircraft finish and reverting back to the German urethane restored the performance. Almost all aircraft today are coated with polyurethanes which are derived from these coatings.

Work on isocyanates was proceeding in the United States as well as in Germany. However, the war time U. S. still retained adequate drying oils and coatings were not a major priority. Rothrock (duPont) obtained two patents regarding the modification of alkyds, one in 1942 and a second in 1944.

Rothrock (duPont) in 1945, patented his work on metallic driers as catalysts in isocyanate reactions. Lichty and Seeger (Wingfoot) added isocyanates to rubber paint. Miller and Pratt (duPont) and Nyquist and Kropa (American Cyanamid) continued modifications of alkyds with isocyanates through the later 40's and into the 1950's.

It was not until the publication of reports made by scientific evaluation teams which explored the German advances, that American industry became interested in polyurethanes. The United States Air Force granted a development contract to Goodyear Aircraft to develop rigid urethane foams with work commencing in September of 1946. In 1947 and 1948, Lockheed Aircraft Corporation did successful work developing two-component foams systems for radome construction and filling components to achieve sandwich effects. During the years 1951 through 1953, seven patents were issued to Simon and Thomas of Lockheed for these systems.

Dupont and Monsanto were developing techniques for manufacturing commercial quantities of 2,4 tolylene diisocyanate and in fact started production in 1951. Preliminary exploration of the TDI-polyol resins was evidenced by their use in casting electrical insulators and phonograph records. Rheineck and Breslow of Hercules were modifying nitrocellulose and polyols with isocyanates to produce coatings.

By the early 1950's the coatings industry in the U.S. had not done much with the two-component or blocked isocyanates although ASTM type I uralkyds were becoming popular in many areas. Germany however, was a different story.

Following the war, Bayer started promoting the isocyanate-based coatings in Europe and elsewhere. These coatings evidenced fuel, water, and corrosive atmosphere resistance equal to or better than other coatings of the day. In general, the early T.D.I.-based isocyanate coatings had weather and chemical resistance equal to amine cured epoxides. Isocyanates seemed to have an edge in acid resistance and epoxies had an edge in alkaline resistance but the differences were not great. Both chalked and yellowed markedly in sunlight, but retained long service life. By contrast, aliphatic isocyanates tended to be transparent to sunlight and while not as chemically resistant as T.D.I.-based isocyanates, had outstanding gloss retention and weather resistance. Germany had both aliphatic and aromatic production facilities to manufacture these isocyanates.

A series of polyesters were formulated with ratios of 1.5 to 3.0 moles adipic acid, 0.5 to 3 moles phthalic anhydride and approximately 2 to 4 moles of triols such as glycerine, trimethyol propane and butylene glycol. To these, Bayer added several isocyanate fractions such as hexamethylene diisocyanate, tolylene diisocyanate (65% 2,4 and 35% 2,6), and naphthalene diisocyanate about 1,5 moles. Because Bayer also recognized that the low vapor pressure of some of the above isocyanates could create health problems in the workplace, they also created modified isocyanates with such combinations as 1 mole of trimethylol propane to 3 moles of hexamethylene diisocyanate, or 1 mole of trimethylol propane to 3 moles of tolylene diisocyanate.

These raw resin materials slowly gained prominence in Europe with both two-component and single component moisture cured polymers being used for coatings applications. By the mid-1960's a large number of working vessels, trucks, railroad stock, and aircraft were being coated by coatings based upon a hexamethylene diisocyanate trimer. A great deal of celebration took place when the isocyanate coated Queen Elizabeth II was launched on September 21, 1967. It was felt that the favorable performance to cost ratio of aliphatic isocyanates had finally gained acceptance.

In the United States, the primary market force was directed at both hard and soft foams. As a result, tolylene diisocyanate was in ready supply much earlier than was hexamethylene diisocyanate. DuPont entered production of T.D.I. about 1951 and remained a large supplier until 1978. They entered production of hexamethylene diisocyanate in about 1961 and phased it out in approximately 1981.

In 1954 Bayer and Monsanto Chemical Company entered into a joint venture which was named Mobay Chemical Company. The purpose of this venture was to produce tolylene diisocyanate and to exploit the Bayer knowledge especially in the field of foams. Mobay also produced polyester resins and T.D.I.-based isocyanates for the coatings field.

Bayer was not satisfied with the Mobay resin production in the coatings field and through Naftone, a United States distributor of goods directed toward the coatings industry and as a distributor for the plastics and coatings division, Bayer started exporting German isocyanates and polyesters. Through this avenue, and supported by a laboratory in New Jersey, European coatings' knowledge was introduced to the American marketplace. In 1963, Naftone started importing Bayer's aliphatic trimer of hexamethylene diisocyanate so light stable isocyanates were available in this country for the first time.

Spencer–Kellogg Corporation was also producing isocyanate coatings resins based upon modified castor oil rather than polyesters. Coatings based upon this system had better exterior weatherability than pure polyesters but lacked chemical resistance to severe environments.

By 1970, Bayer had bought out Monsanto, its partner in Mobay, and Naftone and in that year combined the two companies into the Mobay Corporation of today with its separate Coatings Division.

Elastomeric Coatings

One area of unique and specialized polyurethanes is that of elastomerics. Because of the great usage of polyurethane foams in this country, a great number of polyols both polyester and polyether based have been developed. These polyols have two or more functional groups and molecular weights of up to and beyond 4000. Union Carbide, Shell Chemical, Ltd., the Wyandotte Company, Lankro Chemicals U.K, and Dow Chemical Company were all primary polyol developers.

The polyols are reacted with isocyanates to create both one and two component materials which are used for waterproofing in major construction projects. A typical example might be an automotive parking deck where the coatings must exhibit excellent elongation and durability and weather resistance.

During the early 1960's, considerable development work was being done on polyurethane based caulking materials by a number of manufacturers and interest in urethanes was developed in the construction field. When a need was perceived for a very durable waterproofing membrane that would be able to stretch to cover cracks in a concrete structure, Reichold Chemical Corporation, based upon European elastomeric chemistry, which coupled rubber, caulk and foam chemistries, developed a two-component gun head mixed elastomer and started franchising applicators. The

218

gun head mix materials were very difficult to apply on large construction jobs because of mechanical rather than chemical considerations. Many of the formulators reformulated to create moisture cured variations of the two–component isocyanates. These systems solved the mix problem, but had slower curing and a tendency to foam if applied to damp substrates.

These coatings require great skill to formulate and apply but can give outstanding performance under difficult conditions.

One of the interesting variations of these elastomers has resulted in the development of high performance running tracks in universities and stadiums around the world.

Appendix 1

A Word of Explanation

In order to define some of the time periods as exactly as possible, a patent search of U. S. patents was undertaken covering the approximate thirty years from 1930 to the early 1960's, with emphasis on coatings and especially two or three corporations. Subsequently an excellent review on the overall subject of polyurethanes was discovered which had an extensive patent list covering the entire field during these same years by J. H. Saunders and K. C. Frisch[3]. A number of non–coatings patents from their list were added to the following list to provide perspective. Furthermore, the following list, while fairly complete, is not intended to cover all coatings patents issued. For more complete information on this subject reference to Saunders and Frisch is recommended.

1932

Modification of fibers with isocyanate to improve dyability. Hartman (Society of Chimical Industry in Basle). 1,875,452.

1933

Aminoazo compounds with phosgene gave isocyanates. Hilger (General Analine). 1,916,314.
Carbamoyl chloride from ammonium chloride and phosgene. Theis (I. G. Farben). 1,937,328.

1939

Hexamethylene diamine and diethylurethane from HDI gave filaments. Martin (duPont). 2,181,663.

1941

Resin from alkyleneimines and isocyanates. Esselmann, Dusing (Duisberg). 2,257,162.

1942

Hexamethylene diisocyanate improves alkyd drying. Rothrock (duPont). 2,282,827.
Linear polymers from diisocyanates and diols or dithiols. Catlin (duPont). 2,284,637.
Printing plates by injection molding or casting polyurethanes. Kollek (APC). 2,302,037.

1943

Isocyanate from amine hydrochloride and phosgene, continuously removing HCl. Siefken, Doser (General Analine) 2,326,501.

Coatings from polyesters and, polyester–amides with polyisocyanate; Christ, Hanford (duPont). 2,333,639.
Electrical insulators from HDI and polyamides. Berchet (duPont). 2,333,914.

1944

Linear polyureas and polyurethanes for phonograph records. Mediger (APC). 2,342,679.
Diisocyanates or diurethanes condensed with amine containing bifunctional compounds. Schlack (APC) 2,343,808. .
Formaldehyde resins with isocyanates. Pratt (duPont) 2,349,756.
Cure of drying oils promoted by diisocyanates. Pratt, Rothrock (duPont). 2,358,475. .

1945

Metallic driers catalyze isocyanate or isothiocyanates reactions. Rothrock (duPont). 2,374,136.
HDI from amine hydrochloride and phosgene. Farlow (duPont). 2,374,340.
Glycol–acrylate–styrene polymers with diisocyanate create moldable compounds. Kung (Goodrich). 2,381.063.
Diisocyanates improve adhesion when added to rubber paint. Lichty, Seeger (Wingfoot). 2,388,656.

1948

Drying oils, with silicone and titanium isocyanates give faster drying, harder films. Miller, Pratt (duPont). 2,449,613.

1949

Carbamates from dehydrogenated rosin isocyanates used in varnishes and lacquers, Rosher (Hercules). 2,492,938. .

1950

Unsaturated alkyds and unsaturated isocyanates. Nyquist, Kropa (Am. Cyan.). 2,503,209.
Diisocyanates and diols or diamines. Rinke et al. (U.S. Attorney General) 2,511,544. .
A diethyl isocyanate for a resin compatible with nitrocellulose. Rheineck (Hercules). 2,522,584 .
Coatings from di– and triisocyanates and glycols. Breslow (Hercules). 2,531,392. .

1951

Foams from polyvinyl resins, azo diisobutyronitrile, phenyl isocyanate. Carpentier. 2,576,749. .
Flame resistant foams from alkyd resins, TDI, and unsaturated aryl phosphonates. Simon, Thomas (Lockheed). 2,577,281.

1952

Adhesives from epichlorohydrin, bisphenol, and diisocyanates. Nelson (General Electric). 2,594,979.
Polyesters and diisocyanate as non–volatile plasticizers. Coffey et al (I.C.I.). 2,606,162. .
Polyesters, diisocyanates, and diamines; elastomers. Muller (Bayer). 2,620,516.

1953

Polyesters with diisocyanates give gums, which are cured with diisocyanate. Seeger (Wingfoot). 2,625,531.

Polyesters, and polyester amides cured with 0.7–0.99 mole diisocyanate, then with more polyisocyanate. Mastin, Seeger (Wingfoot). 2,625,535.

Coatings from phenol and unsaturated fatty acid, then addition of polyisocyanate. Hermann (Reichold) 2,645,623.

1955

Urethane tire treads bonded to SBR carcasses using polychloroprene on SBR then urethane elastomer and finally the tread. Schwartz (duPont). 2,713,884.

Nylon coated with polyester–prepolymer, treated with diamine, elastomeric coating. Dacey et al. (U. S. Rubber). 2,721,811.

1956

Elastomers from polyester–prepolymers, cured with glycols. Muller et al. (Bayer) 2,729,618.

Urea–isocyanates from two moles of diisocyanate, one of water. Pelley (duPont) 2,757.184.

Elastomer from linear polyester prepolymer, then diamine. Muller, Petersen (Bayer). 2,778,810.

Articles packaged using foam. Simon, Thomas (Lockheed). 2,780,350.

Cyclic ureas and polyisocyanates for coatings, films, molding compounds. Fraser, Dickey (Monsanto). 2,801,230.

Polyurethanes from glycols and diurethanes with various metal catalysts. Caldwell (Eastman). 2,801,230.

Wood coated with polyisocyanate containing ethyl cellulose. Kelly, Heiser (Dow). 2,804,400.

Prepolymers with unsaturated polyesters, copolymerized with vinyl compounds. Nischk, Muller (Bayer). 2,806,835.

Hydrogenated polyepoxy ethers of polyhydric phenols with polyisocyanates; coatings. Shokal (Shell Dev.) 2,809,177

Tall oil acid glyceride and diisocyanate coatings. Culemeyer (Spangenberg). 2,812,337.

Articles coated by dipping alternately into solutions of prepolymer and diamine. Hess, Dacey (U. S. Rubber). 2,814,834.

Soluble polyurethane obtained from aliphatic diisocyanate and glycol with insoluble polyurethane give baking finish. Altner (Bayer) 2,817,643.

1958

Isocyanates prepared by continuous phosgenation, pressure, 90–180° C.,turbulent flow. Beck (duPont). 2,822,373.

Polyurea–urethanes by heating urea, triamine and glycol. Saunders (Monsanto) 2,824,856.

Linear polyurethanes with ethylene oxide; for plasticizers. Cohen (Polaroid). 2,835,653.

Isocyanates stabilized against color change and polymerization with silicone compounds. Bloom et al. (General Analine). 2,835,692.

Elastomers from OH–terminated polyether prepolymer and NCO– terminated prepolymer, with additional diisocyanate and butadiene–acrylonitrile copolymer. Mason (Armstrong Cork). 2,852,438.

Plates, shells, and shaped elements with rigid foam cores. Hoppe (Bayer). 2,855,021.

Polyisocyanates from diisocyanates and mixtures of triols and glycols. Bunge, Bayer (Bayer). 2,855,421.

Diisocyanate–alkyd foamed with tertiary alcohol and strong acid. Hindersinn, Creighton (Hooker). 2,865,869.

Foamed coatings by applying isocyanates over a catalyst, and water treated polyester. Gensel et al. (Bayer). 2,866,722.

221

1959

Polyisocyanates used to seal porous rock formation in oil well drilling. Mallory, Ayscue (Great Western Drilling Co.) 2,867,278.

Urethane varnish from diethyl alcohol and isocyanate. Leclercq, Paquet (Union chemique Belge.) 2,867,644.

Synthetic paper sheet from batt of polyurethane fibers. Hubbard, Koontz (duPont)., 2,869,973.

Thermoplastics from diisocyanates and mixture of glycols of the types $HO(CH_2)nOH$ and $CH_3CHOH(CH_2)nCHOHCH_3$. Urs (Union Carbide). 2,873,266.

Tires by casting liquid urethane into mold containing textile reinforcing carcass. Cadwell et al. (U.S Rubber). 2,873,790.

Metallic dicyanamide stabilizes of polyurethanes against light or moisture. Burkus (U.S Rubber) 2,877,192. E.

Castor oil prepolymers stabilized by urea. Roussel (duPont). 2,877,193

Crosslinking unsaturated compounds containing blocked isocyanate groups. Bartl, Holtschmidt (Bayer). 2,882,260.

Leather coated with polyisocyanates. Loshaek et al. (Rohm and Haas). 2,884,336

Unsaturated isocyanates as a replacement for linseed oil in paint. Doggett et al. (A. D. Little). 2,886,455.

Salts of organic acids with tertiary amines as catalysts for splitting blocked isocyanates. Bunge et al. (Bayer) 2,886,555.

Latex impregnated paper treated with isocyanate. Hayes et al (W. R. Grace) 2,897,094.

Coating from polyether, and diisocyanate. Croco (duPont). 2,901,467.

Mixing hydraulic cement with prepolymer. Szukiewicz (Allied Chem.) 2,902,388.

Free films created by coating metal with polyurethane, then dissolving away the metal. Engelhardt (Bayer–Mobay). 2,909,809.

Coatings from fluorinated glycol and diisocyanate. Smith (Dow Corning). 2,911,390

Coatings from polyisocyanate, polymerized with vinyl compound. Nischk et al (Bayer–Mobay). 2,915,493.

1960

Coatings from hydroxyalkyl siloxanes and diisocyanates. Speier (Dow Corning). 2,925,402.

Air-drying varnish utilizes tin catalyst, and polyisocyanate. Leclercq et al. (Union Chimique Belge.) 2,926,148.

Resins from diisocyanate and furfuryl alcohol. Leclercq et al. (Union Chimique Belge.). 2,926,157.

Coating textiles with polyester then vinyl chloride polymer containing polyisocyanate. Tischbein (Bayer–Mobay). 2,929,737.

Coatings from diisocyanate and hydroxyalkyl silicones. Clark, Thomas (Dow Corning). 2,931,786.

Coating composition from polyisocyanate and isophthalic polyesters. Fox et al. (General Electric) 2,935,487.

Electrical coating from oil-modified polyester using dimethyl terephthalate. Agens (General Electric). 2,936,253

Ion exchange resins from diisocyanates and certain bis–hydroxy compounds containing acid groups. Fox (U. S. Navy). 2,948,690.

Light sensitive compounds from isocyanates and certain unsaturated ketones. Schellenberg et al. (Bayer–Mobay). 2,948,706.

Crosslinked polyurethane compounds containing plasticizers, and diamino trihydroxy compounds. Khawam (Allied Chem.) 2,956,031.

Polyurethanes from allyl alcohol copolymers and alkenyl–substitute isocyanates. Tess (Shell Oil). 2,965, 615.

1961

Heat cured drying oils from butadiene and diisocyanate. Koenecke, Gleason (Esso Res. and Engr.). 2,968,647.

Drying oil esters treated with polyols and isocyanates. Hauge, Pawlak (Spencer Kellogg). 2,970,062.

Preventing gelation of isocyanate treated oil modified alkyd resin. Rhodes et al. (Allied Chem.) 2,970,123.

Coal tar–epoxy materials cured. Whittier et al. (Pittsburgh Chem.) 2,976,256.

Coatings from polyesters of isophthalic acid and polyhydric alcohols. Sheffer et al. (Schenectady Varnish). 2,982,754

Urethanes stabilized against light and U.V. substituted 2.2'–dihydroxy benzophenomes. Hoeschele (duPont). 2,984,645.

Unsaturated polyester coating with monomeric vinyl compound. Abbott et al. (Sherwin–Williams). 2,993,807.

Modifying drying oils by alcoholizing castor oil with glycol, and reacting with diiocyanate and ethanolamine. Schwareman (Spencer Kellogg). 3,001,958.

Primer coating composition from polyisocyanate, polyester and epoxide resin. Hudson (Mobay). 3,012,984.

Coating from prepolymer of 400–3000 M.W. glycol or triol, with aliphatic diol and another polyhydric polyol prepolymer. Ansul (duPont) 3,012,987.

Electrical insulating films from polyester and polymethylene polyvinyl isocyanate. Flowers (General Electric) 3,013,906.

References

1. Carothers and Arvin., J. Am Chem Soc. 51 2560 (1929).

2. Carothers and Van Natta., J. Am Chem Soc. 52 314 (1930), see also U.S. Patents 2,012,267 and 1,995,291.

3. *Polyurethane, Chemistry and Technology*, 2 parts. J. H. Saunders and K. C. Frisch. Interscience Publishers 1962 & 1964.

4. *Polyurethanes, Chemistry, Technology and Properties*, L. N. Phillips and D.B. V. Parker. Icliffe Books Ltd., London 1964.

5. *Polyurethanes*, B. A. Dombrow, Reinhold Publishing Corp. 1957.

6. Private Communication, A. Pia, Mobay Corp. Retired.

7. Private Communication, W. Griffin, DuPont Chemical Corp.

SATURATED POLYESTER COATINGS

G. R. Pilcher and J. W. Stout
Hanna Chemical Coatings Corp., 1313 Windsor Avenue, Columbus, Ohio 43211.
Mailing Address: P. O. Box 147, Columbus, Ohio 43216-0147

ABSTRACT

The history of saturated, reactive polyester polymers is explored from
the earliest polyester syntheses up through the present work in coatings
science and technology. Emphasis is placed on the three decades following
1960, when the use of saturated polyesters in thin film protective coatings
first became commercially viable. Discussion centers around the composition,
synthesis, and use in coatings of reactive polyesters, silicone polyesters,
polyester urethanes, powder coatings, high solids and waterborne systems,
for a variety of end-uses in multiple marketplaces.

REACTIVE POLYESTERS

The modern history of polyester resins is an intriguing one, beginning
with the crude lactic acid material first made by Gay-Lussac and Pelouze
in 1833[1], and continuing with Berzelius' work with tartaric acid and
glycerol[2], and Watson Smith's synthesis of a polymeric product from
glycerol and phthalic anhydride[3], right into the present, where work in
thin-film coatings made from saturated polyester polymers continues to display both scientific and technological innovation and vigor. Despite the
considerable work reported in the nineteenth century--condensation products
from glycerol and mannitol with tartaric, citric, sebacic, and phthalic
acids, et al., as well as observations regarding condensates from dihydric
phenols, hydroxybenzoic acids, salicylides, etc.[4]--the first significant
use of polyesters may be said to be the glyceryl phthalate (glyptal) coating
and impregnating materials used during World War I[5]. Around 1914, the
term "alkyd" came into use, to differentiate modified resins (e.g., glyptal
modified with fatty acid of oleic) from glyptals[6], and in 1927, Kienle
prepared oil-modified resins using the alcoholysis technique[7]. It was
left to Wallace Carothers and Hermann Staudinger, however, to produce the
applied results which clearly defined the nature of polymeric materials.
Carothers "established conclusively that polymers were just ordinary
molecules, only longer..."[8], defined the groundwork for sequential polymerization, and dealt with the concepts of structure, properties and molar
mass[9]. After the passage of over half a century, Carother's basic
research still stands "as a watershed event in the history of theoretical
chemistry and materials science,"[10] and his early work,[11,12] along
with Flory's,[13,14,15] form a rich and fascinating background for study.

The commercial significance of saturated, reactive polyesters in coatings dates only from the mid-1960's, when heterochain macromolecules with
carboxylate ester groups in repeating units[16] began to find their way,
when suitably crosslinked and thermally set, into a broad variety of
specialty chemical coatings. These commercial polymers are normally
hydroxy- or carboxy-functional, and are synthesized from glycols and dibasic
acids to yield linear polyesters, and from higher polyhydric alcohols and
polybasic acids to give branched polymers. Typical components might be
among the following:

Polyhydric Alcohols

Diethylene glycol
1,3-Butylene glycol

Published 1990 by Elsevier Science Publishing Co., Inc.
Organic Coatings: Their Origin and Development
R.B. Seymour and H.F. Mark, Editors

223

Polyhydric Alcohols

Neopentyl glycol (NPG)
Ethylene glycol
Propylene glycol (PG)
Pentaerythritol
Trimethylolethane (TME)
Trimethylolpropane (TMP)
2,2,4-Trimethyl-1,3-Pentanediol (TMPD)
Dipropylene glycol
1,6-Hexanediol
1,4-Cyclohexanedimethanol (CHDM)

Polybasic Acids

Adipic acid (AA)
Isophthalic acid (IPA)
Terephthalic acid
Phthalic anhydride (PA)
Azelaic acid
Sebacic acid
Trimellitic anhydride (TMA)
1,4-Cyclohexanedicarboxylic acid (CHDA)
Chlorendic anhydride

Synthesis of saturated, reactive polyesters from these materials is via a straightforward condensation reaction, and can be represented by diagrams, modified from Lanson[17], for both linear and branched polyesters:

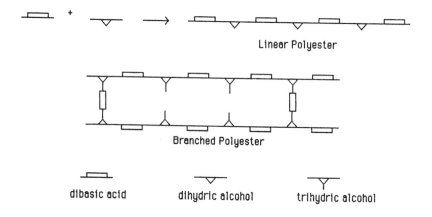

Linear Polyester

Branched Polyester

dibasic acid   dihydric alcohol   trihydric alcohol

Since this is an equilibrium reaction, the final molecular weight for the polymer is significantly affected by the efficiency with which evolved water is removed. The choice of catalyst is critical; e.g., dibutyl tin oxide works fine, but--if the final polyester is reacted with an aliphatic isocyanate in a two-package polyurethane system--the reduced pot-life would render the system unusable[18]. To produce reactive hydroxylated polyesters, the polyol must be in excess, but this is not true for the synthesis of non-reactive polyester plasticizers (which are typically linear, with molecular weights of less than 10,000), which may include a monohydric alcohol or monobasic acid in the formula[19].

During the early work on saturated polyesters, selection of reactants sometimes had as much to do with sheer chance as it had with scientific selection, but empirical work performed in the two decades following the mid-1960's yielded some noteworthy observations. With regard to polyols, it was noted that 1,6-hexanediol imparted both flexibility and chemical resistance to polymers[20], neopentyl glycol contributed to heat stability[21], and TMPD imparted stain- and detergent-resistance. Work done by Finney[22] with melamine-crosslinked polyesters indicated that polyol selection did not greatly affect certain initial physical properties of the tested coatings, such as gloss and hardness, but that other properties—particularly stain-resistance and overbake stability—were significantly affected by polyol selection. Among the polybasic acids, adipic, azelaic and sebacic were found to impart flexibility, and have frequently been used in preparing polyester plasticizers. Chlorendic anhydride and tetrachlorophthalic anhydride have been reported to impart fire-retardant characteristics, and isophthalic acid was shown to contribute higher viscosity and improved weathering performance over orthophthalic, as well as greater flexibility and more chemical-resistant films[23]. Benzoic acid was reported to increase adhesion characteristics, albeit with some decrease in flexibility[24]. It may be possible to produce polymers with improved resistance to hydrolysis by using polybasic acids containing sterically hindered carboxyl groups[25]—work that was being more fully-explored in the late-1980's.

Because of the hydroxyl functionality of the majority of solvent-borne reactive polyesters, they exhibit the potential for reaction with several chemical moieties, among them alkoxy, isocyanate, amino, and carboxyl groups, to produce an impressive array of polymeric products. Commercially, the most common coreactant (crosslinker), which rose to dominance in the late 1960's, is hexamethoxymethylmelamine, which usually yields a balanced set of physical properties—hardness, flexibility, resistance to solvent and salt fog attack, etc.—at the lowest Volatile Organic Content (VOC) and most economical cost. Mixed ether alkylation of the melamine, however, can often impart desirable properties to a polyester film, such as improved flow, sharpened distinction of image and even lower VOC, although often at a price. These products are usually more expensive, and due to the higher molecular weight of their reactive alkoxy groups (which are typically butyl or isobutyl), may offset the VOC advantage of their lower viscosities. The reaction between a saturated polyester and any fully alkylated melamine can be depicted as follows:

$$R-CH_2OR' + R''-OH \xrightarrow{H+} R-CH_2O-R'' + R'OH$$

Where R = Melamine
R' = CH_3, C_2H_5, C_4H_9, etc.
R'' = Hydroxy-functional polyester

Hexamethoxymethylmelamines are predominantly monomeric, and require strong acid catalysts (pKa < 1), such as sulfonic acid. More polymeric melamines, which are both alkoxy-functional and methylol-functional, respond well to weak acids (pKa > 1.9) such as phosphoric acid, and produce films with minimal blistering tendency, and low formaldehyde content.

To achieve systems with maximum detergent, salt spray, and chemical resistance, reactive polyesters may be reacted with alkoxy methyl benzoguanamine, which, following heavy use in the 1960's and early 1970's, fell into disuse for a decade, and enjoyed something of a renaissance in the mid-1980's. Also in the 1980's glycolurils began commanding attention for their ability to achieve the lowest possible releases of formaldehyde during curing, as well as for imparting improved stability and corrosion protection, enhanced adhesion to metal substrates, and—in certain cases—increased flexibility. These polymers are monomeric, may contain alkoxy or methylol reactive groups, require the use of strong acid catalysts, and are generally not used in conjunction with weakly basic reactive materials such as melamine or urea, which can result in inhibited cure response.

A class of crosslinking resins which enjoyed widespread use in the 1960's and early 1970's, but which has seldom been used with exterior-grade polyesters since, are the urea-formaldehyde resins. Nonetheless, they are desirable for their ability to cure at low temperatures, and are used in polyester primers and non-weathering, spray-applied topcoats. The alkylated ureas commercially available may vary in properties from those that are fully methylated, water-soluble, and capable (with controlled acid catalysis) of room-temperature cure, to those that are butylated, hydrophobic, and display very little tendency to self-condense. Basic structures for the various crosslinking resins most commonly used with saturated polyesters in the formulation of chemical coatings are shown on the accompanying table.

Coatings made from saturated, hydroxy-reactive, oil-free polyesters (the term "oil-free alkyd," which came into wide use in the late 1960's, had finally begin to fall into disuse by the mid-1980's) first began appearing in the late 1960's, and were predominantly employed in spray finishes for general purpose applications. Moving into the early 1970's, increasing usage was made of these materials as the basis for coil-coated, post-formed finishes for building products, such as pre-engineered steel building panels and aluminum residential siding and accessories. By the mid-1970's, saturated polyesters were being used in such areas as T-bar coatings for suspended ceilings, spray appliance coatings, and coatings for steel shelving. By the end of the decade, the superior combination of hardness and flexibility of these polymers had allowed them to become the basis for coil-coated appliance coatings, such as refrigerator wrappers, an application which was to have enormous influence and growth potential over the next two decades. In the 1980's, uses extended to automotive underhood applications, and during the final years of the decade, intensive work was underway to produce the prototypes for coil-coated automotive topcoats.

POLYESTER POLYURETHANES

Although polyurethanes have been known since the early 1940's, as a result of Catlin's work[26], the really widespread use of polyester polyurethanes in both two-package industrial spray applications, and in one-package coil coatings applications, didn't occur until the decade which spanned roughly 1975-1985. Although many specialized polyols and "polyol esters" were developed specifically for reaction with isocyanates, it was quite common to react saturated, hydroxy-functional polyesters with isocyanate prepolymers to achieve somewhat differing properties than could be achieved with amino resin crosslinking. Among these, depending upon the polyester and the isocyanate (aromatic or aliphatic), might be improved flexibility and adhesion, greater film "toughness," upgraded mar-resistance, and enhanced chemical resistance. If an aliphatic isocyanate compound was employed, improvements in exterior durability and gloss rentention were often experienced, as well. These reactions were normally thermally-set, in a manner analagous to the curing of polyester-amino systems, although they were frequently called upon to cure under ambient conditions, also. Although isocyanates may react with a number of groups, following is a basic representation of the reaction with a reactive, hydroxylated polyester[27]:

$$R-N=C=O \ + \ R'OH \longrightarrow R-N-\overset{\overset{\textstyle O}{\|}}{C}-OR'$$

COMMONLY-USED CROSSLINKING RESINS

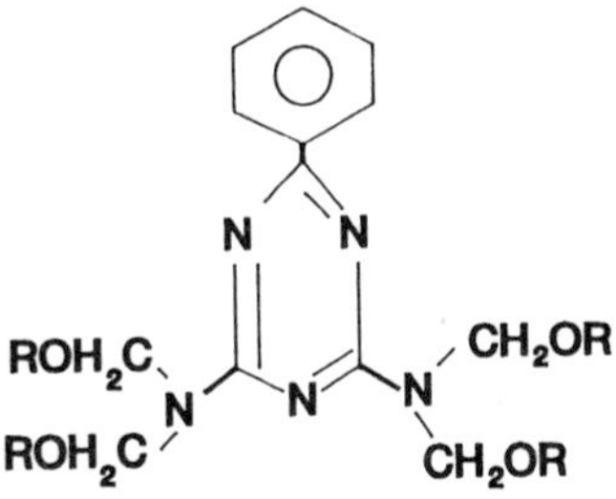

BENZOGUANAMINE

R=CH$_3$, C$_2$H$_5$, et al.

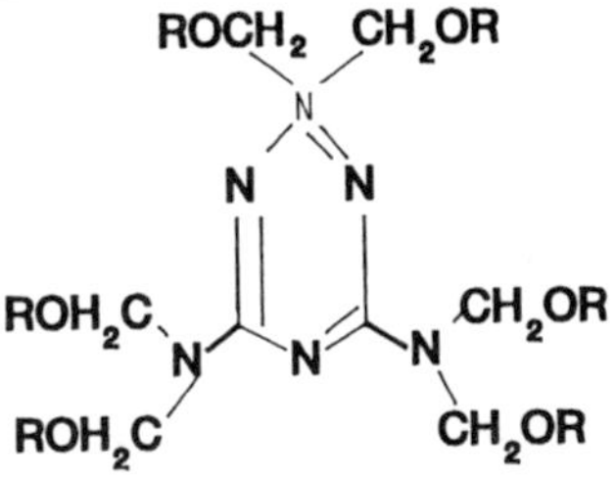

FULLY-ALKYLATED MELAMINE

R=CH$_3$, C$_2$H$_5$, C$_4$H$_9$, et al.

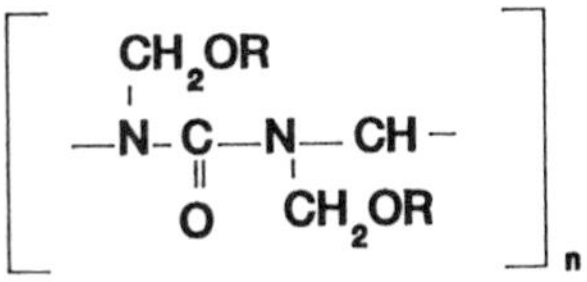

UREA-FORMALDEHYDE RESIN

R and R' = H, CH$_3$, C$_4$H$_9$, et al.

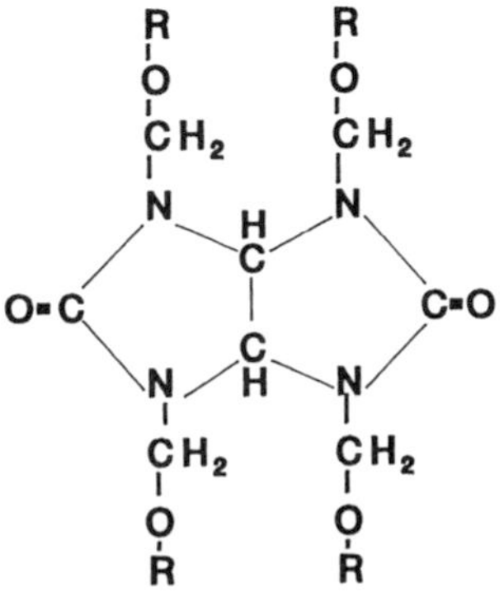

GLYCOLURIL

R=H, CH$_3$, C$_2$H$_5$, C$_4$H$_9$, et al.

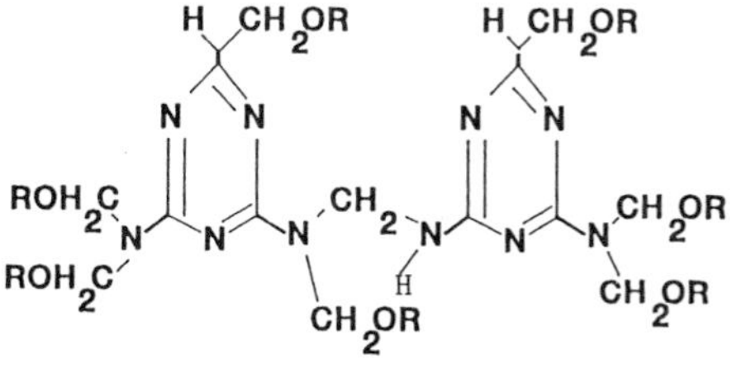

PARTIALLY-ALKYLATED MELAMINE

R=CH$_3$, C$_2$H$_5$, C$_4$H$_9$, et al.

In the decade of the 1980's, significant work was done with one-package, coil coated polyester urethanes, which were destined for a wide variety of end-uses, including appliance primers, automotive underhood primers and finish coats, and topcoat applications requiring a deep draw, such as coatings for oil filters, condenser pans for appliances and air filters.

SILICONE POLYESTERS

Although the unique properties of silicone-modified polyester coatings---resistance to heat and ultraviolet degradation, a high degree of hydrophobicity, and excellent weathering characteristics[28]---have allowed them to find many niches in a variety of industrial coatings applications, they have achieved their highest visibility as thermosetting finish coats in the pre-engineered and agricultural building marketplaces. Beginning in the late 1960's, their outstanding balance of properties caused their use to escalate throughout the 1970's and 1980's to the point where they were the dominant technology in these areas. Most of these coatings were applied by the coil-coating method, and then post-formed into building panels and components, although they were also spray-applied onto pre-formed panels and parts.

In the United States, by the mid-1980's, silicone polyester was the dominant "weather resistant" technology, sharing the marketplace with coatings based upon poly(vinyl)chloride (PVC) and poly (vinylidene)fluoride (PVF$_2$). The demand for silicone polyester coatings was divided into three categories of interest: low-level silicone (less than 20% of the polymer composition), the so-called "thirty percent" silicone level (generally 23-30%), and the "fifty percent" silicone (42-50%) products. Each category reflected the general understanding that the durability of coatings improved with silicone level, although the many approaches which can be used to calculate silicone content (percent Si on polyester portion of resin? On total resin? Percent of reactants?, etc.) led to discrepancies and general blurring of categories. A general understanding, with a well-documented empirical underpinning, has been that a minimum of 20% silicone is required for significant improvement in chalk resistance of weathered coatings, and that 30% is close to optimum for "performance versus material cost"[29]. There is general agreement on this, although unpublished data indicates that marginal, but incremental, improvement in general durability can be seen from levels as low as 2% silicone.

The preparation of silicone polyester polymers involves a straight-forward reaction of the type:

$$R-Si-OCH_3 + HO-R' \text{ ----- } R-Si-O-R' + CH_3OH$$

Where R = Silicone intermediate prepared from chlorosilane monomers

R' = **Any linear or branched polyester,** specifically prepared with excess hydroxyl groups for reaction with a silicone intermediate

A typical structure for a silicone intermediate would be as follows:

$$CH_3O-\underset{\underset{R}{|}}{\overset{\overset{R}{|}}{Si}}-O-\underset{\underset{R''}{|}}{\overset{\overset{R'}{|}}{Si}}-O-\underset{\underset{R}{|}}{\overset{\overset{R}{|}}{Si}}-OCH_3$$

Where R, R', and R'' may be CH_3, OCH_3 or C_6H_5 (phenyl)

The ultimate durability characteristics of any silicone polymer
depended as much on the extent of reaction between the available hydroxyl
groups on the polyester base with the methoxy groups on the silicone intermediate, as on the presence of the silicone itself. A nearly complete
reaction is required in order for the resulting polymer to exhibit optimum
durability. While the base polyester polymer may be engineered in a wide
variety of ways, the ultimate properties required of the finished coatings
film often dictated a specific family of structures. If, for instance, very
flexible films were needed, the base polyester was normally linear, rather
than branched, and application requirements often dictate the molecular
weight. Also, while it is true that silicone polyester coatings exhibit a
high level of hydrophobicity, they may nonetheless be permeable to water
vapor; it was, consequently, fairly common to maximize the flexibility of
the polymers by incorporating a long-chain dibasic acid, such as adipic or
azelaic, into the polyester base. This allowed greater freedom of rotation
in the molecule[30], and helped to prevent microcracking (environmental
stress cracking), and ultimate film degradation. It may also have increased
the solubility of the polymer in hydrocarbon solvents, and rendered it more
HOH/alkali-resistant, as well—probably due to the shielding of the ester
groups by the hydrocarbon chain[31]. Also important, for these reasons,
was the choice of crosslinking polymer used for film formation. This would
commonly be hexamethoxymethylmelamine, but could be extended to include,
depending upon ultimate service requirements, mixed ether melamines and
aliphatic isocyanates. In cases where exterior durability has not been of
primary importance, other desirable properties have been achieved by crosslinking silicone polyester resins with ureas, glycolurils, benzoguanamines
and aromatic isocyanates.

POLYESTER COATINGS IN THE "COMPLIANCE ERA"

Although the largest and most widely-used class of polyesters in the
past has been hydroxy-reactive, solvent-borne coatings, significant work
was being done with waterborne polyesters, beginning in earnest with the
petrochemical shortages in the early 1970's. This work led to anionic
waterborne coatings for conventional applications, which were rendered
carboxyl functional by incorporation of a trifunctional acid, such as
trimellitic anhydride to an acid value (AV) of 70-130. The finished polymers were neutralized with ammonium hydroxide or amine, to render them
water-soluble. Primer and topcoat finishes for various conventional
applications were achieved with increasing sophistication during the decade
following the late 1970's, although these coatings were not suitable as
across-the-board replacements for conventional, solvent-borne polyesters.
There was, during the decade that followed, a certain softening in the
industry-wide reluctance to use waterborne systems. Early objections had
been registered because waterborne coatings frequently required special
handling, demanded more time-consuming clean-up procedures, and frequently
required more exacting metal-cleaning measures. There was also a pervasive,
possibly antediluvian belief, that moisture-resistance of finished systems
would suffer because "once with water, always with water,"[32] but this had
begun to fade by the mid-1980's, by which time the new waterborne technologies had become firmly entrenched.

"If (one) were to characterize the decade of the 1980's in a simple
phrase, few would argue that it might be called the 'Age of High
Solids,'"[33] and reactive polyester coatings dominated this era. Serious
work had begun on saturated polyester polymers for High Solids (H/S) coatings in the early 1970's, and was rapidly accelerated by both the petrochemical shortages which centered around 1974, and the concurrent legisla-
tive concerns with solvent emissions into the environment. Researchers were
driven by the need to meet three major criteria: controlled reactivity, low
viscosity, and low volatility[34]. Because molecular weight needed to be

achieved during crosslinking, rather than being derived from the basic polyester polymer, High Solids polyesters normally supplied a greater number of reactive sites, predominantly hydroxyl groups, available for crosslinking. To avoid the brittleness usually associated with very densely crosslinked systems, it was often found desirable to incorporate longer chain-length polyfunctional acids and/or alcohols[35]. The resultant polymers typically exhibited 80-90% solids when reacted stoichiometrically with isocyanates, but frequently yielded empirical solids up to 12% lower, when crosslinked with melamine. While part of this was understandably attributable to the loss of alcohol resulting from the crosslinking reaction, the remainder was typically the result of unreacted polyhydric alcohol in the polymer; those containing TMPD were often found to be particularly prone to this behavior.

Although High Solids polyester spray finishes could be produced on the same equipment and be employed in many of the same applications as lower solids "conventional" polyesters, many exhibited superior pigment-wetting characteristics, probably due to the presence of the higher number of internal polar groups. In addition to lower VOC, High Solids coatings typically displayed lower energy requirements for curing. It has been calculated that the use of a 75% volume solids coating vs. a conventional 35% volume solids coating, can result in as much as a 52% reduction in gas consumption[36]. As a result of their many strengths—ease of manufacturing and use, low volatile emissions, reduced energy requirements, greater application efficiency, lower handling and storage costs, and excellent physical properties—High Solids polyesters enjoyed spectacular growth throughout the 1970's and 1980's, with outstanding potential for continued growth into the 1990's and beyond.

Although a wealth of interesting polyester work was reported throughout the 1970's and 1980's, including polyesters containing phenoxaphosphine rings[37], metal-containing crosslinked polyester resins based on divalent metal salts of mono(hydroxyethyl) phthalate[38], and intriguing work by Carraher involving the synthesis of titanium polyesters[39], the most interesting, commercially-viable polyester systems were those employed in the production of powder coatings[40]. Although powder coatings had existed as laboratory prototypes since the 1950's, serious work in coatings applications was not undertaken until the early 1970's, and substantial commercial acceptance was not to follow until the mid-1980's, by which time many of the early workers in the field had given-up hope and dropped out. Both thermoplastic and thermosetting polyester resins were employed commercially[41], with the thermoplastic versions typically being linear reaction products of difunctional glycols and phthalic acids, and the thermosetting versions being both hydroxy- and carboxy-functional. The latter could be reacted with a variety of materials, including melamines, epoxies, dianhydrides or blocked isocyanates. By the mid-1980's, the most important polyester thermosetting powder systems were polyester/bisphenol A epoxies, polyester/triglycidyl isocyanurate, and polyester/isocyanate systems[42], which yielded a wide variety of performance properties and parameters. By the late 1980's these systems occupied a small, but dynamic, niche in the chemical coating marketplace. They were nearly ideal from an environmental/legislative point of view, and could be applied electrostatically at significantly higher molecular weights than prevailing High Solids systems. This allowed for the enhancement of certain physical properties, such as extreme hardness and abrasion resistance, not previously considered practical. This technology was expensive, however, and required considerable capital investment in application equipment. It was also difficult to coat all configurations, and contamination was an ever-present concern[43]. Where they could be used and afforded, however, they provided excellent polyester coatings.

The history of polyester coatings, from Gay–Lussac and Pelouze in 1833 to Powder in 1989, has been one of scientific interest, technological challenge, and commercial success. The potential for vigorous, continued technological and commercial growth well into the twenty–first century seems assured.

1. I. Goodman and J. A. Rhys in: Saturated Polyesters, Volume I. (Iliffe Books Ltd., London 1965) p. 5.

2. H. J. Lanson in: Enclyclopedia of of Polymer Science and Engineering, Vol. 1, J. I. Kroschwitz, ed. (Wiley–Interscience, New York 1986), p. 644.

3. ibid.

4. I. Goodman and J. A. Rhys, loc. cit.

5. I. Goodman, Encyclopedia of Polymer Science and Engineering, Vol. 12, J. I. Kroschwitz, ed. (Wiley Interscience, New York 1986), p. 1.

6. J.Boxali, Paint Manufacture 48, 21 (1978).

7. ibid.

8. R. C. Forney, CHEMTECH 18, 180 (1988).

9. I. Goodman, loc. cit.

10. R. C. Forney, loc. cit.

11. W. H. Carothers, J. Am. Chem. Soc., 51, 1548 (1929).

12. H. Mark and G. S. Whitby, eds. Collected Papers of Wallace Hume Carothers on High Polymeric Substances (Interscience Publishers, Inc. New York 1940).

13. P. J. Flory, Chem. Rev. 39, 184 (1946).

14. P. J. Flory, J. Am. Chem. Soc., 63, 3083–3096 (1941).

15. P. J. Flory, J. Am. Chem. Soc., 74, 2718 (1952).

16. I. Goodman, loc. cit.

17. H. J. Lanson, op. cit., p. 645.

18. C. B. Rybny, E. E. Faust, and R. R. Purgason, Paint and Varnish Production 63, 23 (1973).

19. H. Lee, D. Stoffey, K. Neville, New Linear Polymers (McGraw–Hill, New York 1967), p. 214.

20. C. B. Rybny et al., loc–cit.

21. S. Paul, Surface Coatings (John Wiley & Sons, Chichester 1985), p. 131.

22. D. C. Finey, Paint and Varnish Production 59, 27–31 (1969).

23. S. Paul, op. cit., p. 128.

24. K. H. Earhart, Paint and Varnish Production 62, 43 (1972).

25. H. J. Lanson, op. cit., p. 665.

26. W. E. Catlin (assigned to the E. I. du Pont de Nemours Company),
    U. S. Patent 2,284,637 (June 2, 1942).

27. J. J. McLafferty, P. A. Tiglioti, and L. T. Camilleri, J. Coatings
    Tech. 58, 23 (1986).

28. W. A. Finzel, J. Coatings Tech. 52, 55 (1980).

29. H. J. Lanson, op. cit., p. 668.

30. ibid., p. 651

31. ibid., p. 652

32. G. R. Pilcher, Advances in Polymer Tech. 7, 362 (1987).

33. G. R. Pilcher, Modern Paint & Coatings 78, 35 (1988).

34. T. Antonelli, Am. Paint Coatings J. 59, 46 (1975).

35. H. J. Lanson, op. cit., p. 672.

36. ibid., p. 674.

37. M. Sato, and M. Yokoyuma, European Polymer J. 16, 79–83 (1980).

38. H. Matsuda, J. Polymer Science: Polymer Chemistry Edition 15,
    2239–2253 (1977).

39. C. E. Carraher, Jr., J. Polymer Science: Part A–1 9, 3661–3670
    (1971).

40. H. Ades, Powder Coating 5, 2 (1982).

41. R. Chandler, Polyester Powder Coatings (Braintree, Essex, 1976).

42. R. van der Linde, B. J. R. Scholtens, and E. G. Belder, Proceedings
    of the ACS Division of Polymeric Materials: Science and Engineering
    55, 634 (1986).

43. G. R. Pilcher, Advances in Polymer Tech. 7, 363 (1987).

History of Photocuring

Dr. Charles E. Hoyle
University of Southern Mississippi
Hattiesburg, MS 39406-10076

To properly define and understand the historical development of photocuring from its infancy to its present position as one of the fastest growing sectors of the coating industry, it is necessary to first have a general conception of the basic components of photocurable coatings formulations. With the fundamentals of photocuring in hand, one is then ready to appreciate the history of using light to produce viable coatings with properties which not only rival but often exceed those characteristic of other cured coatings. In this brief paper, an annotated overview of photocuring will follow a brief introduction to the principle ingredients in photocured coatings.

Table I shows the basic chemical components in photocurable systems, their approximate concentration range (percentage), and their general function. The primary component in the photocurable coating is the photoinitiator whose task it is to absorb light and convert it to a useful purpose by producing reactive radicals or cations which are capable of initiating chain growth polymerization. The monofunctional and multifunctional monomers are used to lower the formulation viscosity as well as impart important physical properties to the cured film. Likewise, the functionalized oligomers (acrylated epoxies, acrylated urethanes, acrylated ethers, etc.) are selected for their ability to influence the physical and mechanical properties of the cured coating (or adhesive). Finally, the additives are employed, as in any coating application, to enhance the overall integrity of the system. Examples of typical additives in photocurable formulations are fillers, surfactants, adhesion promoters, and glass transition reducers. The history of photocuring involves the parallel development of each of the principal components in Table I. In order to appreciate the growth of photocuring, it is necessary to consider the development of individual components listed in Table I as well as high-intensity light sources.

Before continuing with specific developments, however, Table II lists the journals, preprint collections, etc. which chronicle the rise of photocuring. The preprints of the RadCure/RadTech meetings are in themselves an archival source of the progress in photocuring. As well, the general references (1-16) listed at the end of this chapter are primary review sources for photocuring and photoinitiated polymerization and can provide the interested reader with many of the details of the history of photocurable coatings, which of necessity must be left out of this review. One has but to follow the trends set forth in these documents to become familiar with the photocuring field and its rapid development from infancy to a mature industry in only two decades.

Tables III-X provide very brief highlights of several critical aspects of photocuring of coatings. Radical and cationic initiated photocuring are covered as separate topics since their histories have proceeded along quite different avenues, even though final applications are quite often the same and both have the common requirement of an intense lamp source to initiate the curing process. A brief narrative of each Table will suffice to note some of the selected high points.

The progress and use of photocurable inks, despite getting off the ground in the mid 1940's, experienced its first real commercial venture of any consequence in 1969 (see Table III). In the early 1970's research at Sun, W.R. Grace, American Can, and others rapidly expanded until in 1974 a viable UV curable printing industry emerged based on free radical and cationic curing processes. Numerous refinements in monomer/oligomer composition, red-extended photoinitiators, etc. have helped make photocurable inks a viable alternative

Published 1990 by Elsevier Science Publishing Co., Inc.
Organic Coatings: Their Origin and Development
R.B. Seymour and H.F. Mark, Editors

to conventional printing inks. Table IV lists areas in which UV inks have been successfully introduced during the past two decades.

The photocuring of clear coats (Table V) has its roots in a duPont patent in 1945 by Christ who first indicated that benzoin could be used to induce free radical polymerization. In 1952, Plambeck of duPont formulated a photocurable coating for the production of plastic printing plates. In 1967, Fuhr et. al. at Bayer introduced a highly efficient unsaturated/styrene coating for wood substrates. Following this lead, in the late 60's and early 70's several similar systems were employed in the US. Thus by 1970 the race for development of faster, more efficient photocurable coatings with enhanced physical properties had begun.

As depicted in Table V, the photocuring industry experienced an unprecedented growth in the synthesis and characterization of a multitude of low molecular weight (mono-, di-, tri-, tetra-, and even penta-functional) monomers expressly for the photocuring industry. The monomers/oligomers with acrylate functional groups emerged as the most successful components for fast cure coatings with exceptional physical properties and viscosities conducive to coating application by a number of standard methods (roll coating, brush, spray coating, etc). In addition, acrylated systems were successfully formulated to adhere to a variety of substrates. To assign development in this area to a particular person or group would do an injustice to the efforts of numerous research personnel throughout the world.

One very successful system not based on acrylate chemistry also experienced widespread use in the 1970's as a clear coating (as well as a principal component of pigmented photocurable coatings): thiol-ene based formulations with benzophenone type photoinitiators were employed as clear coats (Table V). But as the acrylate systems continued to develop (as a result of new and improved low molecular weight multifunctional acrylates and the wide variety of functionalized oligomers), the use of thiol-ene systems for clear coats began to wane. Finally, as noted in the last entry in Table V, in the latter years of the 1970 and into the early 1980's several groups began to explore the possibility of dual radical/cationic photocurable systems.

Table VI is of particular importance since it lists several developments which will be critical in leading to new advances and applications in the future. The first entry references the pioneering work of Kloosterboer in describing the inhomogeneties and gelation processes experienced during the photocuring of multifunctional monomers. The second entry refers to the efforts of Decker in demonstrating the viability of using high intensity continuous and pulsed laser sources to initiate photocuring of coatings primarily directed toward imaging systems with implications for other applications. The final entries in Table VI make note of new applications and/or monomers/oligomers which have the potential of greatly expanding the scope of photocuring. Of particular interest is the Mead Imaging lithographic process which involves exposure of an encapsulated photocurable system to light to yield a multicolored image.

By the mid to late 1970's, as indicated in Table VII and reported by Pappas (1) in an excellent review article on free radical cleavage-type and abstraction-type photoinitiators, several effective photoinitiators or photoinitiator packages had been developed and were in use. Since then, significant refinements in speed, oxygen sensitivity, absorbance characteristics, solubility, compatibility, etc. have been made. The last three entries are directed toward totally new classes of photoinitiators which may well have potential in future applications, particularly where the situation calls for absorbance in the visible region of the electromagnetic spectrum. Finally, it should be noted with apology that it is impossible to list all of the excellent developments over the past 30 years in the photoinitiator field.

Table VIII lists critical dates in the history of photocationic curing. Of special note are entries in 1976 and 1977 for Crivello and Lam who discovered and pioneered the development of aryliodonium and arylsulfonium salts as initiators for cationic photocuring of vinyl ethers, cyclic ethers, multifunctional ethers and lactones. This single set of developments, among all of the others in the 1970's and 1980's, stands as a milestone in the history of photocuring. The work on photocationic systems by Crivello and his colleagues at GE, as well as work by others including Pappas, Ledwith, and groups form Ciba Geigy, Union Carbide, and 3M has resulted in the production of viable photocurable cationic systems for application ranging from coatings on optical fibers to numerous clear coats on metal substrates. The next to last entry in Table VIII is for dual epoxy (cationic)/acrylate (radical) photocurable coatings which produce cured films with unique properties.

The next Table (IX) covers developments in light sources. Probably the most notable and widely employed development in lamp sources for photocuring in the last two decades has been the Fusion Systems electrodeless lamps characterized by short initial warm-up periods and long lamp lifetimes. Despite the use of pulsed Xenon lamps on a very limited basis, the electrodeless lamps are the only light sources which have been able to make inroads into the UV curing lamp market which is dominated by the standard medium pressure mercury lamps.

Finally, Table X lists some representative examples of the applications of photocurable coatings which have developed since the late 1960's. The entries in Table X attest to the success of the photocuring industry in formulating coatings for a wide number of applications. Indeed Table X, while not exhaustive, demonstrates the diversity of photocuring as it exists today. The future of the photocuring industry will be shaped by research personnel who will build on the ideas and inventions of the past twenty years while adding fresh, new insight into the overall process of photocuring. It is anticipated that a similar listing compiled two decades from now will yield some astounding surprises.

Literature Cited
1. S. P. Pappas (Ed.) U. V. Curing: Science and Technology Vol. 1, Technology Marketing Corporation (1978).
2. C. G. Roffey, Photopolymerization of Surface Coatings, Wiley-Interscience, New York (1982).
3. K. Hashimoto and S. Saraiya, J. Rad. Cur., 8(1), 4 (1981).
4. J. V. Crivello, Adv. Polym. Sci., Vol. 62, pp. 1-48, Springer-Verlag, Berlin (1984).
5. S. Paul, Surface Coatings: Science and Technology, pp. 601-57, Wiley-Interscience, New York (1985).
6. S. P. Pappas (Ed.), UV Curing: Science and Technology Vol. 2, Technology Marketing Corp., Norwalk, CT (1984).
7. G. A. Senich and R. E. Florin, Rev. Macromol. Chem. Phys., C24(2), 239 (1984).
8. G. E. Green, B. P. Stark, and S. A. Zahir, J. Macro. Sci.-Revs. Macro. Chem., C21(2), 187 (1982).
9. V. D. McGinniss, "Radiation Curing," in Kirk-Othmer Encyclopedia of Chemical Technology, 3rd ed., Vol. 19, Wiley-Interscience, New York, pp. 607-624 (1982).
10. R. Phillips, Sources and Applications of Ultraviolet Radiation, Academic Press, New York (1983).
11. R. Holman (ed.) U.V. and EB Curing Formulation for Printing Inks, Coatings and Piants, SITA-Technology, London (1984).

12.   D.R. Randell (Ed.) <u>Radiation Curing of Polymers</u>, Royal Society, London (1987).
13.   S.P. Papas, Prog. Org. Coat., 13, 35 (1985).
14.   S.G. Wentink and S.D. Kock (Eds.), <u>UV Curing in Screen Printing for Printed Circuits and Graphic Arts</u>, Technology Marketing Corp., Norwalk, Ct (1980)
15.   A.Ledwith in <u>Development in Polymerization-3</u>, Ed. R.N. Hayward, Applied Science Publishers, Ripple Road, Barking, 1981.
16.   D.G. Berner, J. Pulglisi, R. Kirchmayr, and G. Rist., J. Rad. Cur., 6, 2 (1979).
17.   W.C. Perkins, J. Rad. Cur., 8 (1), 16 (1981).
18.   J.V. Crivello and J.H. W. Lam, Polym. Chem. Polym. Lett. Ed., 17 (1979).
19.   J.V. Koleske, O.K. Spurr, and N.J. McCarthy, 14th national SAMPE Tech. Conf., Atlanta, GA, 14, 249 (1985).
20.   J.G. Kloosterboer, Adv. Polym. Sci., 84, 1 (1987) and references therein.
21.   C. Decker, and Polym. Sci, Polym. Chem. Ed., 21, 2451 (1983).
22.   C. Decker and K. Moussa, Polym. Preprints, 29 (1), 516 (1988).
23.   J. Crivello, Polym Matr. Sci. Eng., 60, 217 (1989).
24.   R. Eckberg and K.D. Riding, Polym. Mater. Sci. Eng., 60, 233 (1989).
25.   S. Lapin, Polym. Mater. Sci. Eng., 60, 233 (1989).
26.   G.B. Schuster, S. Chatterjee, P. Gottschalk, and P.D. Davis, SMSE Proceedings, 42nd Annual Conference, Boston 388 (1989).
27.   J.V. Crivello and J.H.W. Lam, Polym. Sci. Symp. No. 56, 383 (1976).
28.   K.B. Wischmann, Proceedings: 31st International SAMPE Symposian, April 7-10, Las Vegas, NV (1986).

TABLE I GENERAL UV CURABLE FORMULATION

Component	Percentage Range	Function
Photoinitiator	1-3	Produces Reactive Initiators
Oligomer Resin	25-90	Primary Component For Estab- lishing Final Film Properties
Monomers	15-60	Determines Cure Rate, Crosslink Density, and modifies formu- lation viscosity
Additives	----	Surfactants, Pigments, Stabilizers, Adhesion Promoters, Fillers

TABLE II. PUBLICATIONS DEDICATED TO RADIATION CURING

Year Initiated	Title	Publisher
1974	RadCure Proceedings	SME
1974	Journal of Radiation	Technology Marketing
1974	Radiation Curing	Technology Marketing
1984	Cure Letter	Captan Associates, Inc.
1987	RadTech International Proceedings	RadTech

TABLE III. HISTORY OF PHOTOCURABLE INKS

YEAR	SOURCE	REMARKS	REFERENCE
1946	Inmont Printing Ink Patent	First UV 2,406,878	US Pat.
1969	----	First Com- mercialization of UV Printing	See ref. 1
1970	Bassemir- Sun Chemical Corp.	Fast Photo- curable Inks	US Pat. 3,551,235 and others
1972	W.R. Grace Co.	Thiol-Ene Chemistry-- Flexographic Printing	US Pat. 3,708,296 and others
1974-	----	Rapid Expansion	----

TABLE IV. REPRESENTATIVE APPLICATIONS FOR PHOTOCURABLE PRINTING INKS

Letterpress	Solder Resist
Intaglio Printing	Plastic Substrates
Offset Lithography	Metal Decoration
Screen	

TABLE V. RADICAL PHOTOCURING-HISTORICAL OVERVIEW: 1945-80

Year	Source	Remarks	Reference
1945	R.E. Christ-duPont	Benzoin employed to initiate radical photopolymerization	US Pat. 2,367,670
1952	L. Plambeck-duPont	Benzoin ethers used in photocurable systems	US Pat. 2,760,863
1967	K. Fuhr, et. al.-Bayer	Highly successful unsaturated polyester/styrene formulation with benzoin ethers photoinitiators-standard wood filler-wood varnish system.	DAS 1,694,142
1969-70	-----	First wood filler and wood varnish photocurable systems began to appear in US--based on unsaturated polyester/styrene formulations	-----
1970-80	-----	Rapid development of efficient photoinitiators, multifunctional monomers, and acrylated oligomers	----
Early 1970's	W.R. Grace. Co.	Thiol-ene photocurable coatings developed	US Pat 3,661,744 and others
Mid 1970's	J.E. Moore-General Electric	Perfected photo-DSC method for assessing progress of photocuring	See. ref. 1
Late 1970's	(1) A. Ketley-W.R. Grace (2) J. Crivello-G.E. (3) W. Perkins-Celanese	Developed Dual Radical/Cationic Photocurable Systems	Ref. 17-19

TABLE VI. RADICAL PHOTOCURING-RECENT DEVELOPMENTS OF NOTE

Source	Development	Reference
J.G. Kloosterboer-Philips	Theory of Inhomogeneties in photopolymerization of difunctional monomers established	Ref. 20
C. Decker and K. Moussa	Demonstrations of Effectiveness of Laser-Initiated Polymerization and Real Time-IR Analysis of Photocuring	Ref. 21-22
F.W. Sanders-Mead Imaging	Microencapsulated System for color imaging	US Pat. 4,399,203
(1.) J. Crivello-RPI (2.) R. Eckberg and K.D. Riding-General Electric	Introduction of Rapid Cure Epoxy-Silicone and Epoxy-Siloxane Systems	Ref. 23 Ref. 24
S. Lapin-Allied Signal	Development of high speed cationic photocurable formulations based on vinyl ether functionalized urethane oligomers	Ref. 25

TABLE VII. RADICAL PHOTOINITIATORS

Year	Source	Remarks	Reference
1945-80	Many Groups	Efficient Photoinitiators available by late 1970's: (1) Benzophenone and thioxanthone types with amine coinitiators (2) Michler's Ketone and Benzil (3) Various Quinones (4) Benzoin ethers, dialkoxyacetophenones, benzoin acetals, benzoyl oximes, and various substituted acetophenones	See ref. 1
1981	A. Hesse et. al-BASF	Development of acylphosphine oxides	US Pat. 4,265,723
1981	D.P. Specht, et. al.-Eastman Kodak	Introduction of ketocoumarin visible active photoinitiators.	US Pat. 4,289,844
1989	G.B. Schuster, et. al.	Characterization of Efficicent Cyananine Borate Photoinitiators which absorb visible light.	Ref. 26

TABLE VIII. CATIONIC PHOTOCURING-HISTORICAL OVERVIEW

Year	Source	Remarks	Reference
1965	Licari, et.al- North American Aviation	Introduction of Diazonium Salt Photoinitiators for Curing of Epoxy Resins	J.J. Licari and P.C. Crepeam US Pat. 3,205,157
1976-77	Crivello and Lam--General Electric	Development of Diaryliodonium Salts for Cationic Photocuring of Olefins, Lactones, Cyclic Ethers.	Ref. 27
1977	Crivello and Lam-General Electric	Introduction of Sulfonium Salt Cationic Photoinitiators	US Pat. No. 4,058,401
1979	See Entry in Table V	Development of Dual Cationic/ Radical Photo-curable System	Ref. 28

TABLE IX. LIGHT SOURCES-HISTORY OF DEVELOPMENT

Year	Source	Remarks	Reference
1970	---	Standard Medium Pressure Mercury Lamp-Workhorse of Industry, Developed well before 1970 and perfected between 1970 and 1980	See ref.1
1975	Fusion-Systems	Introduction of electrodeless lamps for photocuring	See ref. 1
1975	(1.) L.R. Panico, (2.) M. Michalaski (3.) L. Bordzol	Use of pulsed flashlamps for photocuring introduced formally.	1975 RadCure Proceedings

TABLE X. ABBREVIATED LISTS OF SOME PHOTOCURING APPLICATIONS

- Furniture Coatings
- Floor Wear Coats
- Paper & Board Coatings
- Metal Can & Foil Coatings
- Plastic Coatings
- Pressure Sensitive Adhesives
- Filled Coil Coats & Automotive Paints
- Wire-Cable Coatings
- Potting & Conformal Coatings
- Moisture Vapor Barriers

- Optical Fiber Coatings
- Silicone Release Coatings on Paper
- Water-Based Systems for Grafting Modifications
- Laminating Adhesives
- Synthetic Skin
- Compact Disks

HISTORY OF ELECTRON BEAM CURED COATINGS

JAMES F. KINSTLE
James River Corporation, 1915 Marathon Avenue, Neenah,
Wisconsin 54956

ABSTRACT

The history of electron beam utilization in coatings
applications is an interesting composite. It is a
combination of the individual histories of equipment
and hardware (Van de Graaff, scanning gun, large area
filament type), of science (phenomenological, molecular,
mechanistic), of scientists (especially physicists and
chemists; from various countries), and of applications
(bulk systems and coatings, especially on heat-sensitive
substrates). The presentation will cover these intertwining historical threads in the development of this
field.

INTRODUCTION

High energy radiation has been used in chemical processing
for many years. The first sources were x-rays and γ-rays.
X-rays are emitted when rapidly moving electrons strike a metal
substrate; the average emitted energy depends on the energies
of the incident electrons. "Hard" x-rays are high energy,
short wave length, and highly penetrating; they are used in
diagnostic medicine. "Soft" x-rays, those obtained from
conventional x-ray tubes for crystallographic studies for
example, can give higher dose rates (energy per unit time). In
special defocussed configuration, x-rays have been used to
initiate chemical reactions. This is not a common practice but
may reemerge to take advantage of the potentially greater
resolution possible in certain electronic resist applications.
Gamma rays are electromagnetic radiations emitted by the nuclei
of natural or artificial radioactive isotopes. The most
popular source of γ-rays was, and still is, cobalt-60, which
emits γ-rays of 1.17 and 1.33 MeV, along with 0.3 MeV
electrons. [MeV is megaelectron volts, or 10^6 electron volts.]
Gamma rays are very high energy; their intensity is reduced to
1/10 its initial value by passing through 30-45 cm of water
(3-4 cm lead). This has allowed polymerization and/or
crosslinking in thick sections, making possible the formation
of polymer impregnated concrete, wood, etc., by irradiation of
the appropriate monomer saturated compositions. It also means
that extensive shielding is required, especially important
since sources cannot be turned off and on. Other sources of
and mixed high energy radiations include radioactive caesium-
137 (actually its derivative barium-137) and "spent" nuclear
fuel and pile sources. The mixed high energy radiations were
early noted to affect materials, like the color changes
observed in irradiated glass by the Curies about the turn of
the century. These were a few comparisons of β and
γ-rays, see ref. [1] for other early γ-ray related studies.

Published 1990 by Elsevier Science Publishing Co., Inc.
Organic Coatings: Their Origin and Development
R.B. Seymour and H.F. Mark, Editors

245

Work with γ-rays and the natural sources of electrons
suggested that a better controlled source of electrons would be
very useful.

EQUIPMENT DEVELOPMENT

Though the use of artificially generated electrons in
chemistry was described in Science by Coolidge in 1925 [2],
real development of the equipment and its use has been a post-
World War II phenomenon. An early source of electrons was the
so-called Van de Graaff generator [3]. This unit produces
highly energetic electrons by mechanical transport of
electrical charges on a moving insulating conveyor belt or
cylinder, inducing and enhancing a charge that is removed by a
corona discharge, a kind of continuous lightning effect. This
electrostatic generation technique can provide electrons of a
fraction to perhaps 10 MeV, quite smoothly.
Transformer type accelerators, based on resonance between
an incorporated circuit and the supply frequency, were
developed in the 50's. These can produce electrons in the
hundreds of keV to 5-10 MeV, at moderate dose rate. The
electrons, produced during half of the resonant cycle are
intermittent or pulsed. They are rapidly scanned across the
substrate. These resonant transformer type electron guns were
the most commercially significant electron sources for many
years.
Linear accelerators, in which electrons are accelerated
through a wave guide and, like the above types emerge from its
vacuum surroundings through a foil window, produce electrons in
the < 1 to > 60 MeV range. These are rapidly pulsed (hundreds
per second) sources of highly focussed electrons.
There are several variations of "blanket type" electron
radiation, from Energy Sciences (ESI), Nissin, and RPC. Each
has elements in common; a filament from which electrons are
generated (adjust number of electrons per unit time by
current), a vacuum chamber acceleration zone (adjust energy by
potential difference, or accelerating voltage), and a self-
shielded reaction zone through which the substrate to be
irradiated passes (speed adjustable). So each important
process variable is controllable with these homogeneously
distributed, unscanned electrons. All three of the mentioned
companies (or their successors; corporate changes have occurred
in this field) sell units in the 150-500 keV range, which is
the range of principal interest for coatings applications.

PEOPLE AND THEIR SCIENCE

Through the years during which the equipment was being
invented, developed, and refined, there were significant
advances in related disciplines, especially in chemistry.
These were accomplished by a range of scientists. Some of
these were Arthur Charlesby in England [4], Adolphe Chapiro in
France [5], and K. Hayashi and Y. Tabata in Japan. Early
workers in the U.S. included Malcolm Dole at Baylor [6], Vivian
Stannett at Research Triangle and North Carolina State, Sam
Nablo and colleagues at ESI, Alan Hoffman at University of

Washington, Joe Silverman at Maryland, Bill Burlant and
coworkers at Ford Motor Co., and a host of other contributors
from materials suppliers, equipment manufacturers, and
potential end users. Some more of the companies will be
specifically mentioned as we proceed.
     The science that these workers involved themselves in
mostly centered on polymerization and crosslinking. Related
radiation studies continued on equipment, process, and
chemistry of higher energy systems and on the emerging field of
photochemically induced reactions. The combined results of all
these efforts allowed formation of a successive series of
hypotheses that ultimately yielded a mechanistic and kinetic
framework for explanation of results and projection to new
systems. Through this period, products were introduced to the
marketplace. Discussion here will center on results and
developments pertinent to coatings applications.
     Accelerated electrons have high energy; viz about 100X
that needed for bond breakage. Thus, when electrons strike an
assembly of organic molecules, many chemical events can be
induced. These include dissociations electron capture,
electron ejection, and/or excitation, forming ions, radicals,
and various excited states. No specific initiators are
necessary. Under ultra pure conditions, certain monomers can
be polymerized ionically. But most systems are less pure and
less biased toward ionic polymerization, and free radical
reactions maintain. The free radical reactions that take place
during curing of a coating follow much the same kinetic
pathways as the UV induced polymerizations or thermal free
radical systems covered elsewhere in this book. Polymerizable
groups, contained in monomers, multifunctional monomers, and/or
functionalized oligomers or polymers, are consumed according to
the following kinetic expression.

$$R_p = k_p \, [M] \left(\frac{\phi \, I_a}{k_t} \right)^{1/2}$$

where R_p = overall rate of polymerization (rate of
          disappearance of polymerizable groups)

   k_p = specific rate constant for propagation

   ϕ = quantum efficiency

   I_a = amount of absorbed radiation, in turn the amount of
          incident energy times an absorption efficiency times
          the concentration of absorber

   k_t = specific rate constant

One can influence the reaction rate in all the ways suggested
by this equation. For example, by choice of rapidly
polymerizing monomers (high k_p); general polymerizabilities
have proven to be acrylate>methacrylate>vinyl>vinylene>allyl,
with aromatic monomers being sluggish under the electron beam.
The monomer (polymerizable function) concentration is
adjustable; note that the monomer concentration is also
implicit in the I_a term. Systems with high susceptibility

(high ϕ) can be chosen. The k_t term tends to be controlled by bimolecular reactions; these slow as much or more than the propagation reactions as the polymerizing system increases in viscosity and/or vitrifies. Note that several influences are in the square root term; their influences must be quadrupled in order to double the rate. The workers whose efforts led to the mechanistic understanding given in simplistic fashion here also used this understanding to advantage.

APPLICATIONS

The uses of electron beam curing have been varied. The most important ones known to the author will be mentioned here. Wire and cable coatings have been electron cured, at Sequoia, GE, and elsewhere, starting in the 50's. There were early concerns about copper catalyzed oxidations at high dose in air, but the better understanding and formulating that rapidly accumulated has allowed wide spread adoption and utilization of this technique. Polyolefins (predominately crosslinking), and poly(vinyl chloride) plus multifunctional acrylate systems (in which both polymerization and crosslinking occur) are in use. Heat shrink tubing was an early application of electron beam technology. Polyethylene thin tubing is irradiated, causing it to crosslink. It is then heated and stretched laterally, aligning the molecules and placing them in crystallite-locked non-thermodynamic oriented position. Upon heating, the molecules wish to go back to the shape and size that they were when they were crosslinked; thus, upon melting of the "locking" crystallites, they do so. The heat shrink tubing made in this way is an excellent wire splice cover, etc., as we all have become accustomed to. A few years later, W. R. Grace introduced a heat shrink food wrap based on the same principles and practices. In the late 50's Radiation Dynamics was formed. They have designed and built many electron accelerators and have been one of the principal "drivers" of electron beam technology. In the early 60's Sekisui introduced radiation crosslinked poly-ethylene (PE) foam. PE was compounded with a latent chemical blowing agent (carbonamide or other), formed into a sheet, irradiated with electrons to crosslink it, and then foamed. The irradiation limited blow up, since the network formed limited the movement of the molecules past each other. The cell size and homogeneity also appeared to be improved relative to unirradiated PE. Deering Milliken were the first to publish about grafting a monomer onto fiber substrates to impart soil release. This was reported in the mid-60's and appeared to be an example of the simultaneous irradiation grafting technique. There have been many applications involving wood substrates. Gel and filler coats have been electron cured on wood, taking advantage of the fact that the electrons have good penetration power; i.e., they can cure filled monomer/polymer systems even if they get relatively "thick" where they fill a crack in the wood substrate. Electron cured coatings have also been used on coil coatings, on both steel and aluminum. Excellent speed and process performance is obtained, and coatings have been formulated that allow extensive post-forming without delamination or cracking. In the 60's, 70's, and early 80's, the Electrocure process was

used by Ford Motor Company to cure coatings on literally
millions of plastic automobile parts (some small and some very
large). They used scanning guns (resonant transformer type)
with unsaturated acrylic type coatings formulations of their
own design. Very tough coatings were obtained, with good
adhesion to plastic substrates (even polypropylene); all
without harming the heat sensitive plastics. In the late 60's
the self-shielded, low energy electron processor was introduced
by Nissin in Japan; at similar time, manufacturers in the U.S.
introduced them. Another application area involves electron
cured adhesives for tape and other applications; some of these
take advantage of the penetration by irradiating from the "back
side", while others take advantage of the better adhesion (due
to grafting?) of the electron beam systems. Another
application area for electron beam sensitive materials is in
curing highly filled binders for magnetic media like video
tape, floppy disks, etc.

MEETINGS/INFORMATION

There are regular meetings that cover radiation curing.
These include the "International Meeting on Radiation
Processing" (IMRP) series, which started in 1976 as a topical
meeting jointly sponsored by the American Nuclear Society
(Isotopes and Radiation Division), American Chemical Society
(Polymer Division), and the Society of Plastics Engineers
(Electrical and Electronics Division). This first meeting was
in Puerto Rico. In even-numbered years there have been
successive IMRP meetings. There are also "Radcure" meetings
and "Radtech" meetings. Symposia have been conducted within
the programs of the ACS Division of Polymeric Materials;
Science and Engineering, the ACS Division of Polymer Chemistry,
the Australian Chemical Society, the Materials Research
Society, etc. Publications can be found in a broad range of
general and specialized publications, too numerous to list
here. Chemical Abstracts is suggested as a guide to the
literature, searching under "Polymerization (polymn), electron
beam or radiation", "Electron, beam...", "Radiation...", etc.
By following along, you will see and understand the history of
electron beam curing in "real time" as it continues to develop.

REFERENCES

1. S. C. Lind, "Chemical Effects of Alpha Particles and
   Electrons," 1st Ed. 1921, 2nd Ed. 1928, Chemical
   Catalogue Co.
2. W. D. Coolidge, Science $\underline{62}$, 441 (1925).
3. J. G. Trump, R. J. Van de Graaff, J. Appl. Phys. $\underline{19}$, 599
   (1948).
4. A. Charlesby, "Atomic Radiation and Polymers," Pergamon,
   London (1960).
5. A. Chapiro, "Atomic Radiation and Polymers," Pergamon,
   London (1960).
6. M. Dole, "Radiation Chemistry of Macromolecules I, II,"
   Academic Press, NY (1972).

INFRARED SPECTROSCOPY OF COATINGS MATERIALS

CLARA D. CRAVER
Chemir Laboratories, Craver and Craver, Inc.
761 West Kirkham Road, St. Louis, MO 63122, USA

INTRODUCTION

In setting out on this assignment I expected only to bask in the memories of a professional lifetime of IR Spectroscopy – 44 years to be exact. Francis Scofield often quoted a phrase, presumed to come from the Greeks or Romans, but never actually located in the literature by scholars to whom I have posed the question: "All of this I saw, part of this I was". It is pleasant to think of your field in these terms.

Caution made me look behind the great rush of work that occurred during World War II and was published so richly in the following ten years. Significant material was published by 1905. Careful research through the 1920's and 1930's, done without the benefit of commercial spectrometers, prepared us for spectrum interpretation and the age of modern industrial infrared spectroscopy.

EARLY BACKGROUND OF IR SPECTROSCOPY

In 1881 W. de W. Abney and E. R. Festing published "On the transmission of radiation of low refrangibility through ebonite" (1). This work was in the near infrared because 1.2 μm was the limit of sensitivity of their photographic plate. Angstrom used a bolometer and sodium chloride prism to 8 μm in 1890.

A more extensive study of IR spectra of thin films of some of the components of coatings is introduced in the great work of William W. Coblentz, published in 1905 (2). He was exploring the nature of absorption of IR by molecules and he reached conclusions which still impact on the use of IR in coatings analysis and research.

Data

Infrared spectra obtained by Coblentz by his painstaking point-reading of radiometer deflections are shown for some coatings materials in Fig. 1-3.

Fig. 1. Asphalt

The spectrum of asphalt was run as a cast film dried for four months to eliminate solvent bands and then as a melt to make certain that the solvent bands had been removed. Anyone who has run spectra of solvent cast

Published 1990 by Elsevier Science Publishing Co., Inc.
Organic Coatings: Their Origin and Development
R.B. Seymour and H.F. Mark, Editors

252

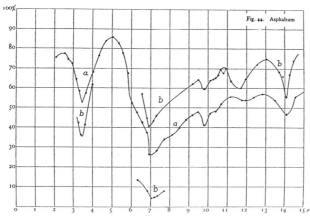

Figure 1.

A 1905 IR
Spectrum of
Asphalt Film.

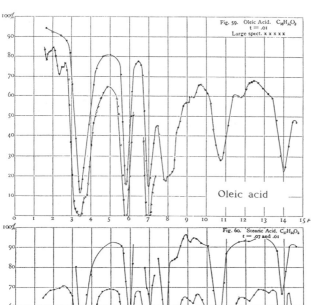

Oleic acid

Stearic acid

Figure 2.

1905 IR
Spectra of
Oleic and
Stearic Acid.

Figures 1-3
reprinted from
Reference 2 by
permission of
the Carnegie
Institution of
Washington,
1962, and
The Coblentz
Society, Inc.
1989.

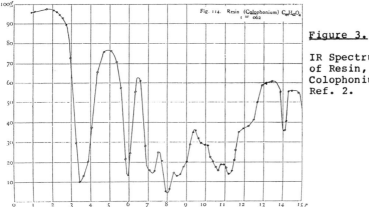

Fig. 114. Resin (Colophonium) $C_{20}H_{30}O_2$
t = 062

Figure 3.

IR Spectrum
of Resin,
Colophonium.
Ref. 2.

films can certainly relate to the need for this caution!
In his interpretation of the asphalt spectrum Coblentz
reported: "as is to be expected, the bands are found in
petroleum distillate".

Fig. 2. Fatty Acids

Coblentz's IR spectra of oleic and stearic acid are
clearly interpretable as long-chain acids. Minor wave-
length errors must be taken into account. Then one needs
to make only a little more allowance for band distortion
than is necessary to interpret data in some of the modern
computer-deresolved spectral libraries, or spectra of
microscopic particles, being used today.

Fig. 3. Resin

This colophonium sample is described in the text as
"violin resin" and is further identified as a residue from
distilling oil of turpentine. Some of the interpretation
of the spectrum was made by comparison with the simpler
terpenes, limonene and pinene.

The similarity in the strong band at 3.5 microns was
noted. It had been seen in too few compounds for him to
assign it to C-H. The "very sharp band of general
absorption" from 7-12 microns was reported as "worthy of
notice in connection with other compounds containing
oxygen".

254

Conclusions

The basic facts were reported or forecast in this work of Coblentz. Coblentz reported that:

> "a study of isomeric compounds shows that the arrangement or bonding of the atoms in the molecule, i.e., its structure, has a great influence upon the resulting absorption spectrum" and " . . . there is something, call it 'particle', 'ion', or 'nucleus', in common with many of the compounds studied, which causes absorption bands that are characteristic of the great groups of organic compounds. . ."

and

> "This ... is in marked contrast with stereomeric compounds, like dextro and laevo pinene, which were found to have identical spectra, showing that the spatial arrangement of atoms, i.e., the configuration of the molecule, had no effect upon the resulting absorption spectrum."

Spectrum interpretation relative to polymers:

> He repeatedly pointed out the observation that absorption characteristics are more dependent upon the arrangement of atoms in molecules than upon the size of the molecule. The low sensitivity of IR spectra of polymers to molecular weight was clearly forecast in this early work.

> Coblentz noted the ease with which we can pick out styrene modification or other pendant phenyl groups. "Benzene and its derivatives, especially phenyl mustard oil, in which we have the characteristic vibrations of the mustard oils superposed upon the vibration of the benzene nucleus, shows that both the group of atoms and their matter of bonding with other atoms, as well as the kind of atom, have a great influence on the absorption curve".

Coblentz made several experimental observations relative to the application of IR. Among these:

> Water has strong absorption bands and hygroscopic materials are difficult to measure.

> Films may retain solvent, which shows in the spectrum.

> Crystalline solids make poor films and scatter the radiation.

INTERMEDIATE YEARS IN THE DEVELOPMENT
OF CHEMICAL INFRARED SPECTROSCOPY

The 1920's and 1930's showed considerable progress in
predominantly university research in the understanding of the
relationships between electromagnetic radiation and the
structure of molecules. The major classes of organic compounds and systematic group frequencies were investigated by
Jean Lecomte starting in the early 1920's (3) and reported in
more than 50 publications covering alcohols (4), aldehydes and
ketones (5), esters (6), amides (7), and most classes of
hydrocarbons and their halogenated derivatives. Lecomte's
1928 summary of "Le Spectre Infrarouge" was 468 pages (8).
Other major summaries and his original publications are found
in nearly every volume of Chemical Abstracts for over 30
years.

G. Kimpflin published on the "permeability of synthetic
resins to infrared radiation" in 1924 (9). Joseph W. Ellis
studied the S-H bond (10), the N-H bond (11), and published
several papers on "polymers and new infrared absorption bands
of water" (12). In 1938 he reported studying "alterations in
near infrared absorption spectra of water and of protein
molecules when water is bound to gelatin" (13).

In 1939 G.B.B.M. Sutherland and W. T. Tuttle published
"absorption spectra of polymolecular films in the infrared"
(14). There had been several papers on rubber at that time.
One by Dudley Williams and R. Tashek was on the "Effect of
stretch on the infrared absorption spectrum of rubber" (15).

The significance of these fundamental research
investigations to coatings is that group frequencies, and
their systematic variations with molecular environments, are
the basis of our interpretation of IR spectra of chemical
structures including polymers and other coating materials.

Bibliographic references for each chemical group
structure and its absorption frequency correlation can be
found in the encyclopedic book of Lionel J. Bellamy, "The
Infrared Spectra of Complex Molecules", first published in
1954 (16).

INDUSTRIAL SPECTROSCOPY

The beginning of industrial IR as we know it today was
the subject of a paper by Norman Wright, in 1941, titled
"Application of Infrared Spectroscopy to Industrial Research"
(17). A more detailed paper by Barnes, Liddel and Williams
appeared in 1943 (18). This paper was expanded into a book,
with a bibliography by Robert C. Gore of 2,701 references
added, and was published in 1944 (19). It relied heavily on
practical work at American Cyanamid in Stamford, Conn., and

was an essential laboratory reference and short course book for many years. It was used by R. Bowling Barnes as a text for a night course lecture series he gave at Rutgers in the winter of 1945-1946. This course was my first introduction to applications of IR beyond the hydrocarbon analyses we were doing at Esso Research Laboratories, at Bayway, N.J.

"Infrared Spectroscopy - Industrial Applications and Bibliography" included a library of 363 representative spectra of organic compounds in the rock salt region of the infrared spectrum. Norman B. Colthup kindly loaned me his copy of this book and said that these spectra were one of his principal sources of information for his famous group frequency table (20).

Common solvents were included in that reference spectrum compilation as well as spectra of naturally occurring vegetable oils, Figure 4, and of phthalates, Figure 5.

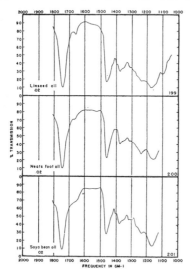

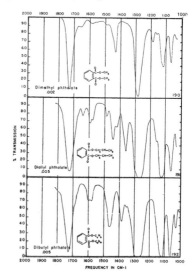

Fig. 4, IR spectra of   Fig. 5, IR spectra of
Vegetable Oils. (Ref. 19)   Phthalates. (Ref. 19)

These spectra serve to emphasize the inconvenient fact that chemically similar groups of coatings materials, such as the vegetable oils, are so similar spectrally that highly detailed spectra are necessary to distinguish among them. The same observation holds for phthalates as a class. It is easy to inspect a spectrum and determine that it is a phthalate. Better spectra and careful interpretation is usually needed to determine which one of the common phthalates it is, or whether it is a mixture.

This practical 1944 book also covered basic information on instrumentation and determining spectra, on the qualitative analysis of mixtures and on quantitative applications of IR.

The preparation of samples as described is very much the same way as it is done today. Resinous materials were prepared as thin continuous films. The suggestion was made for softening rubbery material to a gluey consistency with a solvent, and spreading a thin film on a plate and evaporating the solvent in a vacuum desiccator. Melting specimens to obtain films was recommended if it was certain that the melting did not alter the material. The problem associated with scattering from crystals and the Nujol mull technique were described.

Measurement of reaction rates was demonstrated by styrene polymerization using the 1628 wavenumber band of the vinyl group. Examples of working curves for calibration for quantitative determination of binary and ternary mixtures was described.

The extensive bibliography in reference (19) was intended to be a complete report of all of the infrared work in the scientific literature up through the date of its publication. Many of the references have to do with instrumentation and a great number of them were limited to near infrared work. Those citations which were clearly in the field of rubber, other polymers and coating materials are referenced in the previous section.

COMMERCIAL INSTRUMENTATION

During World War II many companies were involved in the industrial application of infrared. This big step forward was brought about by the pressures in wartime, predominantly for improved aviation gasoline and synthetic rubber. This led to development of the commercial infrared spectrometers. The first Beckman infrared spectrometer, the IR-1, was built originally for the purpose of measuring C-4 olefin streams for the production of rubber. It was followed soon thereafter by the first Perkin-Elmer single beam instrument.

Additional ease in obtaining IR spectra was introduced by the Baird double-beam instrument about 1947. The Perkin-Elmer Model 21, to be dubbed "the workhorse of IR", was introduced in 1950. These advances in instrumentation accelerated the applications of this great new chemical tool. Once the physicists and instrument manufacturers had done their part in increasing the sensitivity and stability of the spectrometers, chemists were ready to put these marvelous tools to the study of their problems.

Only the sales records of Beckman Instruments, Baird and the Perkin-Elmer Corporation could document the number of institutions and industries which began using IR in the late 1940's and 1950's. It would be impossible to document how many of these spectrometers were used for coatings analysis, characterization and research.

IR IN THE ANALYSIS OF COATINGS

The identification of coatings materials was spurred in the early 1950's by this availability of commercial spectrometers and by publication of small useful compilations of reference spectra.

Harry Hausdorff's paper at the Pittsburgh Conference in 1951 combined solubility tests and elemental analyses in a flow chart with a compilation of IR spectra for polymer identification (21). The Perkin-Elmer Corporation made this publication widely available.

The Naval Research Laboratory published a compilation of 92 plastic and resinous materials in 1954 and provided a flow diagram for identification of unknown materials (22). The first breakdown of the spectrum for identification was on the basis of whether or not a carbonyl group was present, and the next by the presence or absence of aromatic bands. The Federation of Societies for Paint Technology published a highly useful book on IR of paints and coatings in 1961 (23).

The technique of analysis of polymeric materials by pyrolysis reported in 1953 enlarged the range of materials which could be analyzed by IR (24). In 1965 a combination of chemical and spectroscopic procedures for deformulations and identifications were published by Haslam and Willis (25).

RESEARCH APPLICATIONS TO COATINGS CHARACTERIZATION

Investigations of rubber and styrene polymerization dominated the literature of infrared spectroscopy in the 1944-1949 period. Drying oils were being investigated at that time, but the study of the complex reactions of natural products moved slowly. The work of Gamble and Barnett (26) was reported in a review by R.T. O'Connor (27) as "the first use of infrared spectra to investigate the drying mechanism of oils". O'Connor's review article cites eight additional articles on this subject through 1954 and reports many other publications on specific functional groups in "fatty-acid derivatives". Mellon Institute research contributed to investigations of the hydroxyl band in autoxidation of linseed oil (28).

Drying oils and the role of driers was presented along with IR of "certain plastics" by Max Kronstein "at various meetings" in the fall of 1948: the American Chemical Society, the National Research Council, and the American Oil Chemists' Society. A summary of this work is in the proceedings of a Symposium sponsored by the Graduate Division, of the College of Engineering, of New York University, that same fall (29). Participants included members of the New York Paint, Varnish and Lacquer Association, the New York Production Club, and a distinguished spectroscopist from Cambridge University, Professor G.B.B.M. Sutherland, who gave an unrecorded, informal presentation on "IR of Polymer Materials". This meeting was one of the early interfaces between the sciences of coatings materials and infrared spectroscopy in this country.

From data reported verbally at conferences in the 1950's and articles and bibliographies in the 1960's, it is apparent that simultaneous methodologies were being developed. Much coatings research was going on in this period. Krimm and coworkers were concentrating on assignments of vibrational modes of halogenated polymers and others from 1956 through the 1960's (30) (31). Publications on over 15 polymer classes were referenced in a 1969 IR chapter in "A Treatise on Coatings" (32).

The combination of spectra of films of coatings and of gaseous products produced in autoxidation was reported by a team of experts doing extensive IR of coatings at the Naval Research Laboratory in the early 1950's (33). This work, like the pyrolysis work (24) was a prelude to our modern hyphenated techniques.

PARTICIPATION BY THE AUTHOR

During this early period of industrial IR I was at Esso Research Laboratories, in Bayway, NJ. There I had the opportunity of applying quantitative techniques of IR to the comparison of very complex mixtures such as crude oils, oil seeps, and other natural hydrocarbon deposits, waxes and asphalts. I moved to Battelle Memorial Institute, in Columbus, Ohio, and installed an IR spectrometer there in January 1949.

Battelle was a leader in coatings research and I soon was applying IR to research on drying oils, alkyds, rosins, and tall oil in a coatings section led by E. E. McSweeney. I had the opportunity to apply techniques of the analytical spectroscopist to that long-standing controversial subject of coatings - the effects of driers on drying oils. We were able to elucidate the "drying" mechanisms. Our work was reported and discussed fully at the 1955 Gordon Research Conference on Coatings. Our formal presentation of these results at an American Chemical Society meeting received the Carbon and Carbide Award (34). The publicity accompanying that award contributed to the momentum that was building toward the use of spectroscopy in the characterization and chemistry of complex coatings systems.

REFERENCES

(1) Abney, W. de W., and Festing, E. R., Proc. Phys. Soc. (London), 4, 256-60 (1881); On the transmission of radiation of low refrangibility through ebonite.

(2) Coblentz, W. W., Investigations of Infra-Red Spectra, Carnegie Institute Publication (1905). Bulletin #35. Abstracted in Phys. Rev. 20, 273 (1905). Reprinted in 1962 under joint sponsorship of the Coblentz Society and the Perkin-Elmer Corporation, Norwalk, Conn.

(3) Lecomte, Jean, Compt. rend., 178, 1530-2 (1924); C.A., 18, 2287.8. Qualitative studies on the infrared absorption spectra of organic compounds.

(4) --, Ibid., 180, 825-7 (1925); C.A., 19, 1661. Infrared absorption of the alcohols.

(5) --, Ibid., 1481-2; C.A., 19, 2299. Infrared absorption spectra of aldehydes and ketones.

(6) --, and Freymann, R., Compt. rend., 208, 1401-3 (1939); C.A., 33, 4872.6. Studies of the infrared absorption of solids by the powder method. Acetates, formates, oxalates, acetylacetonates, sulfates.

(7) --, --, Bull. soc. chim., 8, 601-11 (1941); C.A., 36, 2477.4. I. Infrared absorption of the amides.

(8) --, "Le Spectre Infrarouge." Paris: Press of University of Paris (1928). 468 pp. C.A., 22, 4385.5; C.A., 24, 554.

(9) Kimpflin, G., Compt. rend., 178, 1709-11 (1924); C. A., 18, 2466.7. Permeability of synthetic resins to infrared radiation.

(10) Ellis, Joseph W., J. Am. Chem. Soc., 50, 2113-8; C.A., 22, 3355.7. Infrared absorption by the S-H bond.

(11) --, Phys. Rev., 31, 314 (1928); C.A., 23, 3655.5. Infrared absorption by the N-H bond.

(12) --, Ibid., 38, 582 (1931); C.A., 26, 377.7. Polymers and new infrared absorption bands of H_2O.

(13) --, and Bath, J. D., Bull. Am. Phys. Soc. (3). 13, 7 (1938). Alterations in near infrared absorption spectra of water and of protein molecules when water is bound to gelatin.

(14) Sutherland, G.B.B.M., and Tuttle, W. T., Nature, 144, 707 (1939); C.A., 34, 1251.5. Absorption of polymolecular films in the infrared.

(15) Williams, Dudley, and Taschek, R., J. Applied Phys., 8, 497 (1937). Effect of stretch on the infrared absorption spectrum of rubber.

(16) Bellamy, L.J., Infrared Spectra of Complex Molecules (1954), John Wiley & Sons., Inc., New York, NY.

(17) Wright, N., Application to Infrared Spectroscopy to Industrial Research; Ind. Eng. Chem. Anal. Edit. 13, 1-8 (1941).

(18) Barnes, R. B., Liddel, U., Williams, V.Z., Ind. Eng. Chem., Anal. Ed., 15, 11, 659-709 (1943), Am. Chem. Soc.

(19) Barnes, R. B., Gore, R. C., Liddel, U., Williams, V.Z., Infrared Spectroscopy - Industrial Applications and Bibliography (1944); Stamford Research Labs., American Cyanamid Co., Stamford, CT, Reinhold Pub. Corp., NY.

(20) Colthup, N.B., J. Opt. Soc. Amer., 40, 1950.

(21) Hausdorff, H.H., Analysis of Polymers by Infrared Spectroscopy, The Pittsburgh Conference on Analytical Chemistry and Applied Spectroscopy, March 1951. Published by the Perkin-Elmer Corporation, Norwalk, Conn.

(22) Kagarise, R.E., and Weinberger, L.A., Infrared Spectra of Plastics and Resins, NRL Report 4369, May 1954, Naval Research Laboratory (1954).

(23) Chicago Society for Paint Technology, "Infrared Spectroscopy - Its Use as an Analytical Tool in the Field of Paints and Coatings", Fed. of Soc. for Paint Tech., Phila., PA (1961).

(24) Harms., D.L., Anal. Chem., 25, 1140 (1953)

(25) Haslam, J. and Willis, H.A., Identification and Analysis of Plastics, Van Nostrand, Princeton, N.J. (1965).

(26) Gamble, D.C., and Barnett, C.E., Ind. Eng. Chem., 32, 375, (1940).

(27) O'Connor, R.T., Southern Regional Research Laboratory, New Orleans, La., in J. of Amer. Oil Chem. Soc. Vol. XXXIII, No. 1, Jan, 1956.

(28) Honn, F.J., Bezman, I.I., and Daubert, B.F., J. Amer. Chem. Soc., 71, 812 (1949).

(29) Kronstein, M., New York Univ., Coll. Eng., Symposium on Varnish and Paint Chemistry, 1948, 13.

(30) Krimm, S., et al., Chem. Ind. (London), 1958, 1512; C.A., 52, 17960h (1958); ibid. 1959, 433.

(31) Liang, C. Y., and Krimm, S., J. Chem. Phys., $\underline{25}$, 563 (1956); C.A., 51, 861e.

(32) Myers, R.R., Long, J.S., Treatise of Coatings, Vol. 2, (in two parts) Characterization of Coatings: Physical Techniques, Part 1, Marcel Dekker, New York (1969).

(33) Crecelius, R.E., Kagarise, R.E., and Alexander, A.L., Ind. Eng. Chem., $\underline{47}$, 1643 (1955).

(34) Mueller, E.R., Craver Smith C. D., Battelle Memorial Institute, Columbus, Ohio; Ind. and Eng. Chem., Vol. $\underline{49}$, p. 210 (Feb. 1957).

PLASTICIZERS FROM COATINGS TO PLASTICS, 1938-1948

J. KENNETH CRAVER
Craver & Craver, Inc.
761 West Kirkham
St. Louis, MO 63122, USA

In 1938 the Monsanto Chemical Company, of St. Louis, Missouri, purchased the Fiberloid Corporation, of Indian Orchard, Mass., and a controlling interest in the Shawinigan Resins Corp. at the same location. Monsanto was a producer of fine chemicals, pharmaceuticals, heavy chemicals, resin intermediates and plasticizers (Table I). Fiberloid made nitrocellulose sheets, rods and tubes, cellulose acetate automotive window glass interlayer and cast phenolic resin shapes. Shawinigan made vinyl acetate, polyvinyl acetates and acetals.

In October of that year I went to work for Monsanto in the Applications Research Laboratory, of the Research Department in St. Louis. My job: evaluating plasticizers.

I had graduated that spring from Syracuse University with an M.S. in organic chemistry. I knew nothing at all about resins, polymers, plastics, coatings or plasticizers. But then, no one else in the Research Department did either. My predecessor, who had some little experience, was being transferred to the new acquisition in Massachusetts.

My indoctrination was brief, the general feeling being that plasticizers were passé and that the new polymers would do without them. But there existed a backlog of proposed plasticizers which needed screening for patent filings.

Although not clearly stated then, looking back, I can now define my research objectives as:

A.) Clear up a backlog of proposed plasticizers - some 2-300 of them - by evaluating them in nitrocellulose and cellulose acetate films so that patents could be filed.

B.) Look for a plasticizer which would make cellulose acetate flexible and tough at low temperatures. (This was to stave off the emerging competitive pressure of plasticized polyvinyl butyral safety glass interlayer.)

C.) Find a plasticizer for nitrocellulose to replace camphor. The Japanese had already cut off much of the natural camphor supply from Formosa.

This third objective was a will-o-the-wisp. John Wesley Hyatt had patented the nitrocellulose camphor combination in 1878 and, even today, it remains a unique formulation. No other polymer plasticizer combination produces as distinct a product as "celluloid". We spent a lot of time and effort

Published 1990 by Elsevier Science Publishing Co., Inc.
Organic Coatings: Their Origin and Development
R.B. Seymour and H.F. Mark, Editors

examining liquids and solids whose structure or whose properties resembled camphor, but without success.

The goal of the second objective was also a dead end. We did not know enough about molecular structures then to realize that plasticized cellulose acetate could never exhibit the broad rubbery phase which plasticized polyvinyl butyral and polyvinyl chloride exhibit. Plasticized polyvinyl butyral displaced cellulose acetate safety glass interlayer then and has held on to that market ever since.

I did accomplish the first objective. It took about 18 months to go through the stockpile of acetates, butyrates, propionates, succinates, maleates, fumarates, phthalates, sebacates, citrates, itaconates, glycollates, salicylates, benzoates, phosphates, adipates, alkyl biphenyls, chlorinated biphenyls, sulfonamides and what not. In the end, there was not one new product worth commercialization.

My bible was "Physical Examination of Paint, Varnishes, Lacquers and Colors" by Gardner. My management didn't think it necessary to invest in any fancy coatings or plastics testing equipment so we made do with what we had. Our techniques were simple enough: stock solutions of nitro-cellulose in butyl acetate - toluene and cellulose acetate in acetone were prepared. Plasticizers were added in amounts of 25, 50 and 100 PHR (parts per hundred of resin) to these solutions. Films were then cast on glass plates, air dried in the laboratory, stripped and conditioned in a small constant temperature/humidity cubicle. Measurements for tensile-elongation on a small Scott and flexibility on an old Schopper fold tester were the principal routine tests. Hardness we measured by "fingernail" or Venus pencil; gloss, by inspection; moisture resistance, by placing a drop of water on a coated steel panel and observing rusting, whitening or peeling.

There was a body of arbitrary knowledge as to what constituted a good plasticizer: high boiling esters or sulfonamides, water insoluble, heat stable; and containing halogen or phosphorous or both for nonflammability. But the understanding of polymer plasticizer interaction was still in the future.

After Pearl Harbor, I was transferred to more defense-oriented jobs. By 1945, however, the demand for plasticizers for polyvinyl chloride for military applications had convinced our management to restart the "Plasticizer Application Lab". As I was the only researcher about with any experience, I was told to go to work. This time, however, there was money (grudgingly) for specialized equipment: roll mills for compounding, torsion testers for low temperature flexibility, aging ovens, accelerated weathering units, presses for molding. Most importantly, we had access to our customers, through the Sales Department, an unheard of thing, prewar. Now we could focus on real needs and specifications and could arrange cooperative field trials.

Best of all, there was now the beginning of some theoretical concepts concerning solubility parameters and H-bonding values to help narrow the search.

We could tackle the problem in a focused fashion by using 2nd order transition values as a measure of efficiency; we could screen our museum of experimental plasticizers quickly and select the most promising for detailed study.

In less than a year we had identified and pilot planted three new plasticizers, followed rapidly by others (Table II).

What made the difference? Why was it that in 1938-39 we examined hundreds of plasticizer candidates without a single commercial success while in 1945-46 we identified and commercialized eight, including two new classes, the alkyl diaryl phosphates and the butyl benzyl phthalates?

In 1938, Application Research had no communication channel to sales or the customer. The experimental results were without exception to back up the patentability of new products produced by the Organic Synthesis Group. Beyond that, there was little demand for new plasticizers in the existing markets. Dibutyl phthalate and tricresyl phosphate served the nitrocellulose coatings very well. Camphor was, and still is, the best plasticizer for nitrocellulose sheets, rods and tubes. Dimethyl and diethyl phthalate and triphenyl phosphate were the plasticizers of choice for cellulose acetate coatings and molded products.

By 1944, plasticized polyvinyl chloride had expanded into markets previously served by rubber: electrical insulation, floor tile and raingear, and was moving into coated fabrics and foams. The requirements for low temperature flexibility, permanence, stability and flame proofing, were only partially met by the commercially available dioctyl phthalate and tricresyl phosphate.

Research Management, pressured by its own administration and Sales, allocated the money and space to equip an up-to-date Application Laboratory. Through direct contact with field salesmen and their customers we were able to set up specifications for improved plasticizers which the synthesis chemists could work toward. The most promising were quickly pilot planted and given field trials. The best were then put into full scale production.

What did we learn? We learned about commmercial chemical development, that process of translating market needs into research projects and new products into commercialization. We learned that Research and Development and Sales should not work in isolation if new products were to come into existence.

And along the way, we helped transform plasticizers from flexibilizers for coatings to major components of PVC compounds.

TABLE 1

MONSANTO PLASTICIZERS, 1938

Plasticizer	Use	Introduced
Sulfonamides	NC	1928
Phosphates		1932
Tricresyl	NC, PVC	
Triphenyl	CA	
Glycollates		1935
Methyl phthallyl ethyl-	CA	
Ethyl phthallyl ethyl-	CA	
Butyl phthallyl butyl-	NC, PVC	
Phthalates		
Dimethyl	CA	1935
Diethyl	CA	
Dibutyl	NC	
Chlorinated Biphenyls		1935
Arochlors	NC, PVC	

TABLE II

MONSANTO PLASTICIZERS, 1948

Plasticizer	Use	Introduced
Phosphates		
Cresyl diphenyl-	PVC, NC	1946
Octyl dicresyl-	PVC	1945
Octyl diphenyl-	PVC	1945
Phthalates		
Diphenyl-	PVC, NC	1940
Butyl, benzyl-	PVC, Acrylates	1945
Dioctyl-	PVC	1948
Adipates		
Dioctyl-	PVC	1943
n.octyl, N-decyl-	PVC	1948
Biphenyls		
o-Nitro-	PVC, NC	1946
Hydrogenated (HB-40)	PVC	

History of Alkyd Coatings

Raymond B. Seymour
University of Southern Mississippi
Hattiesburg, MS 39406

Abstract

While it is not possible to produce good yields of graft copolymers by polymerizing styrene, vinyltoluene or p-methylstyrene with a medium oil alkyd in the presence of ditert. butyl peroxide, graft copolymers are obtained where p-tert. butyl perbenzoate is used as the catalyst (initiator). The graft copolymers, which are harder and faster drying than the corresponding blends are readily characterized by methanol precipitation, FTIR and GPC.

Introduction

The term paint has been used since the dawn of history to describe decorative and protective formulations used by primitive artists and builders. The application of crude paint which is the oldest polymer oriented industry is well chronicled by the 25 thousand year old paintings on the walls of caves at Altamira, Spain and Lascaux, France and the recipe for oleoresinous paints supplied by Theophilus in the 11th century.

Prior to the 17th century, the use of paint on houses was restricted to the wealthy. Nevertheless, Thomas Child produced 2 barrels of oleoresinous paint daily in 1927 in Boston, MA, USA, using "ye old paint stone" which was imported from England. The stone has been preserved and can still be seen as a building stone in downtown Boston.

Oleoresinous Paints

In spite of their empirical development and lack of understanding of the "drying process," oleoresinous formulations became the standard for the production of paints, putty, oil cloth, linoleum, artist's color, and printing ink until the early 1920's when they were replaced, to some extent, by synthetic polymers.

Polyesters

Since some of the first synthetic organic compounds were esters which were produced by the condensation of monofunctional alcohols, such as ethanol, and monofunctional organic acids, such as acetic acid, it is not surprising to note that one of the first synthetic polymers was polyglyceryl tartrate which was synthesized from difunction reactants by Berzelius in 1847. Over a half century later, Watson Smith produced thermosetting polyesters (Glyptals) by the condensation of glycerol and phthalic anhydride in 1901.

During the following half century, polymer chemists learned that the curing or "drying" of oleoresinous paints, in the presence of heavy metal salts ("driers"), and air, involved the formation of hydroperoxides on the carbon atoms adjacent to the ethylenic bonds in the unsaturated oils. In the accepted mechanism, the carbon-carbon double bond shifts to a conjugated configuration and the peroxide is transferred to another monoallylic carbon atom. Polymerization proceeds via a radical chain mechanism to produce a crosslinked insoluble film.

Alkyds

In 1931, Kienle utilized Glyptal and oleoresinous paint technology to produce superior polymers which now account for about one half of all coating resins. He obtained

Published 1990 by Elsevier Science Publishing Co., Inc.
Organic Coatings: Their Origin and Development
R.B. Seymour and H.F. Mark, Editors

267

air-curable resins which was more flexible than Glyptals by producing oil-modified polyester resins for which he coined the euphonious acronym, alkyd, after the a̲l̲cohol and a̲c̲i̲d̲ reactants used.

Kienle also utilized Baekeland's concept of functionality which was quantified by Carothers. When the functionality of the reactants is greater than two, crosslinking or "bodying' occurs as the glycerol, phthalic anhydride and unsaturated oil are heated. Kienle reacted the unsaturated oil with glycerol at 230-290° C in the presence of litharge to produce a monoglyceride which was then esterified by reacting with phthalic anhydride. Kienle classified his resins as short, medium and long oil alkyds in accordance with the increase in concentration of the unsaturated oil used in the condensation reaction.

The medium oil alkyds are more versatile than the short or long oil alkyds. Most alkyd resins are produced by the fatty acid process, in which the glycerol, dicarboxylic acid and unsaturated acid are condensed at 200-230° C. These alkyd coating have superior gloss, adhesion, hardness, and chemical resistance and dry faster than oleoresinous coatings.

Alkyd Blends

Selected alkyd resins are compatible with cellulose nitrate, amino resins, phenolic resins, epoxy resins, silicones, acrylics, chlorinated rubber and styrene. When added to cellulose nitrate, short oil alkyds improve gloss, adhesion and flexibility of these commercial coatings. The chemical resistance of short oil alkyds is improved when they are reacted with amino resins. Alkyds also react with phenolics to produce chroman-type alkali-resistant coatings.

The properties of both alkyds and epoxy resins are upgraded when both resins are present in a coatings mixture. The heat resistance, durability and gloss of alkyds are upgraded when they are reacted with silicones. It is customary to add alkyd resins to improve the durability of highway marking paints based on chlorinated rubber. Long oil alkyd resins are also added to latex house paints to improve their resistance to chalking.

Acrylated Alkyds

The degree of hardness of alkyd resins has been increased by blending with low molecular weight acrylic resins. More homogeneous resinous coatings have been obtained by blending acrylic acid copolymers with alkyds. These acrylated alkyds dry more rapidly than alkyd coatings but are not available commercially.

Styrenated Alkyds

The drying rate and hardness of alkyd coatings were improved in 1948 by Bhow and Payne who added polystyrene to the alkyd resins. Xylene solutions of these styrenated alkyds were produced by several different techniques but the polymerization of styrene in a solution of the alkyd was preferred.

In a recipe supplied by Payne, a solution of 4 moles of styrene and 3 percent benzoyl peroxide was added slowly over a period of 30 minutes to a xylene solution of dehydrated castor oil at 115° C these solutions and were heating at 140° C for one hour. The 6.5 percent unreacted styrene monomer was removed by vacuum distillation prior to the traction with phthalic anhydride 232° C for 80 minutes.

However, in the most widely used method, styrene and the benzoyl peroxide are added to a previously prepared resin. Harrison, Tolby and Redkamp maintained that unless long oil alkyds were used, the styrenated alkyds were primarily blends of polystyrene and the alkyd resins.

Precipitation in methanol has shown that many commercially available styrenated alkyds are blends of the alkyd and polystyrene. An attempt to produce copolymers by adding styrene and benzoyl peroxide or ditert. butyl peroxide to a commercial soy-linseed medium alkyd (Cargill 5150) in a nitrogen atmosphere, were also unsuccessful. However, the blends containing at least 40 percent styrene were compatible with the alkyd resina nd the coatings were slightly hazy, fast drying, harder and more chemical resistant than the alkyd resins component.

Graft styrene-alkyd copolymers were obtained when a solution of styrene and tert butyl perbenzoate were added simultaneously to the medium oil alkyd at 120° C. The reaction was continued for 2.5 hours after the addition of the styrene and initiator, but a small amount of unreacted styrene remained in the solution. Precipitation in methanol, and characterization by FTIR spectroscopy and gel permeation chromatography (GPC) showed the presence of the graft copolymer rather than a mixture.

Alpha-Methyl Styrenated Alkyds

Attempts to produce a copolymer by heating alpha-methylstyrene and tert butyl perbenzoate with an alkyd were unsuccessful. However, a graft copolymer was obtained when 50:50 mixture of alpha-methylstyrene and styrene along with tert-butyl perbenzoate was heated with the medium oil alkyd resin.

Vinyl Tolvenenated-Alkyds

Dow produces a mixture of meta (60) and para (40) isomers of methylstyrene (VT) which it calls vinyltoluene (VT). Blends of VT and medium oil alkyds, which are produced when VT is heated at 130° C in the presence of di-tert. butyl peroxide are compatible but the films are hazy. However, graft copolymers are produced when tert. butyl perbenzonate is used as the catalyst at 120° C instead of ditert. butyl peroxide at 130° C. These graft copolymers are soluble in aliphatic hydrocarbon solvents, such as naphtha.

Para-Methylstyrenated Alkyds

A methylstyrene monomer which is a 97% para and 3% meta isomer was produced commercially by Mobil and this production facility was acquired by American Hoechst which sold it to Deltech in 1988. This PMS copolymerizes readily in the presence of ditert. butyl peroxide and a medium oil alkyd at 130° C but the product is a compatible mixture of the two polymers which can be separated in methanol.

Coatings from these mixture are unaffected by water but are pitted and discolored in ethyl acetate and crack in acetone. Their hardness, as measured by a Sward rocker, is much higher than the alkyd of a 40-60f blend of the alkyd and PPMS.

The graft copolymer of PMS (40) - alkyd (60) has a Sward rocker harness of 46 vs. 30 for the corresponding blend. Neither the copolymer nor the blend is adversely affected by water or methanol but both bubble when exposed for 24 hours to ethyl acetate or xylene. These solutions contain about 1 percent residual monomer and this monomer content can be reduced by vacuum distillation.

Conclusion

Styrenated alkyds produced by the copolymerization of the medium oil alkyd (60) and styrene, vinyltoluene or para-methylstyrene (40) in the presence of p-tert. butyl perbenzoate are harder than the unreacted alkyd. The coatings obtained from the graft copolymer are harder, more solvent resistant and more ductile than coatings obtained form blends with similar rations of polymers.

The history of alkyds is of particular interest since it demonstrates considerable ingenuity in combining classic polyester production art with ancient oleoresinous technology to produce a product whose use exceeds that of either component. This technology has been improved by the incorporation of grafting copolymerization techniques. Additional advances can be anticipated and more research and development effort is applied to alkyd resins.

HISTORY OF ELECTRODEPOSITION OF COATINGS

ARTHUR M. USMANI
Boehringer Mannheim Biochemistry R&D Center
9115 Hague Road, Indianapolis, IN 46250

ABSTRACT

The electrodeposition of organic coating compositions has
been commercially practiced for the past 25 years. It
has made a substantial impact on coating application
operations e.g., the priming of automobile bodies and the
coating of aluminum extrusions.

In this chapter, a historical account of electrodeposition and early development will be described.
Additionally, we shall review anodic as well as cathodic
electrodeposition. Other related electrochemically
induced processes and a brief discussion on electrochemical reactions have also been included for
completeness.

INTRODUCTION

The electrodeposition of organic coating compositions onto metallic
surfaces has been known since the early 1930´s, but it achieved commercial
importance only in the early 1960´s. Since then, electrodeposition has
assumed an important role in the priming or one-coat finishing of suitably
pre-treated metallic surfaces, primarily steel and aluminum.

In electrodeposition, charged water-soluble or water-dispersible
macroions are attracted by an electrode of the opposite polarity where they
undergo electrochemical reactions and are deposited. In practice, the part
is immersed into a bath of 7-15% solid water-borne paint that is subjected
to a 50-500V DC electric field. The metal part is rapidly (60-120 s) and
uniformly coated to a self-limiting, reproducible film thickness in the
range of 0.7-2 mils, including edges, recesses, angles, holes, and
projections.

The electrodeposition resins are designed and prepared like other
coating polymers but are required to carry ion-forming groups e.g., amino or
carboxyl. If the electrodepositing vehicle is represented by RX, where R+
is the macrocation, then R+ will be deposited at the cathode (cathodic
electrodeposition).

Conversely, with macroanion R−, deposition will take place at the anode
(anodic electrodeposition). A generalized representation of cathodic and
anodic electrodepositing resins is indicated in Table 1.

TABLE I. Cathodic and anodic electrodepositing resins.

Resin[a]	Solubilizer	Film-forming Macroion	Type Electrodeposition
RNH_2	Acid (HA_{aq})	RNH_3^+	Cathodic
RCOOH	Base (BOH_{aq})	$RCOO^-$	Anodic

[a] R = resin

In this report, we will provide a historical account of electrophoresis and electrodeposition of coatings. Additionally, we shall review both anodic and cathodic electrodeposition. Other related electrochemically induced processes will be discussed, although very briefly. For completeness, we will also include a discussion brief on electrochemical reactions.

Early Historical

The migration of colloidal clay particles to the anode in an electric cell was observed in 1808 by Reuss [1]. He also observed the migration of fluid to the vicinity of the cathode (electro-osmosis). Quincke, interested in electro-osmosis, has referred to the observations made by Faraday and others in his report of around 1860. In water, starch and many other finely divided substances migrate toward the anode, whereas in non-aqueous turpentine most substances migrated to the cathode [2].

The first recorded observation of an electrodeposition was made by Picton and Linder in 1905 [3]. In their experiment, they found that ferric hydroxide was repelled from the anode in the presence of alcohol and that the coagulum deposited on the cathode was horn-like in appearance. Electrophoretic separation of diphtheria toxin and anti-toxin in agar gel was observed in 1907 by Field and Teague [4]. The work on electrophoretic separation continued to progress, resulting in a powerful analytical tool when Tiselius began to publish his findings and results around 1937.

In 1919, Davey of General Electric was granted a patent for a process of making and applying Japan, a product composed of asphalt, oleaginous material, and ammonia or alkali [5]. Anodic deposition at 10 to 200V was claimed in this patent. Anode Rubber Company, with Klein as inventor, obtained a patent in 1923 for electrophoretic precipitation of rubber onto a permeable non-conductive surface surrounding the anode [6]. Between 1936 to 1943, W. Clayton and co-workers at Cross and Blackwell Limited were granted patents for coating of the interiors of food cans using oleoresinous varnishes [7-9].

Recent Historical

Thus, we find that over a 150 year period, the investigations of many researchers in electrodeposition had only limited success without much commercial payoff. The principle reason can be attributed to the lack of suitable polymers that could support current. Also lacking was the know-how about pigmentation of the aqueous system, specifically surface chemistry.

Scientifically sound synthetic polymer chemistry began due to the vision and brilliant work of Carothers in the late 1930's. Soon thereafter, in the 1940's and 1950's, a whole slew of polymers and polymer families was

created. In 1948, styrene-butadiene latex was developed. During the
1950's, aromatic acid monomers for water soluble polymers and coatings
became available. Thus, the stage was set for the "re-discovery" and
commercialization of electrodeposition.

During the late 1940's and early 1950's, we became a more consumer
product-oriented society. The "plastic revolution" had not yet taken
place. Thus, manufacturers of metal products had to intensify ways to
mitigate the corrosion problem. Rusting of automobiles during use was of
much concern to car manufacturers.

During the early 1960's, R&E work on electrodeposition of coatings
began at Ford Motor Company under the leadership of Brewer and Burnside
[10,11]. Several leading coatings and chemical companies also collaborated
with Ford on this project. In a matter of several years, anodic
electrodeposition coating tanks were scaled up in size from 5 gallon to
50,000 gallon and priming of wheels and auto bodies became a reality. For
this breakthrough in paint application, George Brewer has been the recipient
of several awards. Success stories must be told and re-told so that
potential innovators in any organization or industry are not discouraged and
frustrated [12].

The wide acceptance of electrodeposition resulted due to the increasing
number of publications and patents that began around 1965 and continued
unabated until the early 1980's. Three doctoral dissertations significantly
contributed to the chemistry and technology of electrodeposition [13-15].
In a similar vein, voluminous compilations of electrodeposition literature
can be found in several books [16-19].

The commercial success of electrodeposition came when polymers in the
form of colloidal dispersions, ultra-fine particles with molecular weights
intermediate between latex and true solutions containing hydrophilic group
became available. The earliest electrodeposition vehicles, used in 1963 in
the automotive industry, were based on amine and alkali neutralized
maleinized vegetable oils. Such vehicles, when formulated into primers with
strontium chromate, gave marginal performance and corrosion resistance.
Later, phenolic and amino modifications were used with improved properties
due to a greater degree of cross-linking. In the late 1960's and early
1970's, the oleoresinous primer formulations were phased out and replaced
with synthetic polymers e.g., acrylics, epoxies, and polybutadiene.

One-coat finishes based on water soluble acrylics were also introduced
during the early 1970's. These found applications in coating toys and
metallic furniture. Discoloration of whites on steel was a major problem,
however. As an extension of anodic electrodeposition, the cathodic method
was proposed by Brodie and Usmani of Sherwin Williams in 1968 [20]. In the
cathodic process, the article to be coated is made cathode. The resin
binder is required to migrate to the cathode and hence must bear a positive
charge. This can be achieved by incorporating an amine functionality into
the resin and then reacting this functionality with an organic acid to
obtain a positive charge and water dispersibility and/or solubility. The
successful and industrially applicable development of cathodic
electrodeposition was accomplished in the early to mid-1970's by Wismer and
co-workers at PPG [21]. This effort continued at a substantial pace into
the 1980's. For his contributions, Wismer has also received significant
awards.

Cathodic electrodeposition made its debut in the USA in 1976 for the
priming of automobile bodies. Since then, it has been introduced in Europe
and the Pacific Rim. At the present time, cathodic electrodeposition
accounts for more than 60% of total electrodeposition. Because of the

absence of electrode dissolution in the cathodic method, it is possible that anodic electrodeposition specifically on steel may become obsolete in the future.

STATUS OF ELECTRODEPOSITION

Coating is a design concept intended to provide protection from environmental conditions, decoration to enhance appearance, and functionality for specific property. The coating provides protection from adverse environmental conditions by sealing out the undesirable atmospheric conditions. Applications of coating to any metallic substrate always result in cost reduction. Brush, dip, blade, roller, air spray, airless (hydraulic) spray, fluidized bed, and electrostatic spray are some of the commonly used coating methods. Electrodeposition is an application method of rather recent development. The multi-disciplinary nature of electrodeposition is depicted in Figure 1.

Compared to conventional application methods, electrodeposition has the following advantages and features:

1. Deposition of film is extremely uniform over edges, corners, cavities, pores; this results in improved corrosion protection.

2. Utilization of primer or coating is very high, usually > 95%.

3. Coating or priming results are highly reproducible.

4. Coatings deposit at very high solids (92% or better) and can be handled immediately after deposition.

5. Electrodeposition coatings are formulated with the minimum of organic solvents, thus contributing to safety and environmental factors.

6. In electrodeposition, counterions that contribute to water sensitivity are not co-deposited; in conventional coatings, fixed counterions e.g., Na^+, K^+ remain part of the film while volatile organic counterions e.g., amine are partially removed by volatilization.

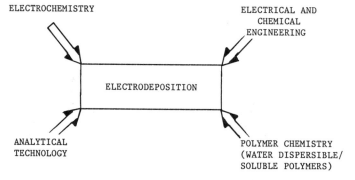

FIG. 1. Multi-disciplinary nature of electrodeposition.

7. Electrodeposition is totally amenable to automation.

8. Material, labor, equipment, and electric energy costs are about 30% less than spray, electrostatic, or dip coating methods.

The disadvantages of electrodeposition are listed below:

1. Formulation latitude is limited due to electrochemical considerations; high pigment loading and use of soluble anti-corrosion pigments and highly ionic simple electrolytes are not possible.

2. Film build-up usually does not exceed 2 mils (50 μ).

3. The substrates to be coated must be electrically conductive; a plastic substrate per se cannot be coated.

4. Color choices are limited.

5. Capital and space intensive.

ANODIC ELECTRODEPOSITION

Mechanism

Commercial anodic systems use various types of water-dispersible or water soluble polymeric compositions, inherently insoluble in water, but capable of combining with inorganic or organic bases. The chemical compositions and properties of such systems are described by Gilchrist [22], Hagen [23], Sullivan [24], Motier and Marion [25], Kita and Kimi [26], Laffargue and Lahaye [27], Subramanian, Sundaram and Patel [28], and Usmani [29]. In addition to the solubilized and dispersed resin system, an electrodeposition bath may contain dispersed pigments [30,32] and cross-linkers [33,34]. Thus, ions and surface charged particles in the broadest sense are involved.

The kinetics, film growth, and mechanism of anodic systems have been studied by many workers [35-42]. In an anionic electrodeposition cell, the significant reactions occurring at the anode can be summarized as follows:

1. Electrolysis of water:

$$2H_2O \longrightarrow 4H^+ + O_2 + 4e^-$$

2. Anodic dissolution of electrode:

$$M^o \longrightarrow M^{n+} + ne^- \text{ e.g.,}$$

$$Fe \longrightarrow Fe^{++} + 2e^- \text{ (steel)}$$

3. Polymer deposition:

$$RCOO^- + H^+ \longrightarrow RCOOH \text{ (Major)}$$

$$RCOO^- + M^{n+} \longrightarrow (RCOO)_n -M \text{ (Minor)}$$

$$RCOO^- \xrightarrow{-e} RCOO^\cdot \xrightarrow{-CO_2} R^\cdot; \ R^\cdot + R^\cdot \longrightarrow R-R \text{ (Minor)}$$

276

A successful continuous electrodeposition operation requires that all the components maintain a material balance; since changes occur unevenly during electrodeposition, these are critical. In a RCOO⁻ B⁺ dispersed electrodeposition resin (anodic), the RCOO⁻ portion will be removed from the bath in the form of an electrodeposit. The solubilizer portion, B, will remain in the bath where it will accumulate and eventually interfere with the deposition process. One method to balance solubilizer B is to utilize leftover solubilizer for solubilization of the feed material (Figure 2). The feed consists of partially solubilized resin RCOOH which is allowed to react with the solubilizer-rich bath. This reaction is facilitated by the use of an ultrasonic or high-shear homogenizer. Another method to balance the electrodeposition bath relies on removal of leftover solubilizer by various methods, e.g., electrodialysis, ion exchange, or dialysis (see Figure 3).

An important feature that is required in the electrodeposition material composition is "throwing power". Throwing power is the ability to form electrodeposits of such uniform thickness that even the most recessed areas and pores of the substrate are covered. Generally, the application of a higher voltage produces higher film thickness and higher throwing power. For every electrodepositing composition, however, there is a maximum applicable voltage; higher voltages cause rupturing of the deposited films resulting in undesirable imperfections. Other factors that improve throwing power are higher bath conductivity and higher equivalent weight of the depositing composition.

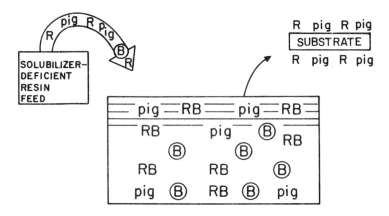

FIG. 2.   Solubilizer-deficient feed schematic: R = resin, B = solubilizer, pig = pigment.

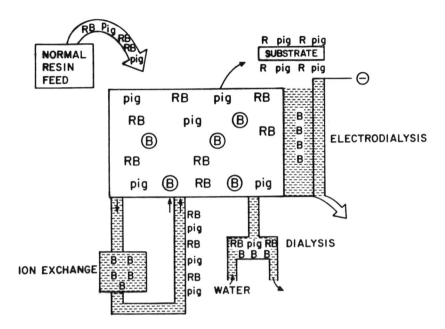

FIG. 3. Schematic of normal solubilized resin feed with removal of leftover solubilizer.

Anodic Resins

For supporting the anodic electrodeposition process, carboxyl groups are required. Table II shows properties that are obtained from specific chemical groups in a resin. A judicious choice of the various groups in a resin can lead to an electrodepositing coating of desired properties.

A large number of proprietary resins lend themselves to anodic electrodeposition. Acrylics and epoxies are by far the most important systems.

Acrylic System: Acrylic monomers can be polymerized to give either thermoplastic or thermosetting acrylic polymers. The various pendant chemical groups present along the polymer chain can be designed to perform the desired function. The acrylic resin may contain a carboxyl-containing acrylic monomer, a hydroxyl-containing acrylic monomer, some neutral acrylic monomers, and certain functional acrylic monomers (see Table III).

The cross-linking reaction that is utilized with the carboxyl-containing acrylic polymer is:

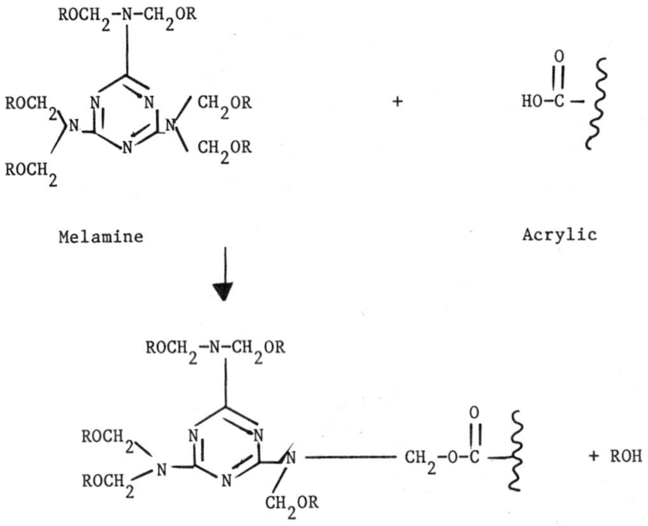

Melamine

Acrylic

+ ROH

TABLE II. Properties versus groups in an anodic resin.

Group	Property Obtained
Carboxyl	Supports anodic electrodeposition; cross-linking with triazine cross-linkers
Aromatic	Thermal and mechanical stability, stiffening effect
Triazine	Thermal and mechanical stability, corrosion resistance
Ether linkages	Chemical resistance, flexibilizing effect, and moisture sensitivity
Amide	Toughening and adhesion; cross-linking with triazine cross-linkers
Hydroxyl	Adhesion, moisture sensitivity; cross-linking with triazine cross-linkers

TABLE III. Typical monomeric components of acrylic anodic resins.

Carboxyl-containing

    Acrylic acid
    Methacrylic acid
    Maleic anhydride

Hydroxyl-containing

    Hydroxyethyl methacrylate
    Hydroxypropyl methacrylate
    Hydroxyethyl acrylate
    Hydroxypropyl acrylate

Neutral

    Methyl methacrylate
    Styrene

Functional

    Butyl methacrylate (flexibilizing)
    Lauryl methacrylate (flexibilizing)

The rate of reaction of the cross-linker can be summarized as follows:

1. In the presence of an acid catalyst

$$-OH > \overset{\displaystyle O}{\overset{\displaystyle \|}{C}}-NH_2 > COOH$$

2. In the absence of an acid catalyst

$$-COOH > > \overset{\displaystyle O}{\overset{\displaystyle \|}{C}}-NH_2 > OH$$

Anodic Epoxy System: Fatty acid esters of hard epoxy resins (when $n > 7$), e.g., Dow's DER 664, DER 664U, and DER 662, are popular electrodepositing resins. Esters are prepared by reacting the fatty acids, e.g., linseed oil with an epoxy resin at about 260°C. About 2 to 3% xylene, as an azeotrope, aids in water removal. Maleic anhydride is then reacted with the unsaturated fatty acid of the epoxy ester to a succinic anhydride derivative. The maleinized resin is thinned with a solvent, e.g., butyl cellosolve and the electrodeposition vehicle is prepared by neutralization with a suitable amine. Because of the excessive flexibilizing nature of the fatty acid, a high level ($\sim$ 40%) of the melamine, e.g., Monsanto's Resimene 745 or 755, is required in the coating formulation.

CATHODIC ELECTRODEPOSITION

Mechanism

    Cathodic electrodeposition takes place at the cathode. The metal surface does not ionize and remains passive during the deposition process. Cathodic polyelectrolytes are polymers containing amines (1°, 2°, or 3°) or

quarternary ammonium, sulfonium or phosphonium groups. Under the influence
of the electric current, such particles or micelles migrate to the cathode
and are deposited. The reactions occurring during cathodic
electrodeposition, as proposed by Wessling et al [43] and Wismer et al [44],
can be summarized as follows:

1. Electrolysis of water:

$$H_2O + e^- \longrightarrow 1/2 \; H_2 + OH^-$$

2. Polymer deposition:

$$R \; (N)_x \; (NH^+)_y + y \; OH^- \longrightarrow R \; (N)_{x+y} + yH_2O$$

Cathodic Resins

The cationic groups required for water dispersibility and deposition
can be introduced into cathodic resins by various methods, e.g., 1°, 2°, and
3° amines neutralized with acids, quarternary ammonium salts and hydroxides,
quarternary phosphonium salts, ternary sulfonium salts, and quarternary
ammonium carboxylates [45-51]. In such resin systems, problems of cure of
highly basic films and corrosivity of acidic bath must be addressed and
solved.

Dow's Wessling has done extensive work on cathodic electrodeposition of
polymer colloids. Among his proposed polymeric systems are the onium-
stabilized acrylic emulsions, onium polyelectrolytes, and onium (quarternary
ammonium, sulfonium, and isothiouronium) functional hydrophobic colloids
[52].

Acrylic: Cationic acrylic or vinyl copolymers are prepared by inclusion of
amine-containing monomers or by inclusion of glycidyl monomers that can be
subsequently reacted with an amine. Acrylics produce a suitable one-coat
system of excellent durability.

Epoxy: Typically, an epoxy resin is reacted with a base. The commercially
significant cationic epoxies are those with quarternary ammonium salts or
amine salts. Due to high alkalinity, these systems cross-link sluggishly
with amino and phenolic resins. Use of blocked isocyanates is an effective
way of cross-linking such cationic epoxies.

ELECTROCHEMICAL REACTIONS: BRIEF DISCUSSION

Table IV shows reactions that can be carried out electrochemically.
The important electrochemical and chemical parameters by which the reactions
can be controlled are indicated in Table V.

Electrochemistry can introduce up to 3.5 eV energies into organic
compounds (for comparison, photochemistry = 0.8-8.0 eV and radiation
chemistry > 8.0 eV), which means that almost any reaction can be carried out
electrochemically. But from a practical viewpoint, one must consider the
variables shown in Table V.

Let us take a closer look at the factors governing electrode reactions
to understand the electroorganic synthesis. Figure 4 shows the potential
distribution in an operating electrolytic cell [53]. The key factor in
determining the selectivity and reaction rate is the electrode potential at
both the anode and cathode. Electrode potential is the difference in
voltage between the solid electrode and the adjacent solution. Figure 4
shows that within the double-layer region a very large electric field exists

281

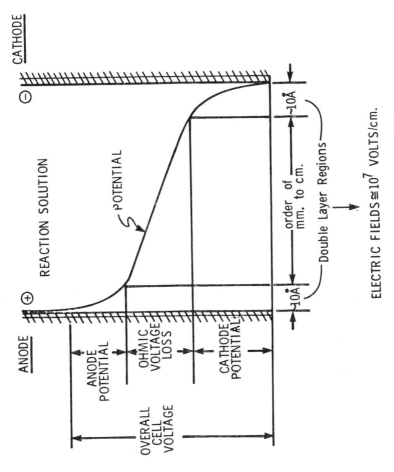

FIG. 4 Cell potential distribution [53].

$(10^7$ V/cm) [53]. This being the case, chemical bonds can be introduced into the molecules at low temperatures [54]. Electrosynthesis and electrodeposition can be done either at constant current or constant voltage. Electrodeposition at constant voltage is industrially popular.

Although electroorganic processes sound very exciting they are not industrially employed to a significant extent. Should energy costs rise again and the environmental problems become more burdensome, electroorganic processes may find new applications.

TABLE IV. Electrosynthesis of organic compounds.

Direct electron transfer to generate

At anode	At cathode
Cation radicals	Anion radicals
Dications	Dianions
Carbonium ions	Carbanions
Free radicals	Free radicals

Indirect electrolysis

To generate redox couples
To generate reactive inorganic species, e.g., $OH^{\cdot}$, $CO^{\cdot}_2$

Generation of acid at anode (anodic electrodeposition) or base at cathode (cathodic electrodeposition)

TABLE V. Electrochemical reactions: Electrochemical and chemical parameters.

Electrochemical parameters	Chemical parameters
Electrode potential	Solvent/supporting electrolyte concentration
Electrode material	Concentration of starting materials
Current density	Concentration of other components of bath
Electric field	
Adsorption	Temperature of bath
Type of cell and design: compartmentalized membrane or porous separated static or flow cell	pH
Solution conductivity	

RELATED ELECTROCHEMICAL PROCESSES

Historically, electropolymerization of vinyl monomers was researched in the early 1950's by Kern and Quast [55], Breitenbach and Srna [56], Parravano [57], and Funt and Williams [58]. Initiators are generated by an electrode process, as described in an earlier section. Ionic or radical propagation takes place with the formation of a polymer solution or dispersion along with the deposition of a porous polymer film on the electrode. By far, the most important industrial process in this class was developed by Baizer of Monsanto, who generated anion radicals from acrylonitrile, leading to the manufacture of nylon.

The electrochemical generation of initiators for polymerization has been reviewed by Silvestri et al [59]. Diacetone acrylamide monomer that has high solubility in aqueous electrolytes forms polymer films of good adhesion upon cathodic electropolymerization [60]. Cvetkovskaja et al have reported the growth of about 10 µm polyacrylamide films by electropolymerization in the presence of zinc electrolytes [61].

The anodic deposition of conducting polymers, an exciting new field, has been initiated by Chiang et al [62], Nigrey et al [63], and Diaz et al [64]. Aromatic or heteroaromatic compounds can be oxidized at the anode to yield polymers with conjugated double bonds. Further oxidation of the polymer backbone to the polymer radical cation, with the parallel insertion of anions to compensate the charge, leads to conducting polymers with low resistivity comparable to metals. Monomers e.g., pyrrole, aniline, and thiophene are typically deposited on inert anodes.

Recently, Salyer and Usmani prepared open pore urea formaldehyde structures from a non-etherified water soluble resin. This is due to a pH decrease at the anode that initiates polymerization and structure formation on and around the anode [65]. Beck et al have electrodeposited carbon black filled systems that, after baking, produce films of good conductivity to which a second coat of electrodeposit can be applied [66-67].

Subramanian et al electrodeposited graphite fibers, resulting in significant improvements in composite strength and toughness when coated fibers were introduced in an epoxy matrix [28]. This novel technique can be extended to boron fibers and may well be used to introduce the epoxy matrix.

CONCLUSIONS

In this chapter, we have given a historical account of electrodeposition along with the working mechanics of both anodic and cathodic electrodeposition.

Availability of improved steel and its low cost indicates that automobile manufacturers may not switch to the more expensive plastics that can withstand 300°F paint baking temperature. Thus, electrodeposition is here to stay. Use of electrodeposition in electronics, current carrying polymer synthesis, and composites are some potential areas of growth.

REFERENCES

1. A. Reuss, Memoires de la Societe Imperiale de Naturalistes de Moscow, 2, 327 (1808).
2. G.H. Quincke, Poggendorffs Annalen, 113, 513 (1861).
3. H. Picton and S.E. Linder, J. Chem. Soc., 87, 1908 (1905).
4. C.W. Field and O. Teague, J. Exp. Med., 9, 222 (1907).
5. W.P. Davey, U.S. Pat. 1, 294, 627; Feb. 1919.
6. Brit. Pat. 223, 189; Nov. 1923.
7. Brit. Pat. 455, 810; 1936.
8. Brit. Pat. 496, 945; 1937.
9. Brit. Pat. 2, 337, 972; 1943.
10. Am. Paint J., 50, 16 (1965).
11. H.N. Bogart, G.L. Burnside and G.E.F. Brewer, SAE J., 81 (Aug. 1965).
12. A.M. Usmani, Guest Editorial, Polymer News, 9, 258 (1984).
13. A.M. Usmani, Ph.D Dissertation, "Electrodeposition of Carboxyl-Containing Polymers", North Dakota State University, Department of Polymers and Coatings, 1968.
14. J.P. Giboz, Ph.D. Thesis, "Mechanism of Electrodeposition of Polymers on Metallic Electrodes", Louis Pasteur University, Strassbourg, France, 1971.
15. C. Laffargue, Ph.D Thesis, "Mechanism of a Technique for Electrodeposition of Anionic Polymer on the Electrodes of Iron and Platinum", Louis Pasteur Unviersity, Strassbourg, France, 1975.
16. R.L. Yeates, Electroplating, 2nd Edition (Draper Ltd., Teddington, UK, 1970).
17. R.H. Chandler, Electrophoretic Painting (1971).
18. M.W. Ranny, Electrodeposition (Noyes Data Corporation, Park Ridge, NJ, 1970).
19. G.E.F. Brewer (Ed), Electrodeposition of Coatings, Advances in Chemistry Series, 119 (Am. Chem. Soc., Washington, D.C., 1973).
20. M.G. Brodie and A.M. Usmani, Sherwin Williams Research Center Report, 1968.
21. M. Wismer and J.F. Bosso, Metal Prog., 71 (1974).
22. A.E. Gilchrist, US Pat. 3, 230, 162, to Glidden; 1966.
23. J.W. Hagen, J. Paint Technol., 38, 436 (1966).
24. M.R. Sullivan, Ibid., 38, 424 (1966).
25. J.F. Motier and D.L. Marion, Proceeding Am. Chem. Soc., Div. Org. Coatings and Plastics Chem., 31, 273 (1971).
26. R. Kita and A. Kimi, J. Coatings Technol., 48, 53 (1976).
27. C. Laffargue and J. Lahaye, Ibid., 49, 85 (1977).
28. R.V. Subramanian, V. Sundaram and A.K. Patel, Soc. Plastics Industry, Composite Institute, 33rd Annual Tech. Conf., Sec. 20F, 1 (1978).
29. A.M. Usmani, Polym. Plast. Technol. Eng., 15, 115 (1980).
30. S.W. Gloyer, D.P. Hart and R.Z. Cutforth, J. Paint Technol., 37, 113 (1965).
31. D.S. Young, Proceeding Am. Chem. Soc., Div. Org. Coatings and Plast. Chem., 31, 288 (1971).
32. F. Beck and H. Guder, Makromol. Chem., Macromol. Symp., 8, 285 (1987).
33. J.N. Koral, W.J. Blank and J.P. Falzone, J. Paint Technol., 40, 518 (1968).
34. W.J. Blank, J.N. Koral and J.C. Petropoulous, Ibid., 42, 609 (1970).
35. L.R. LeBras, J. Paint Technol., 38, 85 (1966).
36. J.W. Hagen, F.N. Orttung and S.E. Chow, Paint Var. Prod., 67, 48 (1967).
37. C.A. May and G. Smith, J. Paint Technol., 40, 493 (1968).
38. A.E. Rheineck and A.M. Usmani, Ibid., 41, 597 (1969).
39. A.E. Rheineck and A.M. Usmani, Adv. Chem. Ser., 119, 93 (Am. Chem. Soc., Washington, D.C., 1973).

40. S.L. Phillips and E.P. Damm, Jr., J. Electrochem. Soc., <u>118</u>, 1916 (1971).

41. P.E. Pierce, Z. Kovac and C. Higginbotham, Ind. Eng. Chem. Prod. Res. Dev., <u>17</u>, 317 (1978).

42. S. Chen and C. Liu, J. App. Electrochem., <u>11</u>, 319 (1981).

43. R.A. Wessling, W.J. Sellineri and E.H. Wagener, Adv. Chem. Ser., <u>119</u> (Am. Chem. Soc., Washington, D.C., 1973).

44. M. Wismer, P.E. Pierce, J.F. Bosso, R.M. Christenson, R.D. Jerabek and R.R. Zwack, Proceeding Am. Chem. Soc., Org. Coatings and Plast. Chem., <u>45</u>, 1 (1981).

45. R.D. Jerabek and J.R. Marchetti, US Pat. 3,922, 253, to PPG; 1975.

46. J.F. Bosso and M. Wismer, US Pat. 4,071, 428, to PPG; 1978.

47. J.F. Bosso and M. Wismer, US Pat. 3,975, 346, to PPG; 1976.

48. R.M. Christenson, J.F. Bosso, M.E. Hartman and W.H. Chang, US Pat. 4,081, 341, to PPG; 1978.

49. M. Wismer and J.F. Bosso, US Pat. 4,066, 592, to PPG; 1978.

50. J.F. Bosso and M. Wismer, US Pat. 4,038, 232, to PPG; 1977.

51. J.F. Bosso and M. Wismer, US Pat. 4,001, 101, to PPG; 1977.

52. R.A. Wessling, Adv. Org. Coat. Sci. Technol., Vol. 4, Chap. 25, 1980.

53. L.E. Eberson and H. Shafer, Top. Curr. Chem., <u>21</u>, 1 (1971).

54. M. Fleischman and D. Pletcher, Platinum Met. Rev., <u>13</u>, 46 (1969).

55. W. Kern and H. Quast, Makromol. Chem., <u>10</u>, 202 (1953).

56. J.W. Breitenbach and C. Srna, Pure Appl. Chem., <u>4</u>, 245 (1962).

57. G. Parravano, J. Am. Chem. Soc., <u>73</u>, 628 (1951).

58. B.L. Funt and F.D. Williams, J. Polym. Sci., <u>2</u>, 865 (1964).

59. G. Silvestri, S. Gambino and G. Filardo, Adv. Polym. Sci., <u>38</u>, 27 (1981).

60. A.F. Bogenschutz, J.L. Jostan and W. Krusemark, Metalloberflache, <u>24</u>, 25 (1972).

61. M. Cvetkovskaja, T. Grcev, L. Arsov and G. Petrov, Kem. Ind., <u>34</u>, 235 (1985).

62. C.K. Chiang, M.A. Druy, S.C. Gau, A.J. Heeger, A.G. McDiarmid, Y.W. Park and H. Shirakawa, J. Am. Chem. Soc., <u>100</u>, 1013 (1978).

63. P.J. Nigrey, A.G. McDiarmid and A.G. Heeger, J. Chem. Soc., Chem. Commun., 594 (1979).

64. A.F. Diaz, K.K. Kanazawa and G.P. Gardini, Ibid., 635 (1979).

65. I.O. Salyer and A.M. Usmani, J. Appl. Polym. Sci., <u>22</u>, 3469 (1978).

66. F. Beck and H. Guder, Makromol. Chem. Symp., <u>8</u>, 258 (1987).

67. H. Guder and F. Beck, Farbe Lack, <u>93</u>, 539 (1987).

History of Epoxy Resins

Raymond B. Seymour
University of Southern Mississippi
Hattiesburg, MS 39406-10076

Most pioneer paint resins, such as oleoresinous paints, alkyds and cellulose nitrate are esters which are characterized by potential hydrolytic degradation. In contrast, there are no ester groups in phenols and amino resins or PVC and hence these resins have better resistance to attack by corrosives. Resins, such as epoxy resins with ether groups in the principal polymer chain, which are also resistance to corrosion, were produced by Blumer by the reaction of phenolic resins and epichlorohydrin in 1930 [1].

Schlack patented the diglycidyl ethers obtained by the condensation of epichlorohydrin ($ClCH_2CH(O)CH_2$) and bisphenol A ((HOC_6H_4)$_2$ $C(CH_3)_2$) in 1933 [2]. The original epoxide resins, (EP) were produced by DeTrey Freres in Switzerland who licensed this "know how" to Ciba AG (now Ciba-Geigy) in the early 1940's.

Casten patented the curing of these resins with cyclic anhydrides [3] and with less than 5 percent alkylies in 1943 [4]. These patents were assigned to Ciba AG which demonstrated the use of these EP's, called Araldite, at the Swiss Industries Fair in 1945.

Greenlee, Preiswerk and Gans of Jones Dabney division of Devoe Raynolds modified these epoxy resins with glycerol in 1944 [5] and patented reactions of epoxy resins with drying oils in 1948 [6] phenolic resins [7] and amines [8]. Shell, which was a producer of epichlorohydrin and Union Carbide who was a producer of bisphenol A joined with Dow and Reichhold to promote the use of these resins under the trade names of Bakelite ERL, EKRA, Epi-Res and EPON.

Chemists at Union Carbide also produced epoxy-type resins by the reaction of peracetic acid on unsaturated cycloaliphatic compounds in the 1960's. 470 million pounds of all types of EP were produced in the US in 1988. The principal end uses for EP are coatings (38%), reinforced plastics (18%), flooring (6%) and adhesives (5%).

Most commercial EP resins are oligomers that can be cured by the room temperature reactions of the oxirane group (-CH(O)CH-) with polyamines such as ethylenediamine ((CH_2)$_2$ (NH_2)$_2$) or the elevated temperature reactions of the pendant hydroxyl groups was cyclic anhydrides such as phthalic anhydride ($C_6H_4C_2O_3$). The structural formula for linear EP is shown below:

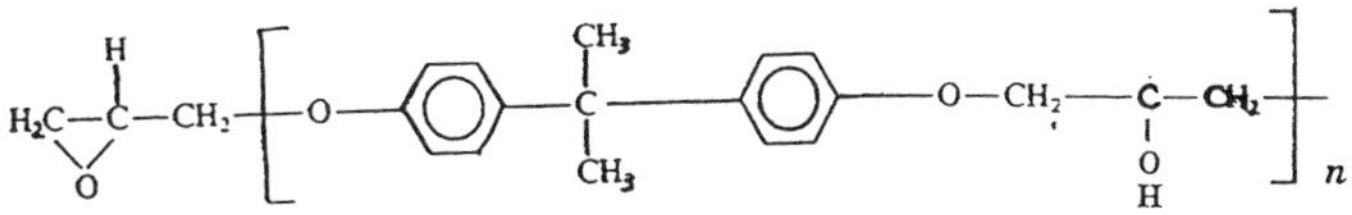

Thermoplastic EP, without epoxy end groups, which are called phenoxy resins are produced by the condensation of bisphenol A and epichlohydrin in a 1:1 ration [8]. Esters of fatty acids are also commercially available. Phillips and Frostick produced cycloaliphatic EP's by the reaction of peracetic acid on cycloolefins [9].

While unmodified bisphenol A-based EP's have a low order of acute toxicity, many amine curing agents will cause dermatitis if the prepolymer mixtures are handled without proper protection [10]. However, cured EP's do not present a carcinogenic or mutagenic hazard. Additional information on EP resins is available in several books 11-16].

Published 1990 by Elsevier Science Publishing Co., Inc.
Organic Coatings: Their Origin and Development
R.B. Seymour and H.F. Mark, Editors

REFERENCES

1. D. Blumer Ger Pat 576,177 (1930).

2. P. Schlack Ger Pat 676,117 (1933).

3. P. Castan US Pat 2,483 (1943).

4. P. Castan US Pat 2,444,333 (1943).

5. S.O. Greenlee US Pat 2,502,985 (1944).

6. S.O. Greenlee US Pat 2,456,408 (1948).

7. S.O. Greenlee US Pat 2,521,911 (1948).

8. Shell Brit Pat 980,509 (1965).

9. B. Phillips, F. Frostick US Pat 2,716,123 (1953).

10. J.E. Berger, K.I. Darmer, C.F. Phillips High Solids Coat 16, Dec 1980.

11. R.S. Bauer "Epoxy Resins Chemistry" ACS Symposium Series 114 Washington, D.C. 1979.

12. P.E. Bruin "Epoxy Resins Technology" John Wiley, New York 1968.

13. H. Lee, K. Neville, "Epoxy Resins and Their Applications" McGraw-Hill New York 1957.

14. C.A. May, Y. Tanaka "Epoxy Resins," Marcel Dekker, New York 1973.

15. W.G. Potter "Epoxide Resins" Iliffe, London, 1970.

16. I. Skeist "Epoxy Resins" Reinhold, New York, 1958.

A HISTORY OF METAL DECORATING CONTAINER COATINGS TECHNOLOGY WITH AN
EMPHASIS ON WATERBORNE TECHNOLOGY, P. V. Robinson, The Glidden Company,
16651 Sprague Road, Strongsville, OH 44136.

ABSTRACT

The idea of using water reducible coatings as liners for metal
containers is quite old - over 50 years. However, commercially
important coatings have only been available for about 12 years.
Containers can be regarded as being used for beverages (beer or soft
drinks) or everything else. There are major differences, globally, in
metal of choice, patterns of container uses and container contents. The
coating used has to reflect these differences. There are many chemical
techniques available for synthesizing water reducible versions of
container coatings. Most rely on the inclusion of acidic groups
followed by neutralization with a volatile amine. Polymers synthesized
via emulsion polymerization also find utility in water reducible
container coatings. This paper will review the various techniques for
obtaining water reducible container coatings and it will compare these
chemistries with each other against their coatings properties.

INTRODUCTION

Through history there has been a need to store and transport food
and any containers of which the local technology was capable of
manufacturing were used. The need to preserve the quality of stored
food was also recognized early, but it was only relatively recently that
progress was made in this area. In the very early 19th century,
Nicholas Appert was awarded a prize by the government of France for the
use of closed containers for the preservation of food, although it was
only 60 years later that Pasteur recognized the need to remove all
bacteria before the food was finally put into storage. The early closed
containers were made of glass and metal containers appeared somewhat
later.

As the Industrial Revolution progressed, various forms of sheet
steel were invented, and some of this was dipped in molten tin in order
to delay the effects of corrosion. This kind of steel plate was known
as tin plate, and gave rise to the so called "tin can". Such metal
containers have been with us for approximately 150 years, although
widespread use started around the turn of the 20th century.

Until fairly recently, the metal container was fabricated from
three pieces of metal. The body of the container was formed from a
sheet of steel in the form of a cylinder by use of solder and then one
end was attached by crimping. After filling, the other end was attached
and the product shipped. In the last 25 years, the solder had to be
removed for reasons of economics and toxicity, and both welded cans and
adhesively bonded cans were introduced. At about the same time during
the last 25 years, the use of aluminum in the form of two-piece cans
(where body and base were formed from one piece of aluminum) experienced
great acceptance in the marketplace for both economic and recycle or
pollution control reasons. Container manufacture is now quite
sophisticated with over 80 billion 12 ounce aluminum cans being
manufactured in the United States alone every year.

When a metal container is used to contain food or beverage, a
coating is usually to be found on it. The coating on the inside of the

Published 1990 by Elsevier Science Publishing Co., Inc.
Organic Coatings: Their Origin and Development
R.B. Seymour and H.F. Mark, Editors

can is used to separate the contents of the container from the metal, while the coating on the exterior of the can is used largely to decorate and to inform. The synthetic polymer industry had its beginnings in the very early days of the 20th century, and prior to the availability of synthetic polymers, metal containers used blends of vegetable oils and naturally-occurring resins to perform the functions of protection and decoration described earlier.

The 20th century has seen a steady and continuous development of new synthetic polymers, beginning with the discovery of phenol/formaldehyde condensates in the early 1900's. FIGURE I shows a highly condensed chronology of the development of the synthetic polymer industry (FIGURE 1). As synthetic polymers became available, the coatings and container industries were quick to utilize the new technology. Coatings were developed for both interior and exterior uses, usually carried in organic solvents at a concentration of about 20-40% in the solvent. As the container industry grew, so also did the amount of solvent being vented into the atmosphere by can coating plants.

THE EMERGENCE OF WATER REDUCIBLE COATINGS

Spray applied container coatings carried in organic solvent were applied at about 30% solids, and during the 1960's it was gradually being recognized that various forms of pollution were becoming unacceptable. An early legislated attempt to control atmospheric pollution was contained in the Rule 66 (issued in 1966) in California. By 1974, the environmental protection agency (EPA) had been established by an Act of Congress, and pollution control became a national issue. Coatings carried in organic solvent became a natural target of the EPA. It was readily recognized that if solvent was to be removed from an organic coating, it would have to be replaced by a less objectionable solvent or by water. This generated a very difficult problem. How to carry a coating in water, yet have it apply and perform like a solvent-borne coating and have it comply with FDA requirements for coatings in contact with food. The effort to solve this problem started in 1970 and waterborne products were introduced to the marketplace in 1976. Currently, spray coatings for metal containers are applied at about 20% solids with the remainder being comprised of volatile organic compounds (VOC) 15%, and water 65%.

Mention has been made of the need to satisfy the requirements of the Food and Drug Administration (FDA). The FDA publishes a Code of Federal Regulations (CFR) which provides approval for acceptable coatings compositions. Most of the coatings compositions within the relevant sections of CFR were grandfathered, in that they were placed onto the list at the time of the establishment of CFR 175.300, largely because they had been used previously without any apparent ill effect. New additions to CFR 175.300 are rare, and are costly and time consuming to obtain FDA approval. A strategy used by most coatings companies is to try to stay within the bounds of CFR 175.300 as they strive to develop new technologies.

A HISTORY OF WATER REDUCIBLE CONTAINER COATINGS

FIGURE 2 shows a brief chronology of the introduction of water reducible coatings for containers. This chronology was obtained by studying the patent literature and it can be seen that, in general, water reducible coatings have emerged during the last 20 years. A surprising exception is the disclosure by Crosse and Blackwell of both oleo and oleo resinous emulsions for use as interior container coatings in the middle 30's. Furthermore, this company disclosed the electrodeposition, both anodic and cathodic, of these emulsions to the interior of cans. This company anticipated technology which was not "invented" again until 20 years later.

An analysis of the patent literature was carried out by chemistry/technology and year of issuance. FIGURE 3 shows the results of this analysis, from which it can be seen that water reducible technology for interior can coatings falls into four broad areas of chemistry, as shown in FIGURE 4. The range is surprisingly narrow, being comprised of various manipulations of thermoset acrylic and epoxy chemistries. In the case of acrylics, polymers are prepared by emulsion polymerization or by solution polymerization in an appropriately selected organic solvent, followed by neutralization and dispersion in water. In the case of the various epoxy technologies, a preformed epoxy resin is modified in some way so it becomes readily dispersible in water or a mixture of water and water miscible solvents. There are many techniques for preparing epoxy and acrylic modified epoxy dispersions in water, as is evident from FIGURE 3. Only a few typical examples will be described in more detail.

CHEMISTRY OF POLYMER DISPERSIONS FOR USE IN INTERIOR CONTAINERS

FIGURE 5 shows the use of acrylamide and its n-butoxy methyl derivative in the synthesis of thermoset acrylic polymers for use in container coatings. If the polymer is synthesized by solution polymerization in a water miscible solvent, then between 5 and 10% copolymerized acid should be used so that neutralization and dispersion in water can be accomplished subsequent to polymerization. If emulsion polymerization is used, then high levels of acid are not necessary, although subsequent formulation with emulsion polymers into a finished coating can be quite difficult for this marketplace. There was a flurry of patent activity describing this kind of polymer for use in interior container coatings from about 1976 to 1980, although the polymers were not successful in the interior container marketplace.

FIGURE 6 shows the chemistry used to produce epoxy-g-acrylic copolymers in which the grafting site is an ester group linking the acrylic moiety to the ends of the epoxy moiety. Again, there are many techniques for synthesizing the so-called ester graft polymers as is evident from the large number of issued patents shown in FIGURE 3. This technology has enjoyed limited commercial success in the marketplace.

FIGURE 7 demonstrates two techniques by which a pure epoxy resin can be dispersed in water. In the first case, the terminal oxirane groups are converted into acidic end groups by reaction with appropriately functionalized acids. In this example, thioglycolic acid is used to open the oxirane ring, although polybasic acids (phosphoric acid) amino acids (para-amino benzoic acid, PABA) and phenolic carboxylic acids (the methyl ester of cellosilic acid is reacted with the oxirane at the phenolic hydroxyl subsequently followed by hydrolysis to remove the methyl group). The range of possible modifications is extensive and limited only by the need to comply with FDA regulations.

Another technique which has been used to provide epoxy dispersions is to make a block copolymer using the diglycidyl ether of polyethylene oxide so that the resin forms sterically stabilized dispersions in water rather than ionically stabilized dispersions. The hydrophobic component

of the polymer backbone should be found within the micell with
thehydrophilic components of the polymer backbone extending into the
water phase. Patents disclosing this kind of technology started to
emerge in the mid 70's, but they have not enjoyed significant commercial
activity in the marketplace.

By far the most successful water dispersed polymers used in the can
interior coatings marketplace have been epoxy-g-acrylic copolymers
formed by carbon-to-carbon grafting. In this case, appropriately
selected acrylic monomers are pumped into a solution in water miscible
solvents of a preformed epoxy resin in the presence of a hydrogen
abstracting addition polymerization initiator. FIGURE 8 shows how this
is accomplished. Hydrogen abstracting radicals derived from the
initiator react with aliphatic hydrogens spaced along the epoxy backbone
to generate a polymer preradical. This then functions as the initiating
site for subsequent acrylic copolymerization. On completion of the
copolymerization, the carboxyl groups derived from the acrylic monomers
are neutralized with a volatile amine and dispersed into water.

Typically, such polymers are comprised of about 80% epoxy backbone
and 20% acrylic. The acrylic component is comprised of ethyl acrylate,
styrene, and methacrylic acid with the acidic monomer dominating the
composition. The outcome of the synthesis, of course, is quite
different from the components going in. About 40% of the final product
is free ungrafted epoxy, about 10% is free ungrafted acrylic copolymer,
and 50% is epoxy-g-acrylic copolymer. The extent of carbon-to-carbon
grafting obtained in this process is of considerable interest to the
author's company. Using carbon 13 nmr analysis, we have concluded that
between 25 and 30% of the aliphatic carbons in an epoxy backbone contain
grafts (FIGURE 9). The chain length of the grafts must be quite small,
based on stoichiometric calculations and partially confirmed by GPC
analysis.

FUTURE

No history of water reducible container coatings would be complete
without some analysis of future requirements. Certainly the reduction
in volatile organic content from 70% down to 15% is a remarkable
accomplishment. Container plants have been certified to comply with EPA
requirements, using coatings that only contain 15% volatile organic
component. Such a plant, however, cannot easily increase its output
because the amount of organic emission is limited. Therefore, the only
way an increased capacity can be accommodated is to further reduce the
volatile organic content of coatings relative to the amount of actual
coating left inside the container. Ultimately, the absolute emissions
from such plants will also be required to be reduced, further
exacerbating the VOC problem. It is also anticipated that lower curing
temperatures will also be required or at least chemistries will be
needed which will allow for faster line speeds. Currently, the
universal technique for application of these coatings is via spray
application at a rate of about 300 cans per minute. Inevitably,
alternative application methods will be required which will deliver some
advantage in either lower VOC or lower cure temperatures.

CONCLUSION

The future for coatings to be applied to metal containers remains
bright. Plastic containers have been proposed (and test marketed), but
significant problems remain, not the least of which is the extreme
difficulty of recycle as compared to the aluminum two-piece can. It is

quite possible that the aluminum two-piece can could be replaced by ysteel two-piece cans, bringing with them different problems relating to corrosion resistance and flavor problems.

An attempt has been made to describe some of the significant developments in the development of water reducible coatings for use inside metal containers. While the chemistry is relatively simple, the technology is very advanced, with the highest level of consistency of product being essential. Human sensory perception of taste or flavor is very well developed. In certain well documented cases, contaminants existing at less that 2 parts per billion are easily detectable in food products. It is the author's contention that if other North American and European industries had addressed quality issues as thoroughly as was required by the container industry (and its suppliers) then we would have nothing to fear from the quality onslaught of the Japanese.

FIGURE 1

HISTORY OF SYNTHETIC POLYMERS

A.) 1900'S - PHENOL-FORMALDEHYDE
CONDENSATES - "BAEKELITE".

B.) 1920'S - ALKYDS - GENERAL ELECTRIC
CO.

C.) 1930'S -
    I. POLYESTERS - ICI
    II. POLYETHYLENE - ICI
    III. ACRYLIC POLYMERS - ROHM & HAAS
    IV. POLYAMIDES, CHLOROPRENE -
        DUPONT
    V. EPOXY RESINS - SHELL CHEMICAL/
        DEVOE REYNOLDS

D.) 1940'S -
    I. EMULSION POLYMERIZATION
    II. URETHANES
    III. AMINOPLASTS
    IV. SILICONES

E.) 1950'S -
    I. THERMOSET ACRYLICS
    II. IONIC POLYMERIZATION

F.) 1960'S

G.) 1970'S - WATER DISPERSIBLE POLYMERS

H.) 1980'S - POLYBLENDS

FIGURE 2

HISTORY OF WATER REDUCIBLE CONTAINER COATINGS

I. 1935 - OLEO EMULSIONS
   - CROSSE AND BLACKWELL

II. 1938 - OLEO - RESINOUS EMULSIONS
   - CROSSE AND BLACKWELL

III. 1972+ - EPOXY-ACRYLIC
    QUATERNIARY SALTS
    - DUPONT

IV. 1976+ - THERMOSET ACRYLIC
    DISPERSIONS
    - PPG

V. 1976+ - EPOXY EMULSIONS AND
    SOLUTIONS
    - MOBIL
    - PPG
    - CELANESE
    - BASF

VI. 1976+ - EPOXY-G-ACRYLICS (ESTER)
    AMMONIUM SALTS
    - DUPONT
    - MOBIL
    - CELANESE
    - PPG
    - GLIDDEN

VII. 1979+ - THERMOSET ACRYLIC LATICES
    - PPG
    - GLIDDEN

VIII. 1980+ - EPOXY-G-ACRYLIC (C-C)
    AMMONIUM SALTS
    - GLIDDEN

FIGURE 3

W.R. CONTAINER RESINS

	30s	70	71	72	73	74	75	76	77	78	79	80	81	82	83	84	85	86	87	88	89
OLEO RESINS	*																				
OLEO-EPOXY-ACRYLIC GRAFT				*		*							*								
ACRYLIC DISPERSIONS								*	*	*	**	*				*		*			
EPOXY EMULSIONS/DISPERSIONS								*	*	**	**		*	*		*					*
EPOXY-G-ACRYLIC (ESTER) AND BLENDS								*	**		*	*	** **		*	** **	**	**		*	*
EPOXY-G-ACRYLIC (C-C)											*	**		*	*	*	*				
ACRYLIC LATEX								*									**				
EPOXY-G-ACRYLIC/LATEX BLENDS													*	*							

FIGURE 4

KEY WATER REDUCIBLE
CONTAINER COATINGS PATENTS

I. 1976-1980 - THERMOSET ACRYLIC
    DISPERSIONS

II. 1976-PRESENT - EPOXY-G-ACRYLIC
    (ESTER) DISPERSIONS - 50-80% EPOXY

III. 1976-PRESENT - EPOXY DISPERSIONS

IV. 1980-1986 - EPOXY-G-ACRYLIC (C-C)
    DISPERSIONS - 50-80% EPOXY

FIGURE 5

THERMOSET ACRYLIC DISPERSIONS/EMULSIONS

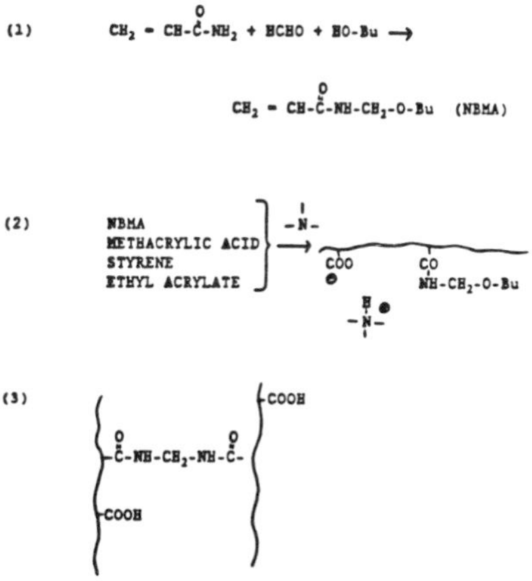

(1) $CH_2 = CH-\overset{O}{C}-NH_2 + HCHO + HO-Bu \longrightarrow$

 $CH_2 = CH-\overset{O}{C}-NH-CH_2-O-Bu$ (NBMA)

(2) NBMA
     METHACRYLIC ACID
     STYRENE
     ETHYL ACRYLATE

(3)

FIGURE 6

EPOXY-G-ACRYLIC (ESTERS)

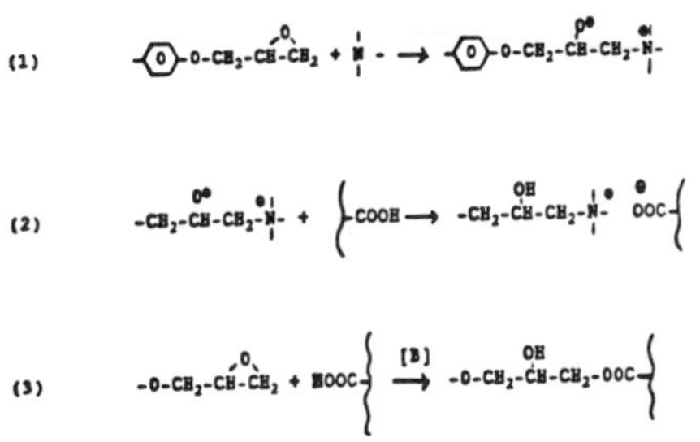

(1)

(2)

(3)

(4) ESTER GRAFTS CAN CONTAIN ESTER GROUPS AND
     QUATERNARY AMMONIUM SALTS

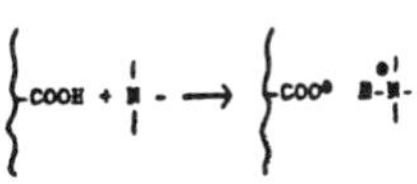

296

FIGURE 7

EPOXY EMULSIONS/DISPERSIONS

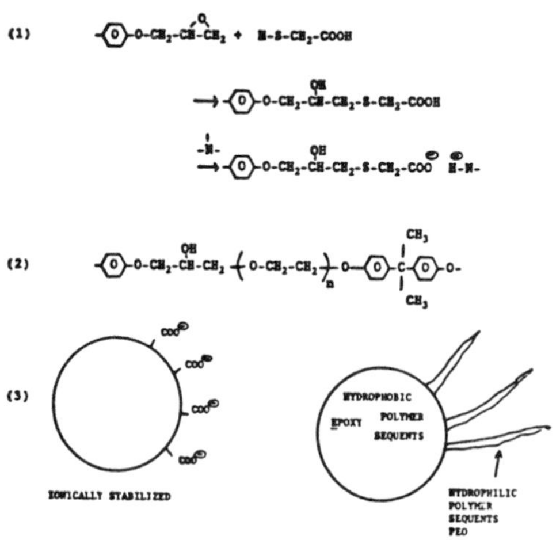

FIGURE 8

EPOXY-G-ACRYLIC (C-C)

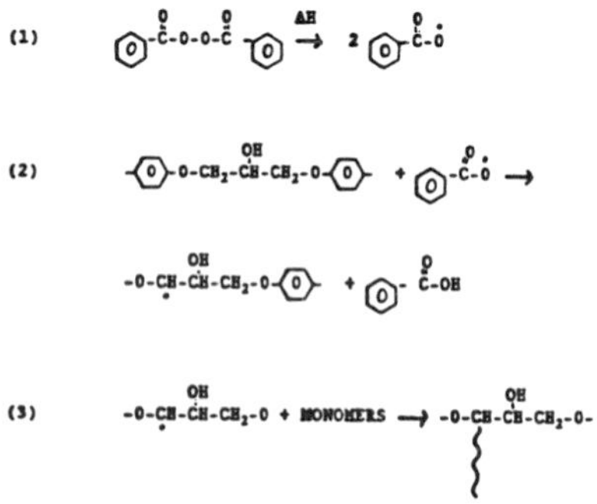

FIGURE 9

EXTENT OF GRAFTING

PEAK, ppm	AREA	EPOXY	GRAFT PEAK
79.1	22016		A
77.0	11523	branched	
73.3	9295		B
72.5	22552*	√	C
71.8	9717		D
71.5	7790	√	
70.3	7745		E

* Estimated 2/3 graft, 1/3 epoxy.
Total graft peak area = 63807, corresponding
to 27.7 % grafting (considering the branching
carbon peak area of 11523 to be 5 %).

HISTORY AND DEVELOPMENT OF ACRYLIC LATEX COATINGS

HARREN, RICHARD E.

Introduction

Prior to 1953 acrylic latex polymers based primarily on poly (ethyl acrylate) had been developed for use in leather and textile finishing. In 1953, Rhoplex® AC-33 acrylic latex copolymer of ethyl acrylate and methyl methacrylate, was introduced for use as a binder in house paints. This polymer had advantages in color retention and durability over styrene butadiene latices and in alkali resistance over the polyvinyl acetate latices which were then used in paints. However, it was deficient in flow and levelling when brushed compared to the larger particle size vinyl acetate latices.

At that time latex paints were inferior to solvent alkyd paints in a number of respects: flow, film build, adhesion, gloss, hardness, and in film formation under adverse conditions. During the period 1953 through 1988 research at Rohm and Haas was directed to eliminating these deficiencies. This paper reviews the following major developments which came from this research.

1. Mechanism of Latex Film Formation

2. Chemical Adhesion

3. Exterior Durability

4. Flow and Film Build by Particle Size Control

5. Flow and Chalk Adhesion by Particle Size Control

6. Associative Thickeners

7. Aqueous Gloss Enamel Latex Polymers

8. Corrosion Resistance

9. Opaque Latex Polymers

10. Rheology Control by Latex Particle Morphology - Multilobe Latex Particles

Figure I shows the sales volume developed by products which resulted from this technology during the period 1953 to 1988.

I. Film Formation

The first acrylic latex polymers used for leather and textile finishing were very soft, low Tg polymers which were applied

Published 1990 by Elsevier Science Publishing Co., Inc.
Organic Coatings: Their Origin and Development
R.B. Seymour and H.F. Mark, Editors

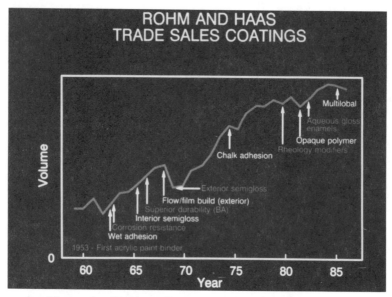

Figure 1 - Volume Growth of Acrylic Binders

under controlled ambient conditions. Film formation was not a problem. The latex polymers for house paints, however, were somewhat harder, Tg 10-15°C; they were applied under variable ambient conditions and the quality of film formation varied from visible cracking and flaking failure in extreme cases to more subtle types of failure whereby apparently well formed films differed greatly in mechanical film and resistance properties.

In 1956 Dr. George Brown proposed[1] proposed that the capillary forces exerted by interstitial water was the major driving force and polymer tensile modulus was the major resistive element in the latex film forming process. His description of the polymer, formulation, and environmental factors affecting film formation was immensely valuable to the coatings scientist.

The concept of minimum film formation (MFT) temperature was developed. This described the lowest temperature at which a latex polymer would coalesce to form a film. The harder the polymer, the higher the MFT. Solvents were used as filming aids and high boiling point and low water solubility were favored. Hoy[2] of Union Carbide Corporation has shown quantitative correlation between the amount of solvent distributed in the polymer and its efficiency as a coalescent.

The film formation process as proposed by Brown is time dependent; if drying occurs too rapidly, there may be insufficient

Figure 2 - Overall View of Test Fence

time for particle fusion to occur as water leaves the film. Figures 2, 3, and 4 vividly illustrate a practical indication of this effect. Half of a wooden fence at the Jersey shore was painted with an acrylic latex paint on a warm, sultry, dry, windy day, conditions which led to very rapid dry. The second half of the fence was painted later latex paint on a warm, sultry, dry, windy day, conditions which led to very rapid dry. The second half of the fence was painted later that day when conditions had changed; the sky had become overcast and humidity increased almost to a drizzle. Initially, the painted fence looked uniformly good. However, two years later the paint applied earlier in the day failed by cracking and flaking. Eight years later the paint on that portion of the fence painted later in the day was still in good condition. Subsequent investigation suggested that increased coalescent, beyond that required to form a film under ideal conditions, should have prevented this type of film failure.

II. Chemical Adhesion

All of the early latex house paint polymers lacked specific wet adhesion to oleoresinous and wood substrates. Adhesion and crack resistance on wood could be improved by softening the binder, but dirt pickup resistance decreased proportionately. Polymer Tg had to be critically controlled to obtain a proper balance of dirt and crack resistance. Modification of the latex with some drying oil or alkyd

300

Figure 3 - Close-up of Fence Painted Under Conditions of Fast Dry
(After 2 Years of Exposure)

Figure 4 - Close-up of Fence Painted Under Conditions of Slow Dry
(After 2 Years of Exposure)

resin was used to obtain wet adhesion to these substrates as well as chalking substrates.

In the late 1950s a major breakthrough in latex adhesion technology came about with the development of ureido functional acrylic monomers which when copolymerized in small amounts with acrylic polymers substantially improved specific adhesion to wood and to oleoresinous substrates. This technology is described in U.S. patent 2,881,171 (1959).[3]

Rhoplex® AC-34 acrylic polymer, introduced in 1963, was the first commercial material with specific adhesion. Rhoplex® AC-34 had greatly improved blister and crack resistance on exterior exposure compared to its non-adhering counterpart, Rhoplex® AC-33 acrylic polymer. It was not surprising that improved wet adhesion led to improved blister resistance, but the large improvement in crack resistance, over dimensionally unstable wood substrates, was not anticipated. This result made it possible to design harder, more block resistant polymers for factory application to wood substrates.

Rhoplex® AC-34 is, like Rhoplex® AC-33, an ethyl acrylate copolymer. Replacing ethyl acrylate with butyl acrylate on a constant Tg basis (Rhoplex® AC-35) led to much improved appearance durability (tint retention and dirt resistance) on exposure.

III. Exterior Durability

General guidelines for the design of acrylic coatings with excellent weatherability from both a film integrity and film appearance point of view were developed during the sixties and seventies. A discussion of these findings was presented by Scott in a paper, *Weatherability of Acrylic Coatings*,[4] at the Pacific Paint Convention in Canberra, Australia (1977). Key conclusions from this paper are summarized here:

First in the selection process for maximum durability is the choice of acrylic comonomers to suit the end use. Methyl methacrylate (MMA) and butyl methacrylate are preferred for rigid substrates of high dimensional stability and for lightly pigmented formulations. Copolymers of MMA with butyl acrylate or ethylhexyl acrylate are more suited for substrates which undergo substantial expansion-contraction cycling and for highly pigmented formulations.

Next is the need to insure good adhesion of the coating to the substrate. This can be accomplished by designating special primers or by building specific adhesion directly into the polymer.

The minimum pigment volume concentration and especially the TiO_2 level that is consistent with hiding, aesthetics, and economics should be utilized. Furthermore, a two-coat system with highly pigmented base coat and clear or lightly pigmented topcoat is ideal, where feasible. Frequently, this type of layering can be induced in a single coat by baking polymers well above their glass transition temperatures.

Polymer glass transition temperature (Tg) must be carefully adjusted to optimize crack resistance and dirt pickup resistance for the particular substrates and pigmentation under consideration. In baking finishes, thermosetting reactions can be utilized to upgrade dirt pickup resistance of the soft, crack resistant polymers.

Finally, maximum weatherability demands the avoidance of disruptive elements. These include: (a) the blending of copolymers which produce grossly heterogeneous films, (b) the deposition of latex coatings that have defects on a microscale due to improper film formation, and (c) the polishing or mechanical abrasion of coatings to remove the durability-enhancing clear layer.

IV. Flow and Levelling by Particle Size Control

Rhoplex® AC-35, like Rhoplex® AC-34, has a small particle size (~0.1μ) and, as a result, paints based on it have good pigment binding ability and a modest degree of adhesion to old chalky paint substrates. It is, however, still deficient in flow and film build in paints. At that time, Brown and Garrett[5] speculated that the poor flow and film build were from the interaction of the small, large surface area particles with cellulosic or acrylic thickeners which result in paints with high yield stress (Figure 5).

Extensive efforts to solve this problem by formulation failed and a major research effort was directed to develop stable, large particle size acrylic latices as well as to search for new types of thickener which would not interact adversely with latex and pigment particles.

In 1964 a 0.6μ particle size, ethyl acrylate/methyl methacrylate copolymer was introduced for interior, flat house paints. This polymer, designated Rhoplex® AC-22, when properly formulated with high levels of propylene glycol to extend wet edge properties

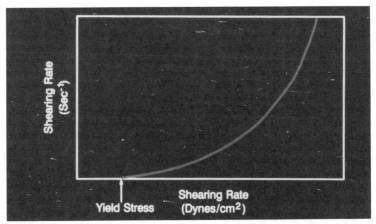

Figure 5 - Typical Flow Curve of a "Plastic" Fluid

and further enhance flow and levelling, also made a superb interior semigloss paint. Rhoplex® AC-22 was the first of a family of related polymers used as binders in interior semigloss paints. Its butyl acrylate analog, Rhoplex® AC-388, followed soon after for use in exterior flat and semigloss paints.

V.   Chalk Adhesion by Particle Size Control

The new, stabilized large particle size acrylic latex polymers just described contained ureido acrylic functionality and had good adhesion to oil or alkyd paints and to new wood. They were deficient, however, in penetration and subsequent adhesion to powdery or chalky substrates compared to their finer particle size analogs.

A partial solution to this problem came about in the early 1970s with the development of bimodal acrylic latex technology. Acrylic latex polymers were developed which contained two distinct particle size modes, one large, the other very small. These polymers had good flow and levelling properties and much improved adhesion to chalky surfaces. However, paints formulated with HEC and related thickeners were deficient in film build compared to their colloid stabilized large particle size counterparts.

Rhoplex® AC-64, a bimodal polymer with flow and chalk adhesion, was introduced in 1974. Despite its deficiency in film build in paint formulations it became widely accepted for use in high quality exterior paints.

VI. Rheology Modifiers

Much of the research throughout the sixties and seventies was directed to modifying acrylic latex technology to accommodate the paint formulation technology which was then available. Latex polymers were modified to adjust to the limitations of the thickeners which were available to the paint formulator. As noted, these thickeners tended to flocculate both the latex and the latex TiO_2, adversely affecting both rheology and gloss potential (Figures 6, 7, 8, 9).

In the early 1980s, a major breakthrough in rheology control was made by Emmons et al.[5] with the development of new types of rheology modifiers called associative thickeners. Associative thickeners, as the name implies, are surfactant like polymers which are capable of hydrophobic interactions similar to surfactants.

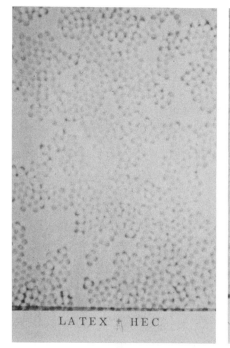

LATEX + HEC

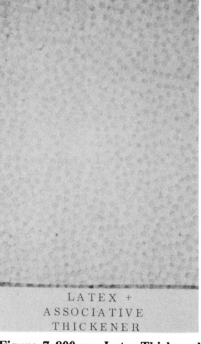

LATEX +
ASSOCIATIVE
THICKENER

Figure 6-800 nm Latex Thickened with Hydroxyethyl Cellulose (via Optical Microscope)

Figure 7-800 nm Latex Thickened with Associative Thickener (via Optical Microscope)

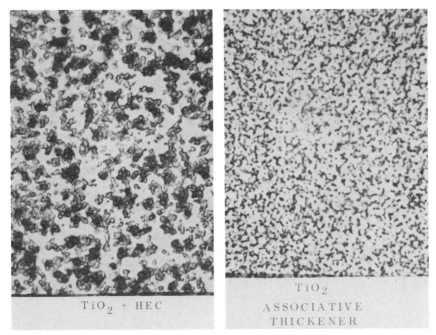

Figure 8 - TiO$_2$ Dispersion
Thickened with Hydroxyethyl
Cellulose (via Optical Microscope)

Figure 9 - TiO$_2$ Dispersion
Thickened with Associative
Thickener

In a simplified representation (Figure 10), they may be considered
to act like double-ended surfactant molecules, with two or more
hydrophobic groups attached to a hyhdrophilic polymer backbone.
Just as surfactants can associate to form micelles, or adsorb on

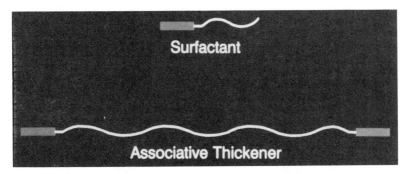

Figure 10 - Schematic Representation of Surfactant vs.
Associative Thickener

306

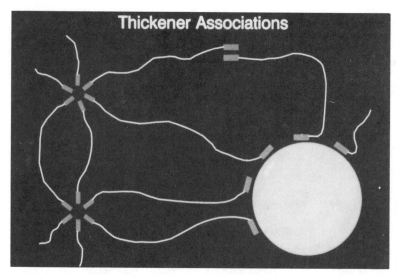

**Figure 11 - Schematic Representation of Associative Thickener
Associating with Itself and with Latex Particle**

particle surfaces (Figure 11), the associative thickener molecules
can link together and adsorb onto latex surfaces to form an
extensive network that can provide greater thickening than would
otherwise be possible from flexible, relatively low molecular weight
polymer chains.

This type of thickening reduces microstructure, thereby enhancing
gloss, rheology and pigment binding. The low molecular weight of
the associative thickeners avoids roller splatter problems that
Glass[7] concluded were principally due to the use of high molecular
weight thickeners. The new thickeners can be designed to have
associative character strong enough to enhance thickening
efficiency, but weak enough to provide the low yield stresses
required for excellent levelling. With associative thickeners we had
in hand the technology to produce latex high gloss and semigloss
paints with more alkyd-like **application** and **appearance** properties.
However, the paint formulation practices established over the past
many years with non-associative thickeners no longer applied.
New protocols had to be formulated to properly exploit this new
technology.

Schaller[8] reviewed this new technology at the 27th Annual OCCAA
convention in July 1985.

VII.   Aqueous Gloss Enamels

Industrial Coatings

In the late seventies, a major advance in acrylic latex technology was made with the development of acrylic latex polymers which could be formulated to give hard, glossy, enamel like coatings. Rhoplex® WL-81 and Rhoplex® WL-91, from Rohm and Haas, and Neocryl® A-601, from Polyvinyl Chemical Corp., were the first of a subsequent family of polymers based on this new technology.

These products appeared before associative thickeners were available, but they were designed for use in industrial coatings which were not applied by brush or hand roller and which generally required little or no thickener. These new aqueous gloss enamel binders are now used in a wide variety of general industrial coatings applications.

When these first generation products were used in state-of-the-art trade sales formulations of the day, flow, film build, and gloss were not good. However, superb gloss, flow, and build could be obtained when they were formulated with the new, non-flocculating associative thickeners. Trade Sales analogs of Rhoplex® WL-81 and WL-91 were developed and introduced in the early eighties. Primal® HG-44 was commercialized first in Europe where high gloss was favored in house paints. Rhoplex® HG-74 soon followed and has been accepted worldwide as a binder for interior and exterior aqueous gloss enamel paints.

This new technology is described in depth in a paper by Scott, Mercurio, and Schaller[9]. Its key technical features are:

1. High Tg for hardness and block resistance.

2. Polymer design to permit rapid release of the high level of solvent needed for film formation.

3. Control of particle size for maximum gloss.

4. Polymer design for the maximum adhesion and water and chemical resistance which is needed in high gloss coatings.

308

VIII. Corrosion Resistance

During the sixties and seventies, Trade Sales acrylic paint vehicles were fine tuned to provide a degree of improved corrosion resistance. When these products, Rhoplex® MV-1, Rhoplex® MV-23, etc., were formulated with corrosion inhibiting pigments and applied in multiple coats, they proved useful as maintenance paint primers for steel structures.

A large further advance in corrosion resistance was obtained in the early eighties when associative thickeners became available. With associative thickeners, non-flocculated paints could be formulated (Figures 12, 13), which gave tighter, less porous films with greatly improved corrosion resistance.

TRANSMISSION ELECTRON MICROGRAPHS

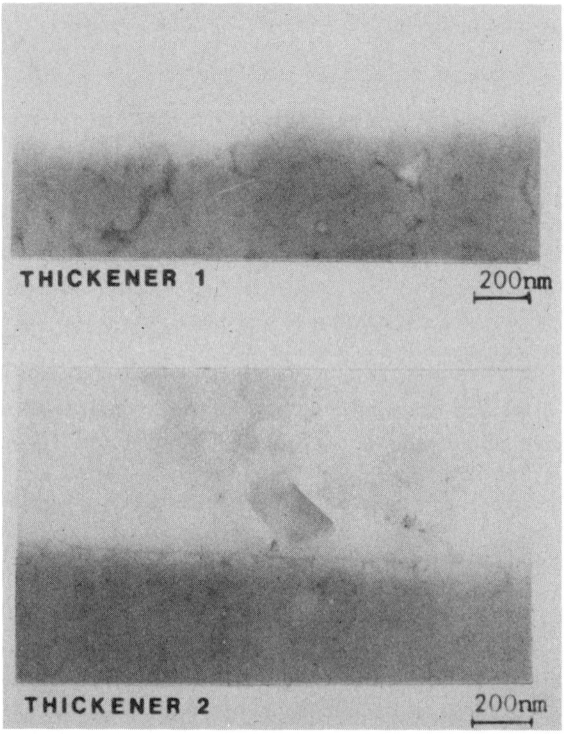

Figure 12 - Freeze Fractured Films (side view) Showing Channels Caused by Thickener 1 (HEC) and Absence of Channels with Thickener 2 (Associative Thickener based on Poly Urethane Chemistry)

SALT SPRAY EXPOSURE

UNPIGMENTED WATERBORNE ACRYLIC FINISH

THICKENER 1 THICKENER 2

1 COAT CLEAN COLD ROLLED STEEL

Figure 13 - Effect of Thickener Type on Corrosion Resistance of
Unpigmented Coatings Over Clean Cold Rolled Steel.
Thickener 1 - Hydroxyethyl Cellulose; Thickener 2 -
Associative Thickener based on Poly Urethane Chemistry

An impressive commercial application of this improved
formulation technology is shown in the next figures. Six years ago,
the badly corroded and pitted cast-iron exterior surface (Figure 14)
of the venerable lighthouse on Hunting Island, South Carolina, was
refinished with a waterborne maintenance coating system. A
three-coat acrylic latex system was applied by airless spray direct to
the cast-iron surface where the old paint had been removed by
sandblasting. After nearly a half dozen years of exposure to the
demanding climate conditions of a coastal environment the acrylic
latex coating system looks almost like new (Figure 15).

As noted, these early acrylic maintenance vehicles were low Tg,
modified trade sales polymers which were deficient compared to
alkyd paints in hardness and water resistance. The development of
Aqueous Gloss Enamel technology removed this limitation. This
was accomplished by fine tuning the first generation trade sales
gloss, enamel binders to maximize corrosion resistance and then
reformulating with associative thickeners.

Maincote® HG-54, aqueous gloss maintenance binder, was

Figure 14 - Steel Light House in Process of Cleaning Prior to Application of Latex Maintenance Primer and Topcoat System

commercialized in the mid eighties and is used worldwide in maintenance primers and topcoats.

IX. <u>Opaque Polymers</u>

Opacity is enhanced by forming air voids or porosity in paint films. The easiest way to accomplish this is to formulate above the critical pigment volume concentration. Opacity increases but paint film quality decreases. The Dulux Paint Co., Australia, avoided this problem by encapsulating air and TiO_2 in a polyester polymer bead (Spindrift). PPG used incompatible solvents to generate air cell voids in the paint film (Pitment).

In 1979 a third approach to polymer encapsulated air voids, Ropaque® OP-42, was introduced (Figure 16). Ropaque® OP-62, a higher efficiency opacifier, was introduced a few years later. The inventors, Kowalski and Vogel,[10] had devised an ingenious emulsion polymerization route to encpasulating water in a hard, non film forming polymer bead. Upon drying in a paint film matrix, the water irreversibly evaporates, leaving behind an encapsulated air void which contributes to opacity without increasing porosity in the paint film.

Figure 15 - Steel Light House 3 Years After Application of Latex Primer and Topcoat System

Opaque polymer is used successfully worldwide as a partial replacement for TiO_2 in paint films. Opaque polymer is also finding commercial acceptance as a component in paper technology to provide opacity, gloss, smoothness, and weight reduction.

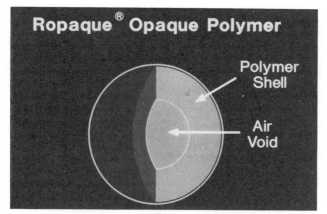

Figure 16 - Schematic of Opaque Polymer Particle

X.   Rheology Control by Particle Size Morphology

One of the most striking deficiencies of early latex paints compared to oleoresinous paints was their lack of film build when brushed. As noted, increasing particle size from 0.1 to ~0.5µ increased the high shear viscosity of formulated paints and improved brush drag as did the replacement of the cellulosic and early acrylic thickeners with associative thickeners.

In the early 1980s a third approach to controlling film build in latex polymers appeared. Acrylic latex polymers were developed which had a multilobal morphology (Figure 17). These lobes were not formed by flocculation or aggregation of latex particles but were formed directly in the latex synthesis process.

A typical multilobal particle has an effective hydrodynamic volume equivalent to a 1µ particle and contains lobes of approximately 0.35µ. The film build achievable in paints reflects that of a 1µ particle while the pigment binding ability is that of a 0.35µ particle. Ultimate film build is controlled by the ratio of multilobal to non-multilobal particles in the latex. Substantially less thickener, associative or non-associative, is required in multilobal latex paints than in non-multilobal ones to achieve rheology control.

Rhoplex® ML-100, multilobal acrylic polymer, the first of what will be a family of new acrylic polymers based on this new technology, was eventually commercialized in 1987.

XI.   Conclusion

A very broad overview of the major developments in acrylic latex technology for trade sales paints has been presented. This overview

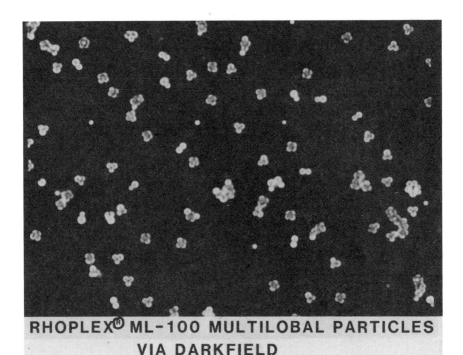

**RHOPLEX® ML-100 MULTILOBAL PARTICLES
VIA DARKFIELD**

Figure 17 - Multilobal Latex Particles via Optical Microscope (Dark Field)

focuses on the contributions of research to the development and maintenance of a strong commercial presence in the market.

Not included here were important developments in dispersant and mildewcide technology nor in those developments in acrylic latex technology for industrial coatings applications. It was also beyond the scope of this review to touch on the developments which took place in the industry with polymers other than acrylics.

Finally, it is impossible in such an overview to acknowledge all those who contributed to the technology covered. Nevertheless, in addition to those mentioned, special mention should be given to B.B. Kine, A.B. Brown, and J.E. Masterson for their contributions to latex synthesis and scale-up, to S. Gusman for his pioneering work on adhesion, to T. Haag for his innovative contributions to paint formulation which first opened up the semigloss market to latex paints, and to A. Kowalski for his major contributions to colloid stabilized latices, to opaque polymers, and to the development of multilobal polymers.

References

1. G.L. Brown, *J. Poly. Sci.*, 22, 423-34 (1956).

2. R.L. Hoy, *J. Paint Technology*, 45, p. 51 (1973).

3. Hankins, Melamed, U.S. Patent 2,881,171 (1959).

4. R.E. Harren, A. Mercurio, J.D. Scott, *Australian OCCA Proceedings and News*, Pacific Paint Conv., Canberra, pp. 17-23 (1977).

5. Emmons, Stevens, U.S. Patent 4,079,028 (1978).

6. G.L. Brown, B.S. Garrett, *J. Appl. Poly. Sci.*, 283-295 (1959).

7. J.E. Glass, *J. Coatings Technology*, 50, No. 641, p. 52 (June 1978).

8. E.J. Schaller, *Surface Coatings Australia*, pp. 2-13 (Oct. 1985).

9. J.D. Scott, A. Mercurio, E.J. Schaller, paper presented at *PRA Fourth Int. Cong., London* (Nov. 13, 1980).

10. Kowalski, Vogel, U.S. Patent 4,427,836 (1984).

A HISTORY OF POWDER COATINGS

Peter R. Gribble,
Associate Director Industrial Coatings
The Glidden Company
16651 Sprague Road
Strongsville, Ohio 44136

ABSTRACT

    Powder Coating represents the fastest growing area in industrial
coatings technology. They provide ecological, performance and economic
advantages to competing liquid technologies. First invented and
commercialized in the fifties, the technology did not progress well until
the development of the electrostatic spray gun and the availability of
improved manufacturing equipment. Ecological pressure on the coating
industry and the demand for high quality coatings has accelerated the
growth of the technology to its present rate of 15-25% annually throughout
the world. The paper will trace the significant factors that influenced
the growth of this technology with emphasis on thin film thermosetting
powder and their application in industry segments.

INTRODUCTION

    Over the past thirty years, powder coatings have become the fastest
growing industrial finishing technology throughout the world. Just in
North America, there are more than 2,000 commercial finishing lines using
powder coating on a variety of articles ranging from aluminum extrusions
and fire extinguishers to major household appliances and automobiles.

    Powder coatings are dry, near 100% solids, baking industrial
coatings. They are finely pulverized plastic compositions which, under
the action of heat, melt and flow to produce a smooth surface coating.
Although there are both "thermosetting" and "thermoplastic" powder
coatings, it is typically the thermosetting systems that are used in
industry to replace liquid paints. This is because of their ability to
produce thin, paint-like film thicknesses in the 1.0 - 2.5 mil range,
whereas thermoplastic powder coatings are typically used for very thick
applications such as the protection of heavy duty pipe or for electrical
insulation. Today, powder coatings are almost universally applied by
electrostatic spray, with some smaller, more specialized operations using
the fluidized bed technique.

    As a mature "compliance coating", powder coatings offer industry a
number of significant advantages. The complete absence of solvent both in
their manufacture and composition translates into the emission of
virtually zero to very low levels of organic compounds on baking. This
makes them fully compliant to today's air pollution legislation.
Similarly, the absence of solvent provides energy savings when compared to
liquid baking industrial finishes. This is because there is no
requirement to exhaust solvent laden air from the application area and in
the oven during the bake cycle. Because powder coatings are solid in
nature, oversprayed powder can be easily reclaimed for reuse. In fact,
nearly all application systems operating today use a powder reclaimation
system. This results in utilization efficiencies of over 98% and the
generation of very little waste. Disposal of finishing system waste
materials is an extremely important issue because of the associated high
cost and increasing legislation.

Published 1990 by Elsevier Science Publishing Co., Inc.
Organic Coatings: Their Origin and Development
R.B. Seymour and H.F. Mark, Editors

315

These aforementioned advantages will contribute to an attractive operating cost package for powder coatings when compared to competing liquid compliance technologies. However, one other remaining advantage is that powder coatings generally produce coating films of high quality, offering an excellent balance of both resistance, i.e. corrosion and weatherability, and mechanical properties such as impact hardness and abrasion resistance. This is again an important aspect to an industry faced with an increasingly quality conscious consumer. Thus, it is perhaps not surprising that this technology is growing at an estimated commercial growth rate of 15-25% worldwide.

EARLY HISTORY

In looking into the history of this technology, the first significant event can be attributed to Erwin Gemmer of Farbwerke Hoechst who was granted a patent in 1953 on the fluidized bed method of applying coating using a powdered polymer. The polymer applied was, in fact, nylon. The fluid bed was not new in 1953, as it had been used in several industrial processes dating back to 1919, but this was the first time it was used for a coatings application of a fairly inert polymer that was not possible to apply by conventional lacquer techniques.

FLUID BED PROCESS

The fluid bed process simply consists of a tank containing a porous plate in the bottom through which air is permeated at a fairly low pressure. Powder above the plate becomes aerated and takes on a liquid-like quality which offers no resistance to objects being immersed in it. The part to be coated is heated to a temperature above the melting point of the powder and then introduced into the bed. This causes the powder particles to quickly melt on its surface and form a continuous plastic film. Upon removal from the bed, the fusion of the particles is completed by either the residual heat in the part or by further heating.

* This process was introduced into the U.S.A. by the Polymer Corporation, under a licensee agreement from Farbwerke Hoechst in 1955. The process was first displayed at an Exposition of Chemical Industries held in Philadelphia in December of 1955.

However, from the mid-fifties, the growth of this process for the application of powder proceeded fairly slowly, being used mainly for the coating of heavy duty parts with a thick layer of plastic coating for protective purposes more than for paint-like decorative purposes. The reasons for this slow growth were numerous:

1. The number of available polymer types available was limited to mainly the thermoplastics and a singular thermosetting epoxy system introduced by Shell Oil Company in the late fifties.
2. Powders were manufactured by the dry blending of the components, ie. resin and pigment, etc. by the use of either a ball mill or a high speed mixer that had the capability of reducing resin particle size. The net result was a fairly uncontrolled broad particle size distribution of non-homogenous particles that produced medium gloss high film thickness coatings. Nevertheless, these coatings did meet a functional requirement.
3. Thin films were difficult to produce from the thermoplastic resin system because of the difficulties in producing fine particle sizes.
4. Early attempts to make homogenous particles by hot melt mixing of the components on Z blade blenders did produce superior

appearances, but in the case of the thermosetting epoxy
systems, great difficulties were experienced in producing
stable powders due to variable cooling times of the mix prior
to grinding.

KEY EVENTS

However, by the 1960's, a couple of significant advancements
occurred that dramatically affected the growth of powder coating.
Specifically allowing the thermosetting materials to be applied at thinner
film thickness approaching those of liquid paint and to produce decorative
paint-like appearances. These advancements were:

    a) the invention of the "electrostatic spray gun", and

    b) the use of a "melt mix extruder" to produce a homogenous
       mix of thermosetting material with a controlled heat
       history.

ELECTROSTATIC SPRAY

The electrostatic spray gun for powder coating as we know it today
was invented by a team of physicists in 1960 consisting of R. P. Fraser
and Marcel Point with secondary input from N. J. Felici, A. W. Bright and
R. A. Coffee. The basic idea for charged powder spraying seems to have
come from Pierre Hampe who developed the first electrostatic crop duster
during the German occupation of France. He published his work in 1946 at
the French Academy and was supported by N. J. Felici, a notable French
physicist. Felici later formed the SAMES (Societe Anonyme de Machines
Electrostatiques) and commercialized the first electrostatic powder spray
gun.

In the operation of an electrostatic powder spray system, powder is
supplied to the spray gun from a feeder unit where powder is stored for
use. Powder is siphoned or pumped from the feeder unit through the powder
feed hose to the gun. The gun directs the powder toward the part to be
sprayed in the form of a diffused cloud. The purpose of the spray gun is
two-fold. First, it charges the powder particles as they pass through the
corona generated by the gun's high voltage electrode. Secondly, it
directs the the pneumtically propelled particles close enough to the part
to be coated so that their image charges can hold them electrostatically
to the part. The part is then subsequently baked to produce a thin film.
This process is very similar to the electrostatic application of liquid
paints, and in 1960 opened up new horizons for powder coating growth.

MELT MIX EXTRUSION

The potential now for electrostatically applied thin films was
quickly coupled with the need to produce homogenous powder particles. Dry
blend powders have poor electrostatic spray properties due to the non-
homogenity of the powder particles produced. This was solved by the use of
melt mix extruders similar to those used in the plastics industry for
compounding the components. The extruders were, however, of specialist
design to provide quick melt and good pigment dispersion with good self
wiping characteristics. The industry quickly found the single screw co-
kneader design to be very suitable and to a lesser extent some twin screw
designs. These events, which transpired in the sixties, lead to a very
aggressive pursuit of thermosetting powder technology in Europe.

It was not until the early 1970's that concern for the environment
caused the U.S. coatings industry to focus on powder coatings technology
as a potential compliance coating. At that time, powder started to

receive a lot of publicity. At the 1st North American Conference on Powder Coatings held by the Canadian Paint and Finishing Magazine in Toronto during February of 1971, the theme was "The Powder Explosion." There were wild predictions at the conference that powder would amount to 20-30% of the industrial finishing market within 3-5 years. That did not occur, but within the following five years, there was certainly an explosion of activity that has in hindsight been responsible for the status to today's technology.

As can already be seen, the advancement of powder coatings technology has required developments in the three quite distinct disciplines of:

a) coatings chemistry,
b) manufacturing equipment, and
c) application equipment.

Thus, in looking at the history of powder coatings, each discipline must be considered for its role in the total advancement of the technology as advancement in one area would not, in all probability, have occurred without the interaction of the others. Thus, for the purposes of this paper, the historical focus will be:

a) examined from the perspective of these disciplines, and
b) specifically focused on the thermosetting materials as they are utilized as thin film decorative coatings competing with liquid coatings technologies and dominate this technology in terms of useage.

Looking first at the decade from 1960-1970:

1960-1970

Coatings Chemistry

-   Thermosetting epoxy systems developed in Europe. (Shell)
-   Accelerated and substituted dicyanimide and Anhydride curing agents introduced for epoxy systems.
-   Hydroxyl-polyester-melamine resin system introduced. (Scado)

Manufacturing Equipment

-   Melt mix extrusion process of premixing, extrusion, and cooling, followed by grinding and sieving established as most prominent method of manufacture.

Application

-   Fluid bed application was the basic method of application.
-   The electrostatic spray gun was introduced.
-   Oversprayed powder was collected by cyclone and baghouse arrangements
-   Manual mixing of reclaimed powder with virgin
-   Film thicknesses were typically in excess of 2 mils

By the end of the decade, powder coatings were established as a viable and commercial decorative compliance technology in Europe. However in the U.S. the reception to this technology was way slow with sales being virtually non-existant. Typical end uses were small metal components (specialities). It was, however, in the 70's that the technology rapidly expanded in all three disciplines and was actively commercialized in the U.S.A.

<u>1970-1980</u>

Coatings Chemistry

- Carboxyl polyester cured with TGIC. Introduced in Europe. (Scado)
- Thermosetting Acrylic. Introduced in Europe. (Bayer)
- Hydroxyl polyester cured with blocked isocyanate. Introduced in Europe and U.S.A. and known as Polyester Urethanes. (Bayer, Cargill, Huls)
- Carboxyl polyester/epoxy system known as "Hybrid" introduced into Europe giving properties of epoxy anhydride systems without the disadvantages. (Note: Most prominent resin system worldwide today)
- Hydroxyl acrylic cured with blocked isocyanate commercialized in U.S.A. known as Acrylic Urethane (Celanese).
- Formulating range of epoxy powders extended by the introduction of specialty solid epoxy resin to provide high flow and superior resistance properties.
- Chemical matting of epoxies introduced to produce low gloss coatings.
- Flow control agents improved and master batches of additives become available for improved processing.

Manufacturing Equipment

- Melt mix extrusion process greatly refined:

  -Twin screw extrusion technology utilized.
  -Grinding equipment improved to give more control of particle size
  distribution by use of internal classification.

- Research effort by U.S. paint companies directed toward alternate manufacturing methods based on liquid to powder techniques, i.e. conversion of liquid paint to powder, by spray drying, thin film evaporation and precipitations.
- Air classification introduced in Europe to reduce fine particle content of powder and thereby improve application properties.

Application

- Electrostatic fluid bed commercialized in the U.S. for special application. (screen wire)
- Improved powder feeding systems for electrostatic spray equipment providing more consistent powder flow
- More control of electrostatic spray pattern through improved aerodynamics.
- New concepts in electrostatic charging were introduced, i.e. tribocharging. (R. A. Coffee patent 1973).
- Improved high voltage generators produce higher KV guns and new designs allow generator to be incorporated in gun, eliminating heavy high voltage cable. (Gema)
- Much improved powder reclaimation systems introduced with automated mixing of reclaimed powder with virgin becoming the norm.
- Twin-air collection system introduced by Gema.

At the close of the decade, U.S. powder coating consumption had grown steadily from 6M pounds in 1975 to 20M pounds in 1980. The General Metals, Major Appliance and Automotive industrials were all using powder on a variety of articles. In the General Metals area for example, powder found high utilization on lighting fixtures, where it replaced both porcelain enamel and liquid paint and on garden products such as lawn furniture and tractors. A slightly unusual use was the coating of insect screen wire, using the electrostatic fluid bed method of application.

In the Major Appliance industry, powder replaced porcelain enamel on the interior of refrigerators and dryer drums, and metal plating on refrigerator racks. It also replaced paint on the cabinets of ranges, microwave ovens, and refrigerators.

In the Automotive industry, powder found numerous uses on under body parts and was used as a primer-surfacer in Japan. There was even some experimental use on automotive bodies by both Ford and G.M. in the mid-70's.

In looking at the next decade, much refinement of the technology occurred.

1980-1990

Coatings Chemistry

-   Expansion of range of both carboxyl and hydroxyl polyester resin by suppliers to extend formulating range.
-   Lower temperature curing potential offered in epoxy and polyester COOH systems.
-   Many proprietary hybridized resin systems utilized to meet specific performance requirements.
-   Much refinement and understanding of use of additives to achieve performance requirements.(i.e. formulation advances)
-   Basic research on the characterization of cure rheology to expand fundamental knowledge base.
-   Research effort started on new resins and crosslinking methods for extended exterior durability.

Manufacturing Equipment

-   Much improved extruder design providing better dispersion, shorter dwell times, and better self-wiping design, i.e. reduced dead zones.
-   Further improvement of powder collection after grinding operation.
-   Laser particle size instrumentation allowing fast on-line adjustment.
-   Color computer system making color measurement fast and matching simple
-   Improved cooling methods from the extruder.
-   Automation for improved consistency and quality.

Application

-   Much computerization of controls and automation of complete system.
-   Improved powder collection system utilizing cartridge filters (compact design).
-   Tribocharging systems became more dominant.

By 1990, annual U.S. sales of powder coatings are expected to top 115M pounds with European sales in excess of 300M pounds. Many new significant uses of powder coatings occurred in the 80's. For example, powder replaced porcelain enamel on the tops and lids of washing machines, an extremely tough environment providing the industry savings. In the automotive area, powder was used as a primer-surfacer for over six years by a major truck manufacturer and it became a dominant finishing system on aluminum wheels and some small exterior body parts.

Beyond 1990, powder is estimated to grow by 1996 to 286 M pounds in North America with European sales expected to be 507M pounds. In terms of the technology, it is expected that powder coatings will penetrate more market areas. Although throughout the 80's no new significant powder coatings chemistries were introduced, it is expected that some new systems will become evident in the next ten years. It is also highly likely that long-term architectural coatings (20-year plus durability) will be available and that powder will find a use on the exterior of automobiles as a clear coat.

Refinements in the chemistry, manufacturing technology, and application technique will result in improved film rheology of the powder coatings, providing higher performance at low cure schedules, while retaining liquid-like film appearance.

ACKNOWLEDGEMENTS

Input from:

D. Heath - Holden U.K.
F. Vissor - DSM Holland
E. Marx - Shell Chemical
R. A. Coffee - ICI U.K.
R. F. Farrell - Glidden Powder Coatings
G. W. Crum - Nordson
E. Meyer - Glidden
P.C.I. Reports
*"Proceedings First North American Conference on Powder Coatings"
 February 25-26, 1971, Toronto, Canada

HISTORY OF STYRENE COPOLYMER COATINGS

V. L. STEVENS*
*The Dow Chemical Company, 2040 Willard H. Dow Center, Midland, MI
48674, U.S.A.

Styrene Copolymer Coatings

There are two main types of styrene copolymer coatings: styrene
butadiene and styrene acrylic. Both copolymer types are well-known in
the latex form and are used in architectural, industrial, and maintenance coatings.

Styrene copolymer coatings were made possible as a direct result of
an urgent need during WWII for large quantities of synthetic rubber. To
meet this need, a program to develop synthetic rubber was financed by
the Federal Government and had the cooperation of industry and universities. This war effort led to the emergence of an industry with the
capability of producing large quantities of styrene butadiene latexes.
This, in turn, induced research and development work, which eventually
resulted in applications for paper, textile, carpet, surface coatings,
etc.(1).

Background

The first styrene butadiene latex was commercialized in 1946 for
paper coating applications(2,3,4) and the first patents on pigmented
paper coatings were issued in 1954 to the S.D. Warren Company (5,6).

In 1948, the first styrene butadiene latex was sold into architectural coatings applications. The prototype styrene butadiene latex
copolymer from which other generations of styrene butadiene latex
copolymers were developed was Dow Latex 546, a 60/40 styrene butadiene
copolymer by The Dow Chemical Company. Dow Latex 512 then came as a
result of improvements in stability and foaming. This was the beginning
of the latex paint technology, which revolutionized the trade sales
paints industry from solvent borne to water borne(7,8). The first
patent on styrene butadiene latex paints was issued in 1950 to The Dow
Chemical Company(9). This evolution from solvent borne to water borne
trade sales paints in the 40's and 50's was not induced by environmental
regulations. Rule 66 came two decades later, and the volatile organic
compounds (VOC) regulations from the Clean Air Act came 30 years later.
Latex paints came into existence as a result of consumer demand: the
consumer wanted a house paint with an acceptable odor such that a
freshly painted room could be occupied within a few hours. Latex
paints also offered safety from fire explosion and asphyxiation
hazards inherent in solvent borne paints. Over and above these was the
fact that after painting the paint brushes and equipment could be
cleaned with water instead of solvent.

The styrene butadiene latexes, after leading the move into the
architectural coatings market in the 1940s and 1950s, were displaced by
vinyl acrylic and acrylic latexes during the 1960s. This displacement
of styrene butadiene latexes from the market was due to color and
chalking upon exposure to UV light. The styrene acrylics, namely
copolymers of styrene with soft acrylic monomers such as n-butyl

Published 1990 by Elsevier Science Publishing Co., Inc.
Organic Coatings: Their Origin and Development
R.B. Seymour and H.F. Mark, Editors

acrylate and 2-ethylhexyl acrylate, subsequently entered the market due to improved color and chalk stability. The first commercial styrene acrylic latex, Lytron 680, was developed by Monsanto in 1953 and later sold to Union Carbide. Lytron 680 latex is a styrene/acrylonitrile/ 2-ethylhexyl acrylate terpolymer modified with methacrylic acid. It is often considered the prototype styrene acrylic latex from which subsequent styrene acrylic latexes were developed.

Monomer & Polymer Considerations

Styrene monomer is 100% hydrocarbon in nature. Styrene was discovered by Newman prior to 1781(10). The first successful polymerization of styrene monomer is credited to E. Simon in 1839(10). The first commercialization of polystyrene was not until 1930 by The Dow Chemical Company(11,12).

Styrene is presently made by dehydrogenation of ethylbenzene, which is made by alkylation of benzene, which in turn comes from crude oil.

Styrene is a very reactive monomer and readily undergoes homopolymerization. It can be copolymerized with a variety of unsaturated monomers with different physical and chemical properties. Copolymers of styrene have a wide range of performance properties depending on the comonomer. Table I lists the common acrylic and methacrylic comonomers as well as other common functional monomers. Table II lists the typical physical properties of styrene vs. selected common comonomers. The polymerization characteristics of styrene and common comonomers are summarized in Table III. Table IV shows the relative homopolymer properties of styrene and selected comonomers. By comparing the properties of the homopolymer listed in Table IV, a general idea about a specific copolymer property can be obtained. Polystyrene is a very hard polymer, which will not form a good continuous dry film at room temperature unless it is plasticized. Polybutadiene, on the other hand, is a very soft gummy and rubbery polymer, which will form a film of poor strength. Thus, butadiene can be and is used as the plasticizing component in styrene butadiene copolymers. Acrylate homopolymers, especially n-butyl acrylate and 2-ethylhexyl acrylate, tend to be soft and tacky. These acrylic monomers are also used as the plasticizing component in styrene acrylic copolymers.

The behavior of styrene, butadiene, and other comonomers in copolymerization is best described with reference to the Alfrey-Price Q-e scheme. Table III and Figure 1 compares the Q-e value of styrene vs. common comonomers. Monomers falling near each other in this plot tend to react uniformly. Monomers with similar Q values, but widely differing e values, tend to form alternating copolymers. Monomers with large Q values (resonance stabilized radicals) do not react easily with monomers of low Q value and similar polarity. Styrene and butadiene monomers, when copolymerized, will form random styrene butadiene sequence in a chain. Monomers such as acrylic and vinylidene chloride, which fall into the middle of the Q-e map, tend to react reasonably well with styrene as well as a wide variety of other monomers.

Styrene butadiene copolymers for coatings are most widely made via emulsion polymerization using an emulsifier such as sodium alkyl benzene sulfonate and an initiator such as potassium persulfate. Latex recipes can be quite complex and involved(21).

Styrene forms a linear polymer; however, butadiene can undergo 1,2 or the 1,4 addition. The 1,4 version normally makes up approximately 80% of the butadiene structures in a styrene butadiene copolymer. However, both modes will result in a double bond per butadiene unit. This residual double bond is capable of being further reacted giving rise to a growing three dimensional polymer network. Depending on the polymerization variables, latexes with significantly different properties will be produced even if the styrene butadiene ratios are kept constant. Thus, styrene butadiene latex producers can have significant control over the design of the final product by changing not only the styrene butadiene ratio but also such variables as:

1. Temperature
2. Pressure
3. Initiator type and level
4. Chain transfer agent type and level
5. Surfactant type and level
6. Agitation
7. Mode of addition (batch and continuous)
8. Etc.

Which, in turn, will lead to changes in properties such as:

1. Molecular weight
2. Backbone configuration (degree of random or block)
3. Degree of cross-linking
4. Degree of branching
5. Ease of coalescence
6. Degree of stability (mechanical, ionic)
7. Particle size
8. Glass transition temperature, minimum film forming temperature

Styrene butadiene latexes can be further modified if other backbone monomers such as acrylonitrile, isoprene, and/or unsaturated carboxylic acids are copolymerized with styrene and butadiene. Presently, the most widely used modifying monomer types are the unsaturated carboxylic acids such as acrylic, methacrylic, itaconic, fumaric, etc. The carboxylic acid functionality imparts excellent strength and stability properties to the latex.

The ability of the polymer chemist to modify and control the morphology of the latex particles is another tool in designing unique products. This flexibility made possible latexes with surface modifications, uniform particle sizes, grafted latexes, structured latexes having core shell, microdomain structures, interpreting polymer networks, etc.(22,23,24,25).

Styrene butadiene copolymers oxidize or degrade by free radical mechanisms. The unsaturations in the styrene butadiene copolymer are vulnerable to attack by atmospheric oxygen, ozone, and by heat and UV light. This results in chain scission, cross-linking, and hardening of the film, which leads to cracking and chalking(13,26,27). A slight yellowing of the copolymer film also results as a consequence of oxidation. Antioxidants, hindered amines, and UV absorbers are being employed in some cases to minimize oxidation and degradation.

Styrene butadiene latexes are not as widely available as are acrylic and vinyl acrylic latexes. This is due to the fact that most latex suppliers are not equipped to handle the butadiene monomer. Butadiene monomer has high vapor pressure, a large heat of polymerization, and is highly flammable (See Table I). Butadiene monomer readily forms violently explosive peroxides when in contact with air or oxygen. Because the utmost care for safe, efficient operation is required in handling and using butadiene, the styrene copolymers of butadiene are only available from a few suppliers.

Styrene acrylic copolymers in coating applications are also made by emulsion polymerization. Unlike emulsion polymerization of styrene and butadiene, the production of styrene acrylate ester copolymers do not require high pressure or special equipment. Styrene and acrylate monomers readily copolymerize in water in the presence of a surfactant and a water soluble initiator such as ammonium or sodium persulfate(15, 28,29).

Most styrene acrylic copolymer latexes are promoted and sold as "modified acrylic" latexes. It is difficult to determine the popularity of styrene acrylic in the coatings market, because they are grouped under the category of "modified acrylic" latexes even though in some cases "modified acrylic" may contain close to 50% styrene by weight of copolymer.

Styrene Copolymers in Coatings

Styrene butadiene and styrene acrylic latexes usually have lower particle sizes than the vinyl acrylic and acrylic latexes. On the average, styrene butadiene and styrene acrylic latexes have particle sizes of approximately 900-2,000Å, whereas the vinylacrylic and acrylic latexes range from 4,500Å to 8,000Å. Although poor shelf life and lower film build is predicted for lower particle size latexes, present day technology in emulsion polymerization and rheology modifiers such as surfactants and associative thickeners have produced styrene butadiene and styrene acrylic latexes that have overcome, to a significant extent, the stated stability and rheology deficiencies.

The performance of architectural coatings formulated with styrene butadiene and styrene acrylic latexes have the following unique and differentiating properties vs. the architectural coatings formulated with acrylic and vinyl acrylic latexes(31,32,33,34,35,36):

1. Superior moisture vapor barrier
2. Superior alkali resistance
3. Superior acid resistance
4. Higher binder efficiency
5. Ease of coalescence
6. Ability for higher gloss
7. Corrosion resistance
8. Tannin blocking potential
9. Lower cost per dry gallon due to lower density

On the other hand, styrene butadiene and styrene acrylic latexes are not as good as the acrylic and vinyl acrylic latexes in the following:

1. Color retention
2. Resistance to chalking
3. Hiding power

     Some of the above properties can be induced via additives or blends
for those latexes not possessing them. The above properties both
positive and negative are inherent in the styrene butadiene and styrene
acrylic latexes due to a large extent to the contribution of the styrene
moiety. The following paragraphs give a more detailed account of some
of the above properties.

Moisture Vapor Barrier

     One of the unique characteristics of styrene butadiene latexes is
the ability to form tight films, which allows the formulation of specialty primer sealers with excellent moisture vapor barrier properties*.
The function of a moisture vapor barrier primer sealer is to keep
moisture from migrating into or away from a structure. For example, in
the humid summer season, it is desirable to keep the moisture outside
from entering the house. This not only increases the comfort level, but
also lowers air conditioning costs since less moisture needs to be
removed from the house. In the cold, dry winter season, a moisture
vapor barrier primer sealer assists in keeping the moisture in the
house, thus providing more comfort and keeping inside moisture from
freezing onto insulation or outside walls. This, in turn, reduces the
heating bill. A moisture vapor barrier primer sealer paint is usually
applied to the interior of the house.

     A commercially-available modified styrene butadiene latex was used
in a series of studies to determine its relative moisture vapor barrier
characteristics vs. commercially-available acrylic, vinyl acrylic, and
styrene acrylic latexes formulated by paint companies in an attempt to
obtain moisture vapor barrier of less than 1 perm. The modified styrene
butadiene latex at 60% PVC and 50% PVC (with no coalescing solvent)
outperformed the specially formulated moisture vapor barrier paints from
the different paint companies (PVCs less than 30%). See Table V(32).

     Moisture vapor barrier comparison was also made by keeping the
moisture vapor barrier primer sealer formulation constant with the
latexes as the variables. Table VI shows the moisture vapor barrier
of the modified styrene butadiene latex outperformed the acrylic,
vinyl acrylic, and styrene acrylic latexes(33).

     Not only did the modified styrene butadiene show superior moisture
vapor barrier, it also gave more efficient moisture vapor barrier
properties from the standpoint of film thickness, ambient drying,
and under forced drying (Table VII and Figures 2-3). Table VIII
shows no loss to moisture vapor barrier performance after 500 hours
of QUV exposure.

---

*Moisture vapor barrier tests are done according to TAPPI
(T4480M-84). A paint film of 6 mils wet is placed on a support
substrate and dried for seven days. The moisture vapor transmission is measured in grams of moisture per day per 100 square
inches of surface area at 73°F and 50% RH. U.S. perms are
calculated by multiplying the rate by 2.439. A perm of less than
one is desirable for acceptable moisture vapor barrier.

Hydrolytic Stability

Another inherent benefit of styrene butadiene, and to some extent styrene acrylic latexes, is their ability to withstand hydrolysis and, thus, provide stability to caustic and acid. Acrylic and vinyl acrylic latexes, as well as alkyd binders, all contain ester groups throughout the polymer network (Figure 4). The ester groups are known for their susceptibility to hydrolysis. Although hydrolysis is, to some extent, controlled in acrylic latexes by the use of hydrophobic units such as the butyl group in butyl acrylate, it is a limitation for paints used on very alkaline, high pH substrates. Modified styrene butadiene latex is composed mainly of carbons and hydrogens and does not have ester groups in its structure. As such, it is very resistant to hydrolysis(32).

The importance of alkali resistance in architectural primer sealers comes from the fact that some substrates such as green plaster and improperly cured concrete are very caustic. This causes short-term and long-term paint performance problems, such as cracking and peeling due to hydrolysis in latexes containing ester groups. One new problem area reported recently is the use of alkaline sizing on wallboards which causes paint performance problems also due to hydrolysis(37).

Binder Efficiency

Both styrene butadiene and styrene acrylics have shown superior binder ability as compared to acrylic and vinyl acrylic latexes. The binder ability is expressed in measurement of critical pigment volume concentration for properties such as enamel holdout and wet scrub.

Figure 5 compares the gloss of a standard semigloss topcoat over the different primer sealers made with vinyl acrylic, alkyd, and modified styrene butadiene latexes. The gloss is compared as a function of % pigment volume concentration. The styrene butadiene latexes showed higher critical pigment volume concentration and did not show dramatic loss of the gloss of the topcoat even at 70% pigment volume concentration(32).

Studies also showed styrene acrylic latexes to be superior binders as compared to polyvinyl acetate homopolymers, vinyl acrylic, acrylic, and ethylene vinyl acetate copolymers(31). Figure 6 plots the weight of film eroded by 500 scrub strokes (gm) vs. the pigment volume concentration % of the different types of paints.

Styrene acrylic latexes exhibit slightly lower opacity compared to acrylic, vinyl acrylic, vinyl acetate ethylene, copolymers and polyvinyl acetate homopolymer latexes(31). Figure 7 shows the relationship between contrast ratio vs. the pigment volume concentration % of the different types of paints.

Tannin Blocking Resistance

Another property of a modified styrene butadiene latex is the ability to block water soluble tannin from staining wood when formulated with certain reactive pigments such as calcium barium phospho-silicate(33).

Corrosion Resistance

Styrene butadiene and styrene acrylic latexes have also been reported to contribute corrosion resistance to metal coatings (34,35,36).

Conclusion

Styrene butadiene latex binders initiated and styrene acrylic latexes continue the conversion of the architectural coatings industry to water borne coatings. Because of the special properties of the styrene moiety, styrene butadiene and styrene acrylic copolymer latexes continue to be significant participants in the coatings industry.

Acknowledgments

I wish to thank John A. Gordon of Pacific Technical Consultants, Art Mees of Union Carbide, Ronald R. Brown of Unocal, and Ben Kine (formerly of Rohm & Haas) for their thoughts and comments on the history of styrene-containing copolymer coatings.

REFERENCES

(1) Thompson, S. S., "The Styrene Butadiene Latex Story," Pendell Publishing Co., 1980.

(2) E. K. Stilbert, R. D. Visger and R. H. Lalk, Tappi, vol. 33, 16 1950.

(3) D. A. Taber and R. C. Stein, Tappi, vol. 40, 107 1957.

(4) L. H. Silvernail and E. J. Heiser, Tappi, vol. 36, 172A, 1953.

(5) U.S. 2,685,571. J. C. Stinchfield and F. Kaulakis (to S.D. Warren Company) August 3, 1954.

(6) U.S. 2,685,538. J. C. Stinchfield and F. Kaulakis (to S.D. Warren Company) August 3, 1954.

(7) L. L. Ryden, N. G. Britt and R. D. Visger, Official Digest, vol. 303, 292 (April, 1950).

(8) W. W. Burr and P. R. Matvey, Official Digest, vol. 304, 347 (May, 1950).

(9) U.S. 2,498,712. L. L. Ryden (to The Dow Chemical Company) February 28, 1950.

(10) A. J. Warner, in R. H. Bounty and R. F. Boyer, eds., "Styrene, Its Polymers, Copolymers, and Derivatives," Reinhold Publishing Corp., New York, pg. 3, 1952.

(11) D. Whitehead, "The Dow Story," McGraw-Hill Book Co., New York, pg. 145, 1968.

(12) J. L. Amos, Polym. Eng. Sci., vol. 14, 1, 1974.

(13) Encyclopedia of Polymer Science and Technology. New York:
      J. Wiley & Sons, 1971.

(14) Kirk-Othmer Encyclopedia of Chemical Technology, Third
      Edition. New York: J. Wiley & Sons.

(15) Kine, B., R. W. Novak, "Acrylic & Methacrylic Acid Polymers,"
      Encyclopedia of Polymer Science & Engineering, Vol. 1, 2nd
      Edition, John Wiley & Sons, pp. 234-299, 1984.

(16) Brandru, J., et al., Polymer Handbook, 2nd Edition, Wiley,
      1975.

(17) Van Krevelen, W. "Properties of Polymers," Second Edition.
      New York: Elsevier Science Publishing, 1976.

(18) Alfrey, T. and C. C. Price. J. Polymer Sci., vol. 2, 101,
      1947.

(19) Alfrey, T., J. J. Bohrer, and H. Mark. "Copolymerization."
      New York: Interscience, 1952.

(20) Ham, G. E. (ed). "Copolymerization." New York:
      Interscience, 1964.

(21) Poehlein, Gary, W., "Emulsion Polymerization," Encyclopedia of
      Polymer Science & Engineering, Vol. 6, pp. 1-51, 1986.

(22) Lee, D. I., "The Control of Structure in Emulsion
      Polymerization," presented at the IUPAC International
      Symposium on Molecular Design of Functional Polymers, June
      25-28, 1989 (Seoul Korea).

(23) Lee, D. I. and T. Ishikawa, "The Formation of Inverted Core
      Shell Latexes," Journal of Polymer Science: Polymer Chemistry
      Edition, Vol. 21, 147-154, 1983.

(24) Lee, D. I., "Morphology of Two Stage Latex Particles
      Polystyrene and Styrene Butadiene Copolymer Pair Systems," ACS
      Symposium Series 165, 1981.

(25) Lee, D. I., T. Kawamura, and E. F. Stevens, "Interpenetrating
      Polymer Network Latexes: Synthesis, Morphology, and
      Properties," NATO ASI Series E: Applied Sciences No. 138
      (1987).

(26) E. M. Fettes, "Chemical Reactions of Polymers," New York,
      Inters Publishers, 1055, ff, 1964.

(27) E. M. Bevilacqua, Rubber Age (N.Y.,) vol. 80 (2), 271, 1956.

(28) "Emulsion Polymerization of Acrylic Monomers," CM-104, Rohm
      and Haas Co., Philadelphia, PA.

(29) G. L. Brown, Off. Dig. Fed. Soc. Paint Technol., vol. 28, 456,
      1956.

(30) Wessling, R. A. and V. L. Stevens, "Vinylidene Chloride, A
      Monomer for 1980's," Modern Paint & Coatings, pg. 148ff,
      October 1982.

(31) Safe, K. A., "Emulsion Polymer Types--Is There A Best Buy?"
      J. Oil Col. Chem. Asso., vol. 53, 599-614, 1970.

(32) Lednicky, R., V. L. Stevens, "Moisture Vapor Barrier Latex for
      Primer Sealer & Undercoats," Modern Paint & Coatings, June
      1988.

(33) Stevens, V. L., "Latexes Offer Alkali Resistance, Moisture
      Barrier, and Tannin Block," Modern Paint & Coatings, February,
      1989.

(34) Gruber, W., "Waterborne Anticorrosion Latex Paints," Proc. 2nd
      International Polymer Latex Conference (London, 1985), RAPRA
      Abs., Vol. 24, No. 8, 1987.

(35) Peene, R. A., "Optimal Pigmentation of Emulsion Coatings for
      Steel in Corrosive Environments," J. Coatings Technology, vol.
      54, No. 690, July, pp. 51-56, 1982.

(36) Tolmachev, J. A., et al., "Water-Thinner Primers Based on
      Butadiene-Styrene Latexes for Coatings Metals," Lakokras,
      Mater, IKh Primen., vol. 5, 4-5, 1980.

(37) Richardson, J., "New Alkaline Sizing Causing Wallboard
      Painting Problems," American Pt. & Coatings Jour., 6/13/88,
      pp. 53-54, 6/13/88.

TABLE I

Common Comonomers & Abbreviations

Monomer	Abbreviation	Formula
Ethylene	ET	$CH_2=CH_2$
Butadiene	BD	$CH_2-CH=CH-CH_2$
Styrene	STY	$CH_2=CH-C_6H_5$
Methyl acrylate	MA	$CH_2=CHCO_2CH_3$
Ethyl acrylate	EA	$CH_2=CHCO_2C_2H_5$
n-Butyl acrylate	BA	$CH_2=CHCO_2C_4H_9$
2-Ethyl hexyl acrylate	2EHA	$CH_2=CHCO_2C_2H_3(C_2H_5)C_4H_9$
Methyl methacrylate	MMA	$CH_2=C(CH_3)CO_2CH_3$
Ethyl methacrylate	EMA	$CH_2=C(CH_3)CO_2C_2H_5$
Butyl methacrylate	BMA	$CH_2=C(CH_3)CO_2C_4H_5$
Vinyl acetate	VA	$CH_2=CHO_2CCH_3$
Acrylonitrile	AN	$CH_2=CHC=N$
Vinyl chloride	VC	$CH_2=CHCl$
Vinylidene chloride	VDC	$CH_2=CCl_2$

TABLE II

Comparison of Properties of Common Comonomers(13,14,15)

Monomer	Bp. $^{\circ}$C	MP $^{\circ}$C	Density g/cc (c)	Flash Pt. $^{\circ}$F COC	Refractive Index nD20
ET	-103.7	-169.19	0.566(-104)	-350	
BD	-4.4	-108.9	0.6211(20)	-155	1.4292
STY	145.5	-30.6	0.9059(20)	88	1.54
MA	80	--	0.9535(20)	50	1.4040
EA	43	-75	0.9234(20)	48.2	1.4068
BA	35	--	0.8998(20)	120	1.4190
2EHA	85	--	0.8852(20)	195	1.4365
MMA	101.1	-48.2	0.944(20)	50	1.4120(25)
EMA	118	--	0.909(25)	70	1.4116(25)
BMA	162	--	0.889(25)	150	1.4220(25)
VAc	72.5	-93.2	0.9317(20)	18	1.3953
AN	77.3	-83.55	0.806(20)	23(TOC)	1.393(20)
VC	-13.83	-153.71	0.9016(25)	-108	1.398(15)
VDC	31.56	-122.56	1.214(20)	-19	1.4249

TABLE III

Comparison of Polymerization Characteristics
of Selected Monomers(13,15,16,18,19,20)

Monomer	Heat of Polymerization Kcal/Mol	Q	e
ET	25.6	0.012	-0.02
BD	17.4	2.39	-1.05
STY	16.68	1.0	-0.8
MA	18.8	0.44	+0.60
GA	18.7	0.41	+0.46
BA	18.5	0.30	+0.74
2EHA	14.5	0.14	+0.90
MMA	13.8	0.75	+0.38
EMA	13.8	0.73	+0.51
BMA	13.5	0.74	+0.39
VA	21.3	0.031	-0.84
AN	18.4	0.48	+1.24
VC	27	0.056	+0.2
VDC	18	0.26	+0.4

334

TABLE IV

Comparison of Selected Homopolymer
Properties(13,14,15,17)

Monomer	Density (Amorphous) g/cc	Glass Transition Temperature °C	Sol. Parameters (Cal/cc)	Refractive Index
ET	0.855	-120	7.9	1.49
BD	0.913	-85	8.3	1.51
STY	1.04	100	9.1	1.59
MA	1.22	6	10.1	1.479
EA	1.10	-22	9.4	1.47
BA	1.08	-55	8.8	1.474
2EHA	--	-70*	--	--
MMA	1.18	105	9.5	1.49
EMA	1.119	65	9.1	1.485
BMA	1.055	20	8.7	1.483
VA	1.19	31	9.4	1.47
AN	1.18	104	11.4	1.52
VC	1.18	81	9.7	1.55
VDC	1.77	-17	10.1	1.63

*Private communication with supplier (Union Carbide).

TABLE V

**Relative Moisture Vapor Barrier Performance Of
Commercial Latexes Using Optimized Moisture Vapor
Barrier Formula For Primer Sealers (23)**

Type	Source	MVB in Perms
Acrylic	Rohm & Haas	8.65
Acrylic	Rohm & Haas	2.88
Acrylic	Unocal	7.31
Vinyl acrylic	Unocal	5.03
Vinyl acrylic	Hoechst	3.41
Styrene acrylic	Rohm & Haas	2.15
Styrene acrylic	BASF	1.63
Styrene acrylic	Unocal	1.52
Modified styrene butadiene	Dow Chemical	0.46

TABLE VI

**Moisture Vapor Barrier – General Comparison*
(MVB Paints From Paint Companies)
Formulations Optimized for Individual Latexes Listed(32)**

Primer Sealer Latex Paint	MVB (U.S. Perms)
Acrylic A (<30% PVC)	2.22
B (<30% PVC)	2.31
C (<30% PVC)	4.14
Styrene Acrylic A (<30% PVC)	3.92
Vinyl Acrylic A (<30% PVC)	5.19
B (<30% PVC)	3.92
Modified S/B (60% PVC)	0.46
Modified S/B (50% PVC With No Coalescing Solvent)	0.66

TABLE VII

**Modified Styrene Butadiene 60% PVC Latex Primer Sealer Paint
Effect Of Accelerated QUV Exposure
On Moisture Vapor Barrier (MVB) Performance(33)**

	MVB (Perms)
No exposure	0.46
500 hrs.	0.38

TABLE VIII

**Effect Of Force Dry On
Moisture Vapor Barrier (MVB) Performance**

Drying Conditions: 30 Min. at 50°C

Latex Paints	(MVB Perms)
Acrylic	5.03
Vinyl Acrylic	6.29
Modified Styrene Butadiene	0.42

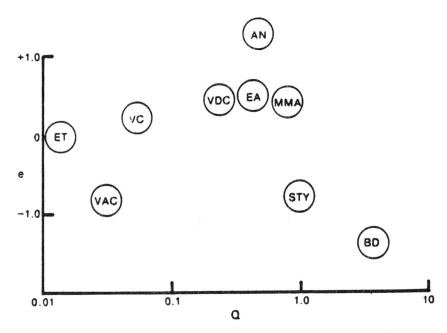

Fig. 1. Comparison of copolymerization characteristics of various monomers (Q-e map).

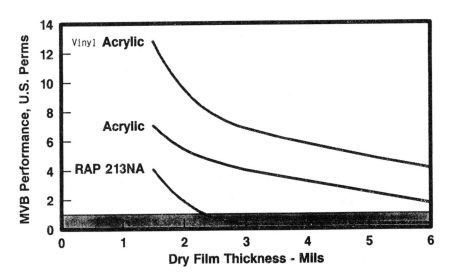

Fig. 2. Effect Of Dry Film Thickness On Moisture Vapor Barrier (MVB) Performance.

338

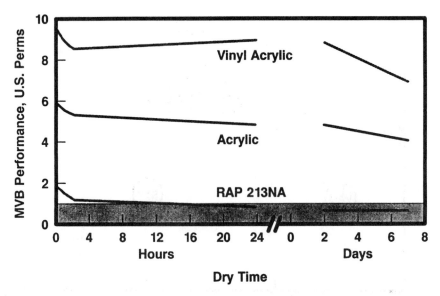

Fig. 3.   Effect Of Dry Time On Moisture Vapor Barrier Performance.

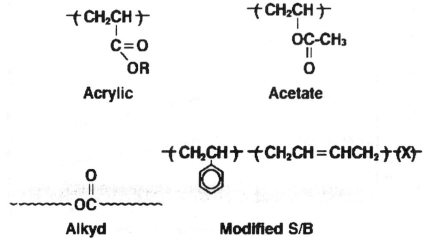

Fig. 4.   Ester Containing Binders And Modified Styrene Butadiene Binder.

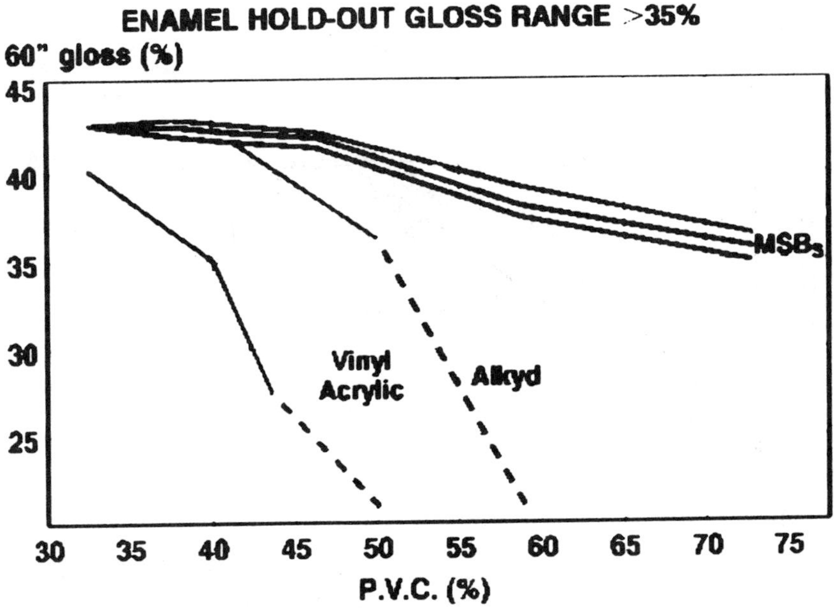

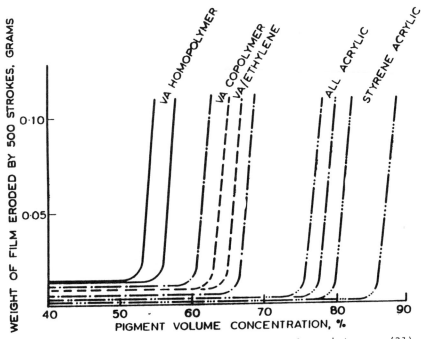

ENAMEL HOLD-OUT GLOSS RANGE >35%

60" gloss (%)

MSB₃

Vinyl
Acrylic   Alkyd

P.V.C. (%)

Fig. 5.   Enamel Hold Out PVC Ladder.

Fig. 6.   Effect of polymer type on wet scrub resistance. (31)

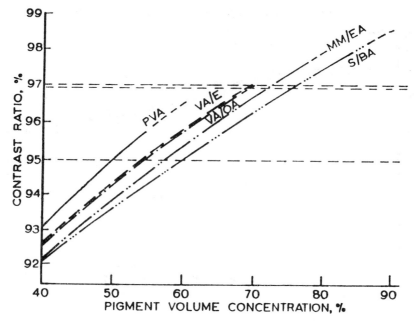

Fig. 7.   Opacity vs. pigment volume concentration (20% titanium dioxide in film volume). (31)

INDEX

342